Creative Couples in the Sciences

LIVES OF WOMEN IN SCIENCE

Founding Editor: Pnina Abir-Am
Series Editor: Ann Hibner Koblitz

VOLUMES IN THE SERIES

Creative Couples in the Sciences

HELENA M. PYCIOR,
NANCY G. SLACK,
AND PNINA G. ABIR-AM,
EDITORS

Rutgers University Press
NEW BRUNSWICK,
NEW JERSEY

Library of Congress Cataloging-in-Publication Data

Creative couples in the sciences / Helena M. Pycior, Nancy G. Slack,
and Pnina G. Abir-Am, editors.
p.   cm.—(Lives of women in science)
Includes bibliographical references and index.
ISBN 0-8135-2187-4 — ISBN 0-8135-2188-2 (pbk.)
1. Scientist couples—Biography. I. Pycior, Helena M. (Helena
Mary), 1947– . II. Slack, Nancy G. III. Abir-Am, Pnina G., 1947–
IV. Title: Creative couples. V. Series.
Q141.C68   1995
509.2'2—dc20
[B]   94-41059

British Cataloging-in-Publication information available

Contents

Series Foreword

Given the paucity of biographies and memoirs of women scientists, it may come as a surprise that we are, so to speak, using up a volume in this series, Lives of Women in Science, on scientific *couples*. Do we really need to focus on women in the company of those most likely to overshadow them, when we could be investing our energy in the urgent task of incorporating more women into the collective memory of science?[1]

The question applies, of course, to women as members of couples in other fields—the arts, literature, politics[2]—wherever two individuals of different gender, linked in a prolonged relationship of formalized intimacy, engage in a shared creative endeavor. In a world long convinced that individuals are the sole units of creativity, these couples incessantly face the social complexities of the "two-body problem."

The practice of collaboration clearly has profound implications for the work and careers of both partners, but historically it has been the women in such partnerships who have borne the brunt of society's asymmetric evaluation of the genders. It is no longer news that, by and large, creative work by women has been underestimated. The gender hierarchy of a patriarchal society cannot, however, so easily ignore women who have been coauthors with men or whose contributions have outlasted those of their male partners. Because it is no longer possible in such cases to render the women's share of the work invisible, society's solution has been to regard the women's contribution as merely derivative. Influence could flow only one way, from men to women, not vice versa. Rectifying such distortions through in-depth historical case studies would be sufficient justification for including this collection in a series about scientifically creative women.

Yet, collaborative marriages also deserve our attention because, aside from women's colleges, they were historically the single most important avenue for recruiting women to science and retaining them as active participants. Because the institution of marriage had a monopoly on permissible intimacy, it was the inevitable solution for those men and women who wished to collaborate within socially acceptable boundaries. Yet, marriage posed a real—but rarely articulated—risk that an individual spouse's work would not get independent recognition. In practice, this almost always meant the wife: legally, economically, and socially, marriage made her dependent upon her husband's name, position, and resources. We cannot account for a woman scientist's achievements and honors—or lack of them—unless we understand that inexorably the social condition of being her collaborator's wife spilled over into every assessment of her work as a scientist.[3]

Valuable as these studies of collaborative couples are for redressing the place of women in science, they go beyond that crucial goal to look directly at the pivotal role of collaboration in the scientific enterprise. The nature of scientific collaboration has become a critical issue, especially in the twentieth century when the increasingly transdisciplinary character of scientific innovation makes collaboration highly advantageous, if not absolutely essential. Historians of science now must include in their repertory of analysis not only the lone geniuses and "founding fathers" of scientific disciplines but also a variety of social groups ranging from research schools to pairs of scientists working together. Innovation in twentieth century science—it has become clear—is as much a property of social relations as of individual talent. The close inspection of a male scientist's work may reveal not only his dependence on formal and informal collaboration but his outright plagiarism of his wife's intellectual property (quite often with her "rational" consent). Both simple justice and the need to avoid erroneous epistemological assumptions about individual industry or genius make it imperative for historians to reexamine the credit due to each member of the collaboration.

Studying cross-gender collaboration has the benefit of highlighting features of scientific collaborative work that might otherwise be taken for granted. For example, the superimposing of the gender hierarchy upon the positivist ranking of the sciences has led to one of the most persistent myths about the collaboration between Marie and Pierre Curie. Marie Curie is often seen as the chemist, and Pierre, as the physicist of the pair. Why? Because physics is regarded as a more basic, more theoretical, more important science than chemistry; so it follows that the husband must have been the physicist, and the wife the chemist. As this collection amply demonstrates, such assumptions do not hold up to the scrutiny of historical analysis.[4]

Cross-gender collaborative couples occupy, by definition, a joint public and private space. That means not only a blurring of personal and professional spheres but an expansion of their scientific and social opportunities.

As they confront the traditional gender-bound uses of time, for example, both partners gain new flexibility. Because women are often required to devote disproportionately more time and attention to social affairs, they have less control over their professional time; a husband-collaborator, who benefits from that joint social life, is more likely to tolerate the absences of his collaborator-wife from both home and lab. Moreover, the collaborative marriage provides both spouses with the social cover for unencumbered professional interactions with other people of whatever age, gender, or marital status. In short, the social and scientific resources and opportunities available to a collaborative couple are apt to be both richer and more diverse than those of same-gender collaborators.

Appreciation of that diversity again steers us away from simple, one-way, one-dimensional models of relationships between the genders or between science and gender. But it also leads us to ask where egalitarian rapport is easier to reach: in the sphere of intimacy, or in the sphere of creativity? How do these two critical aspects of human existence become convertible currency in collaborative work? How do the rates of exchange, or the balance of power, vary over the course of the spouses' lives, together or apart?

Finally, these studies of cross-gender collaboration force us to ask to what extent science lives up to its own ideal of rationality. If, in judging the achievements of these creative couples, science cannot bring itself to reject gender stereotypes, how can science use its claims to objectivity to justify its unique epistemological authority in society?[5]

The studies in this collection contain plenty of new, surprising material to help us start responding to these questions. Through its rich representation of different scientific disciplines, national cultures, institutional settings, historical periods, personalities, and outcomes for the collaborative partners—and especially through its selection of "quasi-egalitarian" couples in which the woman's accomplishments match the man's—this collection fulfills a central aim of this series: the integration of women and gender into the historiography of science. At the same time, this book enlarges the audience for Lives of Women in Science by recognizing that men as much as women readers share a keen interest in the fortunes of collaborative couples. We deeply hope that our readers, regardless of their own collaborative choices, are inspired by the synergy of intimacy and creativity shown by the scientific couples portrayed in this volume.

Pnina G. Abir-Am

Boston, Massachusetts
January 1995

Creative Couples in the Sciences

HELENA M. PYCIOR, NANCY G. SLACK, AND PNINA G. ABIR–AM

Introduction

In 1903 Marie and Pierre Curie became the first spousal collaborators to share a Nobel Prize; in 1922 Albert Einstein alone received the Nobel Prize in physics but, following his divorce agreement, gave the prize's monetary award to Mileva Marić, his ex-wife who had shared in his scientific studies and early scientific work. By 1935 another generation of modern scientific couples had come of age, symbolized once again by the receipt of a Nobel Prize by a married couple, Irène and Frédéric Joliot-Curie. Around the same time, the American geneticist Frieda Cobb Blanchard fondly remembered how she and Frank Blanchard, her recently deceased husband-herpetologist and sometime collaborator, had "consciously enjoyed the unusually close companionship . . . allowed" them. The Curies, Einsteins, Joliot-Curies, and Blanchards were "creative couples" in the sciences, according to the liberal interpretation of the term adopted here. At one time or another, the spouses in these couples evidenced significant scientific potential or, in many of the cases, actual scientific achievement. Spouses—some more consciously, some less so—bound themselves to one another in the promise of a special blending of creativity and intimacy. Most, but not all, of the creative couples collaborated on scientific work, at least to some extent, even if their joint efforts fell short of "formal collaboration," or that leading to coauthored publications.

As a historical study of creative couples in the sciences, this collection attempts to enrich not only the history of women in science but also the history of scientific collaboration. In both respects, its concerns are relevant to science of the late twentieth century, for there are now more women scientists than ever before and contemporary science is more collaborative than individualistic.[1] Like other recent scholarship in the history of science that has sought to correct the long-standing myth of the solitary (usually

3

male) genius of the past,[2] the collection pursues connectedness as well as gender diversity to the very core of scientific innovation. Its primary concern lies with neither individual women scientists nor their collaborator-husbands, but with the creative couples that they constituted.

Historians have already stressed the opportunities that women scientists found in the "family firms" of nineteenth-century science, in which many family members, if not entire households, engaged in the enterprise of science.[3] The present collection positions these family firms as early collaborative units, and suggests links between such domestic collaboration in nineteenth- and early-twentieth-century science and the more public, spousal collaboration that came into its own in the first half of the twentieth century. More essentially, the collection's essays document and analyze the high as well as low points of spousal collaboration, as only detailed case studies can. In explaining synergistic couples, many authors emphasize the complementarity—drawn along the lines of disciplinary commitment, socially constructed gender, personality, or scientific style—that seems to have permitted some couples to do scientific work that surpassed what either the husband or wife alone would have been able to accomplish or the wife alone would have been allowed to pursue.

As another major feature of the collection, authors, where possible, carefully reconstruct the actual division of work between spousal collaborators and their decisions about division of credit. The collection uncovers credit sharing ranging from that which was meritocratic and gender-neutral through plagiarism, by which a husband, often with his wife's complicity, signed his name alone to joint work. Although this study explicitly concerns scientific creativity in the context of cross-gender, formalized intimacy, some of its conclusions hold the promise of general significance in the study of scientific collaboration, since, like scientific couples, unisexual research teams or those pairing unmarried women and men were (and are) controlled by power relationships, personalities, and institutional factors as well as accepted scientific methods and norms.

As a subgroup of scientific research teams and of dual-career couples, the creative scientific couple is an important phenomenon in the history of science and gender. The earliest Nobel laureate women scientists— Marie Curie (physics, 1903, and chemistry, 1911), Irène Joliot-Curie (chemistry, 1935), and Gerty Cori (medicine and physiology, 1947)— shared Nobel Prizes with their husband-collaborators. Another laureate, Maria Goeppert-Mayer, was an episodic collaborator with her husband, who did not however participate in the research leading to her sharing the Nobel Prize in physics in 1963. Some men scientists, too, have been the beneficiaries of spousal collaboration on research that led to Nobel Prizes for them but not for their wives.[4] More basically, as Marilyn Bailey Ogilvie and others have noted, "collaboration with a mate traditionally has been a way through which women could enter science unobtrusively."[5]

In her pioneering turned classic work on American women scientists,

Margaret Rossiter provided the table "Notable Couples in Science before 1940" as well as comments on the possible effects of scientific marriage. She observed that there was a greater proportion of married women among the women scientists newly singled out for "stars" (or special recognition) in the *American Men of Science* (*AMS*) during the 1920s and 1930s than among all women included in the *AMS* during the same period. For women scientists, not marriage alone, but rather marriage to a scientist, seemed to be positively linked to the recognition symbolized by a star in the *AMS*.[6] Rossiter noted, too, that the *AMS* suggested that men zoologists had benefitted from marriages to women zoologists, who "would probably be unemployed and thus able to assist . . . [their] researches."[7]

Despite the developing awareness of the significance of scientific marriage to the history of women in science, there has been no systematic study of the creative couples of the scientific past. In *Uneasy Careers and Intimate Lives* of 1987, several authors explored collaborative marriages in various scientific disciplines and historical periods, but as part of a more general focus on "how the interplay between career and personal life has affected the participation of women in science."[8] In *Despite the Odds* of 1990, Marianne Gosztonyi Ainley reflected especially on the plight of married Canadian women scientists who had been forced into a "two-person single career," while Joan Pinner Scott reported on contemporary experiences of women engaged in collaboration with other women, their husbands, and other men.[9] More recent scholarship has begun to probe the uneven rewards for spousal collaboration in science, documenting the extent to which wives have been victims of the "Matthew Effect," according to which the more distinguished scientist receives the bulk of credit for joint work. Stressing that women's underrecognition in science is a "sex-linked phenomenon," Rossiter has challenged historians and sociologists of science to explore her new "Matilda Effect," according to which a man scientist (e.g., a husband) receives the bulk of credit for joint work with a woman scientist (his wife). There has also been a detailed case study of Marie and Pierre Curie's distinctive publication policy, which minimized the Matthew/Matilda Effect on Marie.[10]

As historians of science were beginning to identify scientific couples as a unit of analysis, scholars across the disciplines were developing a general analytical perspective (called here "intimate, creative connectedness") that rests on a heightened appreciation of the complex relationships—those involving immediate or extended families, close friendships, or heterosexual or homosexual pairings—that nurture, constrain, and convolute the emotional and intellectual sides of women and men. This perspective, which has already deepened the understanding of literary and artistic genius, was embedded in Phyllis Rose's *Parallel Lives* of 1983, an examination of the marriages of five Victorian literary couples.[11] Along somewhat similar lines, *Mothering the Mind* took the basic tack of "experiment[ing] with the idea that certain writers depended upon particular other persons to help create the conditions, the inspiration, the atmosphere for their work." The essays of the latter collection found four different types of largely "silent partners" of major writers: women muses of men writers, "lovers as nurturers," nurturing (biological and surrogate) mothers, and other supportive family members and friends.[12]

In 1989 the anthropologist Mary Catherine Bateson poignantly told the stories of five contemporary women, all of whom had experienced "the joys and benefits of real partnership" with husbands or lovers. What distinguished Bateson's *Composing a Life* was its emphasis on the creativity embodied in the "composing" of women's lives—lives that do not so often proceed in careful steps toward one set goal but rather are "lives in which commitments are continually refocused and redefined." Bateson led her readers to understand that these five women, subject to the "physical rhythms of reproduction and maturation" and (at times) to subordination of their professional aspirations to those of husbands and lovers, succeeded precisely because they actualized "the creative potential of interrupted and conflicted lives."[13] Recently, *Significant Others* pursued the perspective of intimate, creative connectedness through a collection of essays on artistic and literary couples who shared a sexual relationship and who each enjoyed artistic or literary recognition. Restricted to successful women and men artists and writers, the latter collection, too, emphasized "not the limitations of partnership . . ., but the innovations" by means of which these couples transcended the constraints of gender-based "stereotypes, myths, and images."[14]

The present collection of essays is to varying degrees informed by these earlier studies, and yet different from each. The collection studies only married couples, and hence spouses who experienced the social advantages and disadvantages of patriarchal marriage. The collection, moreover, focuses on couples in which both partners had at least the potential for scientific reciprocity. If some essays probe silent partners, the wives were silenced at least partially by the circumstances of their married lives or by the pens of their husbands or later historians, and this silencing itself becomes an object of study. *Significant Others* included a lesbian literary couple and a gay artistic couple. As lesbian and gay couples are recovered from the scientific past, their scientific relationships can be compared with the cross-gender collaboration that is central to the present study.[15]

The collection offers analytical reconstructions of the shared lives of two dozen scientific couples. Focusing on past rather than contemporary couples, the collection includes only couples of whom both partners are deceased. In order to suggest meaningful comparisons and contrasts across time periods, disciplines, scientific ranks, and nations, these couples are drawn from the nineteenth and twentieth centuries, the "soft" and "hard" sciences, the ranks of the not so distinguished and distinguished, successful and unsuccessful marriages, and Europe and North America. Diverse in many ways, the couples however had predominantly middle-class roots, with just a few partners coming from working-class families.[16]

By its very choice of the creative couple as its unit of analysis, the collection probes not simply scientific relationships between women and men but relationships involving a promise of interlocking intimacy and creativity. The creative collaborative couple occupies joint private and public spaces. The creative couples of the collection crossed and mediated the boundaries between the private and public spheres,[17] intimacy and creativity, and the reproductive and productive social functions that their various societies imposed upon them. Some couples, including Margaret and William Huggins, whose observatory was located in the garden of their home, shared physical space that was both private and public. Other couples, where one or another of the spouses worked in the public domain, regularly imported and exported ideas, manuscripts, and even scientific equipment to and from their residences. Pierre Curie thought of his laboratory as a home. At their collaborative best, couples transformed intimacy into a resource for creativity and vice versa. Some spouses (e.g., Marie Curie)[18] regarded joint intellectual fruits as children, and some couples (Mead-Bateson) made their biological children the subjects of joint scientific inquiry.

Analyzing and, as far as possible, categorizing the work or "professional" relationships that prevailed among scientific spouses, the collection's authors uncovered patterns ranging from relatively egalitarian partnerships (those of the Nobelist couples) through husband-creator/wife-assistant relationships (Clementses) or wife-celebrity/husband-assistant relationships (Berkeleys and Lonsdales) and, finally, relationships where marriage seemed to end the wife's involvement in science (Einsteins and Whitmans). Many of the marriages typified not so much one steady relationship as changing and negotiated relationships. Some authors faced the additional challenge of making historical sense of conflicting interpretations of a couple's relationship (Clementses, Hugheses, and Russells). All authors wrote with the understanding that categorizations are tentative, dependent on presently known sources, which in some cases (Whitmans) are limited and, in others (Youngs), extensive. Too, primary sources can reflect one period of a creative marriage more than another or one spouse's viewpoint more than the other's.[19]

Authors also uncovered strategies, embedded in cross-gender collaboration, by means of which women and men scientists worked around the feminine versus masculine, private versus public, and amateur versus professional boundaries that prevailed even in supposedly objective science. The essays suggest that collaboration—certainly spousal but also familial in a larger sense, involving parents, siblings, children, and even more extended family members—is as important a strand of the history of women in science as professionalization. When the collection's authors describe the "careers" of scientific wives, even those of the early twentieth century, they are as often as not describing "lives of science" that fell short, temporarily or permanently, of professional careers. A few wives lacked appropriate credentials, while paid and progressive appointments, formal titles, and honors eluded many.

The term "collaboration" was left undefined for the collection's authors—its meanings to be fixed in the different contexts of the couples' lives. The result is that in the following seventeen essays "collaboration"[20] ranges from adjunct or "behind-the-scenes" support to formal collaboration that ends in joint publication. Husbands more often than wives benefited from adjunct support (including typing, editing, proofreading, slide and lecture preparation, and entertaining for professional purposes), although a few couples (Blanchards and Hoggs) perhaps sought some degree of equity even in adjunct work. The essays recognize, too, that spousal collaboration, as a process that is both emotional and professional, has as much to do with each spouse's creating a nurturing environment for the other's scientific work as with sharing that work itself. The successful scientific wives and husbands collaborated by constructing together a lifestyle that offered each partner "the time in which to accomplish the work at hand" and by sustaining each other's "achievement drive" in the face of the stresses associated with the practice of science.[21] Spouses (Blanchards, Campbells, Goulds, Pickford-Hutchinson, and Mead-Bateson) encouraged each other under the extreme conditions of fieldwork in a foreign country. On a more daily basis, supportive husbands (Blanchard, Brandegee, Curie, and Lonsdale) helped to design creative strategies to balance family and work or, at a minimum, agreed to their wives' strategies.

Some of the collection's essays argue for spousal intimacy as the basis for a special kind of creativity, which neither spouse alone would have realized or, as in the cases of Mead and Bateson and the Myrdals, perhaps one for which a cross-gender and, indeed, marital relationship was essential. Some couples (Blanchards, Mead-Bateson, Myrdals, and Russells), as couples, created new transdisciplinary spaces that were required for innovative scientific activities. Wives, as perhaps no one else would have been able to have done, led husbands into new scientific fields (Berkeleys and Curies), rejuvenated old studies with new methods (Hugginses), and, in a few cases, added an aesthetic component to their husbands' more traditional scientific work (Comstocks and Goulds). Alva and Gunnar Myrdal's early work, especially, highlights the role of gender in bridging otherwise gender-segregated forms of expertise, via the special medium of marriage or legalized intimacy.

The collection documents, moreover, that scientific marriage more frequently enhanced a woman's prospects for carrying on significant research than for securing credit for that research, especially if collaborative. Although intimacy could foster spousal collaboration, and collaboration could enhance intimacy, dividing credit involved separation more immediately than closeness. Defending the decision to publish some joint research in William Henry Young's name alone (and perhaps unwittingly high-lighting the patriarchal bias of a family's assuming the father's surname), Grace Chisholm Young reminded her sister that wife and husband are one, and their work, like their children, is simply theirs.

The collection's essays are divided into five clusters. The essays of the first cluster study the Nobelist couples (Curies, Joliot-Curies, and Coris), whose intense collaboration resulted in not just significant scientific contributions but also shared publications and rewards (a Nobel Prize in each case) as well as personal satisfaction. In contrast to the Nobelist couples, the scientific relationships of couples of the second cluster began in the subordinate one of wife-student to husband-instructor. Other couples met as near-equals and realized great emotional and professional satisfaction without necessarily engaging in the kind of intense, public, and mutually rewarded collaboration that characterized the Nobelist marriages. These couples, in their stunning variety, are the subjects of the third cluster. Couples of the fourth cluster fell short of not only their collaborative potential but, in most cases, mutual emotional satisfaction as well. Each marriage of this cluster seemed to check the wife's professional involvement in some significant way, and most ended in estrangement or divorce. While treating additional couples (who can be divided according to type among the preceding clusters), the essays of the fifth cluster are essentially comparative studies along specific disciplinary or transdisciplinary lines.

Together, the five clusters of essays argue, perhaps above all, for a thesis of creative adaptability as a key ingredient in successful scientific marriages. A few couples (Curies and Youngs) seemed to lay explicit foundations for scientific partnerships; many wives (Campbell, Clements, Comstock, Hughes, and Lonsdale) became more self-conscious about their shared lives as they lived out roles different from their mothers and many of their contemporaries. But, most of the time, the scientific wives and some of the husbands, too, simply tried to make the best of varying circumstances that repeatedly introduced new challenges to the couples' emotional and professional reciprocity. At one time or another, wives (Cori, Hogg, Hughes, and Lonsdale) moved at the dictates of their husbands' careers, and with each move renegotiated, with more or less success, for the space, instruments, or other support required for their scientific work. A few husbands (Berkeley and Lonsdale) proved especially resilient in adjusting their lives to accommodate wives whose scientific accomplishments had come to overshadow theirs.

Peaks of Collaborative Success: The Nobelist Couples

The fact that Marie and Pierre Curie, Irène and Frédéric Joliot-Curie, and Gerty and Carl Cori first won Nobel Prizes not as individuals but as couples challenges historians to find the most illuminating balance between the study of these creative couples as couples and as pairs of individual scientists—a challenge taken up in the essays of the first cluster. Together, the essays uncover some striking patterns running through the lives and careers of the wives, husbands, and indeed couples. First, all the Nobelist spouses entered marriage on somewhat equal terms. Pierre Curie was eight years older than Marie and, at the time of their meeting, already a well-published scientist, but he received his Ph.D. only months before their marriage in 1895. Irène Curie was older than Frédéric Joliot and more advantageously positioned within French science when they married in 1926, but the two had become acquainted as *préparateurs* at the Radium Institute in Paris two years earlier. Gerty Radnitz and Carl Cori, who married in 1920, had met as medical students. Second, none of the Nobelist wives was exclusively her husband's collaborator, not even in the early years of marriage. Marie Curie and her daughter Irène both engaged in publishable research prior to collaborating with their husbands. Although the Coris published jointly as medical students before publishing independently, the couple was separated in employment and hence in research and publications immediately following completion of their medical studies.

If independent research and publications helped to protect the women Nobel laureates from absorption into the personae of their husband-collaborators, it was still the joint work of these creative couples that sealed the scientific reputations of the husbands and wives alike. In all these cases, one or both partners seem to have been predisposed to spousal collaboration by early exposure to what can be called "familial science"—that is, a scientific practice in which coworkers were family members or others who came close to being regarded as such (see chapter 1). Pierre Curie, for example, first pursued science in the "family firm" run by his father, then in collaboration with his brother, and only later with Marie Curie; Irène Joliot-Curie collaborated with her mother before doing so with her husband. Seeming to imbibe familial science from a distance, young Frédéric Joliot cut pictures of Marie and Pierre Curie out of a magazine and hung them in his makeshift laboratory.[22] It is perhaps significant that Carl Cori, as a boy, shared in some of the boat expeditions that his father ran for the Marine Biological Station in Trieste.

Furthermore, all the Nobelist couples embodied a certain degree of complementarity. As Helena M. Pycior (chapter 1) shows, however, this complementarity, at least for the Curies, was not primarily gender-based. By education and temperament, Marie Curie more than Pierre had been acculturated to the institutionalized science of turn-of-the-century western Europe. The Curies' success was due to a complex balance of their personalities, scientific styles, and, to a certain extent, differing disciplinary commitments to physics and chemistry. Whereas Pierre reflected slowly on scientific problems, shunned competition, and cared little for priority, Marie moved quickly from ideas to experiments to bold, published hypotheses. Bernadette Bensaude-Vincent (chapter 2) alludes to complementarity between the Joliot-Curies, with Irène resembling Pierre Curie and Frédéric, Marie Curie. Mildred Cohn (chapter 3) agrees with Carl Cori's proclamation that his and Gerty's efforts had "been largely complementary, and one without the other would not have gone as far as in combination."

None of the Nobelist spouses wavered from commitment to their scientific partnerships, at least not at crucial times in their shared scientific practice. The Curies rigidly followed what they called the "anti-natural path," according to which they divided their time between science and family almost exclusively. Only after the Joliot-Curies' Nobel Prize and especially as Frédéric pursued his hectic pace through World War II did Irène become increasingly absorbed in family matters, especially the care of the couple's children. Together, the Coris maneuvered again and again around obstacles that scientific administrators placed in the way of Gerty's career and, in particular, their collaborative efforts.

Finally, the essays of the first cluster suggest that the primary historical images of the Curies, Joliot-Curies, and Coris as "scientific couples" need to be qualified. Although the collaboration of the Curies and Joliot-Curies was especially fertile, it was intense for a relatively short period of about four to five years in each case. The Coris, for their part, collaborated with each other for the entire duration of their marriage, but still each had multiple collaborators.

Couples Beginning in Student-Instructor Relationships

The essays of the second cluster (and sections of some of the comparative essays of the fifth cluster)[23] feature couples in which the wife was initially an informal or formal student of her scientist-husband. In the nineteenth century, Elizabeth Gould and Margaret Huggins seem to have come to a serious study of ornithology and astronomy, respectively, largely after marriage and under the inspiration and guidance of their husbands. As the century advanced, opportunities for formal education for women increased as did the likelihood of marriages between students and instructors. Three of the creative couples of this collection—Anna and John Henry Comstock, Elizabeth and Wallace Campbell, and Grace Chisholm and William Henry Young—evolved from formal student-instructor relationships. Analogously, Edith Clements, who originally studied German, began work toward a Ph.D. in ecology "under the stimulation and enthusiasm" of her husband, an already published ecologist and botany instructor.

The stories of the Goulds, Hugginses, Comstocks, and Youngs document the potential for individual development as well as intellectual synergy within couples in which the wife was originally her husband's student. Related but somewhat different, the stories of the Campbells (chapter 16) and the Clementses (chapter 15) are those of formal and informal student-instructor relationships leading to modified intellectual synergy. Each of the latter couples did better science than the husband alone could have, and more science than the wife alone would have been permitted, but the wife always remained a subordinate scientific partner. Still, in the course of their marriages, all six wives grew into larger private scientific roles, if not public ones. Taken together, the essays begin a serious reexamination of the role of the wife-assistant in the scientific past and of the "invisibility" of these women. Whereas Anna Comstock and Grace Chisholm Young managed to establish somewhat enduring scientific reputations, Elizabeth Gould, Margaret Huggins, Elizabeth Campbell, and Edith Clements have had, to varying degrees, to be rescued from oblivion by late-twentieth-century historians. Yet, in their lifetimes, Gould and Huggins were visible spousal collaborators, sometimes coauthors with their husbands. Discussions of their invisibility, therefore, are at least partially discussions of the writing of women's accomplishments out of the history of science. The invisibility of Elizabeth Campbell and Edith Clements raises additional historical issues, including their internalization of gender-segregated roles.

Elizabeth Gould (chapter 4) seems to have brought artistic talent but no scientific training to her marriage in 1829. At John's instigation, she soon became his chief lithographer, providing artistic renderings of birds, based on his sketches and stuffed specimens. But, as Janet Bell Garber documents, the distinction between lithographer and ornithologist began to erode when (as Elizabeth put it) she caught some of her husband's zeal. By the time of their expedition to Australia, she was nearly acculturated to the male-dominated practice of nineteenth-century ornithology, with its heavy reliance on the robbing of nests and the shooting of specimens, and she was drawing her own birds and plants.

If John Gould married an artistic partner early in his life, the astronomer William Huggins (chapter 5) remained a lone observer until his late marriage in 1875 at the age of fifty-one. Using major unpublished sources, Barbara J. Becker carefully recovers Margaret Huggins as William's scientific partner and eventually formal collaborator rather than mere "able assistant."

The three couples whose initial relationships were formally that of student and instructor, as well as the Clementses, faced a more professionalized world. Whereas William Huggins's personal wealth supported the Tulse Hill Observatory, where he and Margaret worked side by side, and the Goulds' successful volumes of ornithological lithographs granted them a measure of economic and scientific independence, couples of later generations were usually beholden to universities for scientific appointments and resources. The Comstocks (chapter 6)—the most professionally successful of the couples beginning as student and instructor—met at Cornell University, one of the first of the major American universities committed to coeducation. Anna was John Henry's pupil in an undergraduate zoology course but then left her college studies and married him in 1878. With John Henry's support, as Pamela M. Henson documents, Anna moved through the roles of university wife and assistant to her entomologist-husband, returning undergraduate student (earning her bachelor's degree seven years after marriage), her husband's collaborator and coauthor, instructor of nature-study at Cornell, independent author, and, just two years before her retirement, full professor.

In contrast, Elizabeth Campbell (chapter 16), who met her husband as a student in one of his mathematics classes and married him in 1892, never assumed, and did not appear to desire, any roles other than university wife and invisible assistant to her husband. Formal titles aside, as Marilyn Bailey Ogilvie establishes, Elizabeth's collaboration with Wallace was essential to the success of what were publicly *his* eclipse expeditions.

Somewhere between the dual-career Comstocks and the two-person, single-career Campbells on the continuum of creative couples of the second cluster are the ecologists Edith and Frederic Clements (chapter 15), who married in 1899. Although the first woman to earn a Ph.D. at Nebraska, Edith seemed never to evolve significantly beyond the role of an "extraordinary helpmate." If present historians are hard-pressed to make sense of her scientific career, Nancy G. Slack implies that it is at least partially because Edith—a victim of gender-based socialization—had a split self-image, seeing herself as a wife who tagged along with her husband-scientist and as one of "two plant ecologists" who worked, traveled, and consulted together.

Grace Chisholm first knew William Henry ("Will") Young as her tutor at Girton College, Cambridge, but theirs was a complicated relationship (chapter 7). Grace, and not Will, received a Ph.D. in mathematics before their marriage in 1896, and she mentored him in research mathematics during their first shared year. Unlike many other scientific couples, the Youngs (as mathematicians) did not require laboratory or observatory facilities, and early in marriage decided to form their own home-based research team. By so intensely linking their personal and mathematical fates, as Sylvia Wiegand shows, they were able to achieve a synergy that led to mathematical results of the first order. But they paid a price for their intellectual freedom: despite their decision to publish the bulk of their joint mathematical results in Will's name alone in order to enhance his employment prospects (a distinctive aspect of their relationship that Wiegand analyzes in some detail), Will remained underemployed for most, if not all, of his career.

The backgrounds of the partners, the circumstances of their common lives, and their personalities played roles in assuring the varying success of these six creative relationships. In three of the marriages marked by some degree of publicly acknowledged scientific success for both spouses (Hugginses, Comstocks, and Youngs), husband and wife seem to have come to marriage with a preexisting interest in a common area of science. Family and socioeconomic backgrounds, as well as personalities, helped to shape less or more egalitarian relationships. William Huggins, the only surviving child of an affluent couple and himself a solitary astronomer for many years before marriage, came to the writing of his first joint paper with Margaret only slowly and under very special circumstances. In contrast, John Henry Comstock, who had worked as a cook for a couple who were surrogate parents to him, willingly shared household chores, scientific work, and credit with Anna.

Wallace Campbell (who was emotionally fragile), Frederic Clements (who had a disdain for "using his hands"), and Will Young (who was unable to completely formulate "his" mathematical proofs) were extraordinarily dependent on their wives not only for the running of the couples' households but also for crucial assistance with their science. The husbands' deficiencies, which may have been exaggerated in order to justify an excessive reliance on a spouse turned professional helpmate (Elizabeth Campbell and Edith Clements) or an often silent partner (Grace Chisholm Young), both enabled and constrained the wives.[24] They justified the wives' participation in science at the same time that they, actually or rhetorically, reduced that participation to the more practical, and what Elizabeth Campbell saw as feminine, aspects of science.

A more traditional version of gender complementarity may have facilitated the evolution of Elizabeth Gould and Anna Comstock from their subordinate roles to those of coworkers with their husbands and, in Comstock's case, beyond. Their scientific abilities aside, the hallmarks of Gould and Comstock were artistic talent and sensibility. As women-artists, they—like Eliza Sullivant (mentioned in chapter 15)—could perceive themselves and be perceived as doing "women's work" in science, and so pose no professional threats to their husbands.

But even when there were no obvious rationalizations of gender complementarity available, a few of the husbands condoned, if not actually fostered, the evolution of their wives beyond subordinate roles. Will Young indirectly explained why this was so, at least in his own case. He made it clear that he had never completely let go of the relationship of instructor to his wife. This relationship evoked emotions of the instructor's authority but also of the instructor's pride in the student's accomplishments, for example, the independent research that Grace did later in marriage. Since Grace had mentored Will during their first year of marriage, the dual spouse-mentor relationship was possibly a major ingredient in their remarkable synergy. The mentor's pride in a protégée's work seems to have informed even those marriages that remained asymmetrical (Clementses), but in these cases perhaps enhanced the husband's self-esteem more than it broadened the wife's professional prospects.

A Spectrum of Mutually Supportive Couples

The essays of the third cluster and parts of chapters 15, 16, and 17 analyze eight creative couples who negotiated mutually supportive marriages that enabled both partners to engage in significant scientific activities and that yet, in only half the cases, included, or came close to including, the kind of intense and public collaboration enjoyed by the Nobelist couples. All eight couples were symmetrical or near-symmetrical. Three of the couples met as graduate students, and four of the remaining five met in other professional settings: Elizabeth and Nathaniel Lord Britton, as members of a botanical society; Kate and Townshend Brandegee, as a paid curator and member of the California Academy of Sciences, respectively; Annie and Walter Maunder, as coworkers at the Royal Observatory at Greenwich; and Edith and Cyril Berkeley, as research students in Sir William Ramsay's laboratory. Only the Myrdals met in a nonprofessional setting, as Gunnar and two fellow students stopped during a summer trip at the farm owned by Alva's family.

In three of the couples—Helen and Frank Hogg, the Brittons, and Kathleen and Thomas Lonsdale—the partners maintained distinct specialties in a shared scientific discipline, and so avoided direct competition. Although both Brandegees worked in plant taxonomy, each concentrated on a different geographical group of plants. Of these mutually supportive couples, only the Maunders, Blanchards, Berkeleys, and Myrdals did their best, or at least some of their best, scientific work in formal collaboration with one another.

Although Helen and Frank Hogg (chapter 8) each held a Ph.D. in astronomy, Helen fell into the common pattern of the late-nineteenth-and-early-twentieth-century scientific wife dependent on her husband and his associates for research support. Upon marriage in 1930, she followed Frank to the Dominion Astrophysical Observatory in western Canada, where he held a formal position and she, with special permission, used the powerful telescope. As Ainley stresses, the Hoggs cooperated in close, collegial ways, but they stopped short of formal collaboration. Helen typed Frank's manuscripts, and the couple spent long nights in the Dominion Observatory as she photographed star clusters and he changed the exposure plates. This professional mutuality, continuing through their years at the University of Toronto, helped them to establish themselves as authorities in different astronomical specialties.

Enjoying what Slack describes as a "companionable and supportive" relationship (without children), Elizabeth and Nathaniel Lord Britton (chapter 15) traveled widely and collaborated on establishing the New York Botanical Garden, for which Elizabeth was a leading fund-raiser. Nathaniel became the first director of the Botanical Garden, and Elizabeth—who before her marriage in 1885 had been a paid instructor at Hunter College—served as an unpaid but influential curator at the Garden. They carved out solid, independent reputations: she for the study of lower plants and he, higher plants.

The Brandegees' relationship (also chapter 15) was more symmetrical, possibly because of the paid curatorship that Kate continued to hold after her marriage in 1889. The couple, too, had no children, and Kate, perhaps more so than any of the other creative wives, refused to constrain herself by accepting prevailing conventions about housekeeping and cooking. Their marriage was egalitarian and collaborative, but not entirely so. Kate eventually resigned her curatorship, and the couple, relying on Townshend's independent income, devoted their lives to botanical exploration (together and separately), research, and writing.

As Ogilvie shows, nepotism and a certain amount of creativity in circumventing institutional restraints combined to sustain the largely collaborative partnership of the Maunders (chapter 16). In 1890, unhappy with his subordinate position at the Royal Observatory at Greenwich and concerned about advancing professionalism in British astronomy, Walter Maunder formed the British Astronomical Association (BAA), which was open to amateurs and women as well as more established men astronomers. As Walter's wife from 1895 on, Annie played a major role in the BAA; and by using the association for institutional support, the couple pursued a number of collaborative studies, publishing their results jointly and independently.

The other couples of the third cluster either had originally separate career paths and then worked their way into significant collaborations or vice versa. Upon their marriage in 1922 Frieda and Frank Blanchard (chapter 9) held doctorates with specialties in plant genetics and herpetology, respectively, as well as independent appointments at the University of Michigan. Although both published in their disciplines, totally separate careers would never have satisfied Frieda. Frieda (like Pierre Curie and Iréne Joliot-Curie) had been raised in a tradition of familial science, and aspired to the intimacy and creativity of spousal collaboration in the fullest sense. Shortly after their marriage, as Sylvia W. McGrath explains, the Blanchards designed a transdisciplinary project on the genetics of the garter snake, which ran for twenty years, on a track parallel to their individual disciplinary commitments.

The Berkeleys (chapter 8) moved toward close collaboration later in marriage than the Blanchards; the Berkeleys' collaboration was more absorbing; and Edith determined their specialty of polychaete taxonomy. As captured by Ainley, the Berkeleys provide a case study of a wife whose scientific career was delayed by marriage (in 1902) and motherhood, but then flourished to such a degree that it largely displaced her husband's original career (in bacteriology). For the final thirty years of their shared life, they collected polychaete specimens together, Edith did most of the dissections and all of the drawings, and Cyril assisted her in the writing of the ensuing papers, which were published jointly with Edith as lead author.

Like the Berkeleys, the Lonsdales (chapter 10) stand as models of the hitherto unheralded scientific couples in which husbands turned from careers of their own to support those of more successful wives. In the first stage of their marriage, which began in 1927, Thomas Lonsdale, a physicist turned applied scientist, was the family's principal breadwinner. During this period, as Maureen M. Julian documents, Kathleen managed to earn some independent income and to do the major research that led to a chair in crystallography at University College, London, in the late 1940s. A little over a decade later Thomas retired early from his career so that he was able to take over the couple's extensive correspondence on behalf of pacifism and prison reform and later to assist Kathleen in her duties as the first woman president of the International Union of Crystallography.

Showing yet again the diversity of the successful scientific couples of the past, Alva and Gunnar Myrdal (chapter 17) collaborated most intensely early in their marriage. Trained in separate areas (Alva in sociology and psychology, Gunnar in economics), the Myrdals married in 1924. As Pnina G. Abir-Am stresses, over the next decade the couple took full advantage of their joint, transdisciplinary spectrum of expertise, which they skillfully integrated in *Crisis in the Population Question*, their coauthored book that helped to define the modern welfare state. If the couple's shared creativity and intimacy declined over the course of marriage, and they later enjoyed distinguished but separate careers, it was partially due to the self-confidence and experience that Alva—who lacked an advanced degree—had gained through the early collaboration.

Couples Devolving from Creative Potential to Dissonance

The couples of the fourth cluster, as well as Dora and Bertrand Russell (chapter 17), Grace Pickford and G. Evelyn Hutchinson (chapter 15), and Margaret Mead and Gregory Bateson (chapter 17), fell short of what seemed to be great potential for mutually supportive and intellectually creative marriages. Four of the marriages (Einsteins, Russells, Pickford-Hutchinson, and Mead-Bateson) ended in divorce; the Whitmans' included a long period of separation and estrangement; the Jacobis' devolved into an emotionally and professionally strained relationship; and the Hugheses' left Helen pondering whether all that she had done had added up to a career. Faced with such lost potential, then, the authors of the essays of this cluster and chapters 15 and 17 try especially to explain what went wrong with these couples.

One problem was some husbands' lack of stable, compatible employment. During the first decade of their marriages, Charles Otis Whitman and Albert Einstein worked to raise themselves from underemployment to full-time academic positions that were commensurate with their interests and talents. Although Everett Hughes found an academic appointment at McGill University as he was finishing his dissertation, he remained an assistant professor there for a decade. Having been stripped of a lectureship in mathematics by Trinity College, Cambridge, because of his public stand against conscription during World War I, Bertrand Russell struggled to support Dora and their children. As the men threw themselves into research, writing, and lecturing under the pressure of establishing their careers and/or the patriarchal pressure of making ends meet, they failed to take the time to cultivate the collaborative potential of their marriages.

During the early years of marriage, neither Charles Whitman, Albert Einstein, nor, for that matter, Bertrand Russell enjoyed the kind of academic or institutional affiliations that permitted Nathaniel Britton, Carl Cori, Pierre Curie, and Frank Hogg, for example, to help further their wives' careers. Perhaps, as John Stachel suggests, another problem was an imbalance in the talents of the spouses. Furthermore, some husbands (Jacobi and Russell), already recognized as leaders in their fields, cast long shadows, which obscured the professional personae of their new wives. In other cases, the reputation of a scientific wife (Mead) or a husband (Einstein) solidified following marriage, and even a spouse who had served as a respected collaborator or sounding board during the couple's early years together became dispensable as the emerging partner's new positions guaranteed wider circles of colleagues, assistants, and downright admirers.

To a much larger extent than was true of most marriages of the other clusters, children seriously complicated these relationships. Mileva Marić and Dora Black were pregnant before marriage. The first of the Hugheses' two daughters arrived a year after Helen received her Ph.D., but then Helen stayed home for five years to care for the girls. Emily Nunn Whitman bore the first of her sons when she was forty-four years old. Illnesses of children (Hans Albert Einstein and Francis Nunn Whitman), with the burden of special care that fell invariably on their mothers, and the premature death of Ernst Jacobi, helped to erode some of these potentially creative marriages.

Unlike the other couples, the Jacobis (chapter 11) enjoyed stable, dual careers from the very beginning of their marriage in 1873. Maintaining their separate medical practices and teaching positions, Mary Putnam and Abraham Jacobi initially cooperated in other professional ways, including her editing a book, *Infant Diet*, taken from his lectures but rewritten in a popular style for young mothers. As interpreted by Joy Harvey, however, the Jacobis' shared life devolved from initial personal and professional reciprocity to estrangement. Among the factors that went into the couple's unmaking were, perhaps, competition from working in the same medical specialty (pediatrics) and, certainly, the death of their son, which struck at the very core of his parents' intersecting personal and medical personae.

Marrying late in life, the American biologists Emily Nunn and Charles Otis Whitman (chapter 12) formed a couple that was creative potentially but not actually. According to Linda Tucker and Christiane Groeben, the once "defiantly independent" Emily remained involved in science for only a few years after her marriage in 1883 and mostly peripherally. As Charles slowly worked himself from underemployment to a professorship at the University of Chicago, she moved into the role of the wife dutifully following her husband-scientist and then the role of wife-mother. Plagued by continual financial difficulties and seemingly bereft of mutual emotional and professional support, the Whitmans' marriage exacted a heavy toll on both spouses and on their biological research.

The relationship shared by Mileva Marić and Albert Einstein (chapter 13), from their meeting in 1896 through their marriage in 1903, separation in 1914 (when Einstein became openly involved with his cousin Elsa Löwenthal), and divorce in 1919, was also one of lost potential. But, unlike the Whitmans, Marić and Einstein shared intimacy and science for perhaps a decade. Their union helped to sustain his creativity, seemingly more than hers, as she was driven from the partnership of their student days into the almost exclusive roles of wife and mother. Arguing that there is insufficient documentation to support the claim that Marić contributed significantly to Einstein's major papers of 1905, John Stachel sketches the disintegration of the Marić-Einstein relationship and begins to explain the reasons behind it. Marić, failing her final examinations at the Swiss Federal Polytechnical School twice, bearing a daughter before marriage and afterward two sons, and encouraged by Einstein to play the role of hausfrau, seemed to lose the drive and confidence that had sustained her through her early scientific studies.

Like the Einsteins' relationship, the Russells' (chapter 17) joined a talented woman with (what Dora herself called) "a dominant genius." Although Bertrand Russell was older and already recognized as brilliant if eccentric at the time of their marriage in 1921, Dora, a graduate of Girton College, seemed eager to hold her own in their creative partnership. As Abir-Am explains, however, Bertrand's efforts to support his family ran counter to the Russells' development as a creative couple and Dora's professional growth. Dora seemed to submerge herself in their collaborative work. But some joint books by the couple appeared in Bertrand's name alone, arguably to enhance sales, and, when off on his lecture tours, Bertrand left Dora behind and burdened with their two young children and the running of the alternative school that was supposed to be among their greatest collaborative ventures.

Whereas divorce took a great toll, economic as well as emotional and professional, on Mileva Marić and Dora Black Russell, the divorce of Grace Pickford and G. Evelyn Hutchinson (chapter 15) may have had the positive effect of freeing both partners from competitive and mutually inhibiting scientific collaboration.[25] Having met as zoology students at Cambridge University, this rather symmetrical couple married in South Africa and collaborated on the ecology of shallow lakes. Zoology, and not ecology per se, was Pickford's first field, Slack reminds readers, and, as a divorced woman without children, Pickford concentrated on her own scientific interests, which led ultimately to a professorship in biology at Yale University. Hutchinson, like Albert Einstein, subsequently married a woman with no scientific interests, and through a very long career at Yale he became the foremost American ecologist.

Margaret Mead—probably the most visible American in the social sciences throughout her long career—and Gregory Bateson (chapter 17) met "in the field" (New Guinea of 1932), as Abir-Am stresses, and married in time to do significant anthropological research in Bali before the outbreak of World War II. The marriage offered them intimacy amidst the isolation of field research as well as the opportunity to pool insights from the different anthropological traditions they represented (resulting in a joint book, *The Balinese Character*). But their "field marriage" did not seem to transfer to other places and times, especially the disruptive period of World War II, when they were physically separated, and the marriage ended in 1950.

Helen and Everett Hughes (chapter 14) were graduate students in sociology at the University of Chicago under the same mentor when they married in 1927. But Helen soon fell into a pattern of mobility according to the dictates of her husband's career and underemployment, and she seemed never to be able to climb back to the kind of career track enjoyed by men graduates of Chicago or even some of the women graduates who remained single. Her many academic positions, including editorships with low wages and subordinate titles, were typical of those reserved for talented university wives at least into the 1960s. Although in many ways the Hugheses' marriage was a happy one, Everett (like other husbands of the fourth cluster) seems to have contributed to Helen's professional marginalization by, for example, publishing *French Canada in Transition* in his name alone although the fieldwork for the influential book was a joint venture. Both self-reflective sociologists, Helen and Everett constructed separate, subjectivist views of their marriage—which Susan Hoecker-Drysdale describes and tries to reconcile.

Those involved in less successful creative marriages than the Hugheses' tended to ignore, if not actively obliterate, the scientific accomplishments and memories of their deceased or divorced spouses. As a widower, Abraham Jacobi wrote an article on women physicians and never mentioned Mary Putnam Jacobi by name or her earlier article on the same topic. When Charles Otis Whitman died, Emily proved at best ambivalent in her attempts to see that his research on pigeons was continued. Although Dora ran the Beacon Hill School for a decade after the Russells' separation, Bertrand remembered the school as a mere mistake. Moreover, wives of marriages that ended in divorce or serious estrangement were routinely written out of autobiographies and even biographies of their husbands (Einstein, Hutchinson, Jacobi, and Whitman).[26]

Comparative Study of Couples along Disciplinary and Transdisciplinary Lines

The present collection, covering two dozen creative couples, permits, indeed challenges, editors, authors, and readers alike to begin the comparative history that will build toward a "subtle, sympathetic understanding"[27] of scientific couples in their great diversity. As the collection's first four clusters demonstrate, useful groupings for comparative studies can be made on the basis of the levels of marital and professional success that couples achieved as well as on the basis of the couples' initial relationships (subordination through reciprocity). Discipline, national affiliation, time period, lifestyle, and commitment to childrearing could also be used as key factors for comparative studies of the couples of this collection.[28]

Case studies of an even more diverse group of couples would permit the consideration of other factors, including race, class, and (for contemporary couples) coresident versus commuter marriage. For example, according to Darlene Clark Hine, many nineteenth-century African-American women physicians—the majority of whom came from socially privileged African-American families—married African-American men physicians, ministers, or educators. Hine's work thus suggests the need for historical studies of the creative marriages of African-American women physicians as well as a comparative study of medical couples of the past along race and class lines.[29]

In the final cluster of essays, three authors compare the lives and creative work of couples in similar fields—including botany and astronomy, two of the disciplines most responsible for women's early presence in science. Slack (chapter 15) explores a "continuum" of types of botanical and ecological marriages, ranging from the husband-creator/wife-executor type (Clementses) through the "not entirely equal partnership" (Brittons) and finally to the egalitarian type (Brandegees). She isolates the wife's education as a key factor of comparison. Kate Brandegee, Elizabeth Britton, and Grace Pickford—each of whom was successful in her own career— were well educated in their fields before marriage. Edith Clements, who was featured above as a second-cluster scientific wife, and Eliza Sullivant, whose career is briefly sketched in chapter 15, had no scientific training prior to marriage, and were largely trained by their husbands. The latter women made excellent helpmates but did not have independent careers. A striking feature of all the marriages covered in chapter 15, including that of Pickford and Hutchinson while it flourished, was happy companionship: even where a husband and wife pursued separate research, their interests were mutual, and they were able to spend much of their time together. As the chapter also shows, despite the fact that women were active earlier in botany and plant ecology than in most other fields, the professional positions open to women botanists, especially the married ones, lagged behind those open to men with similar training right into the mid–twentieth century.

The two astronomical couples, the American Campbells and the English Maunders, whose contemporary lives are explored by Ogilvie in chapter 16, provide considerable contrast. Elizabeth Campbell, like Edith Clements, was a dependent scientific wife, whereas at the time of their meeting Annie Russell Maunder had more university training than Walter Maunder. In each case, however, the marriage had a synergistic effect on the work that was accomplished. Elizabeth Campbell became, after marriage, her husband's able—perhaps indispensable—helpmate. Annie, who (despite her education at Girton College) had occupied a low-paid, uncertain position as a "lady computer" at the Royal Observatory at Greenwich when she met Walter Maunder, became his scientific partner. In further contrasting the Campbells and the Maunders, Ogilvie stresses that the Campbells had three sons and the Maunders no children together, and that the Campbells believed in a version of gender-complementarity that posited different and unequal scientific roles for women and men and the Maunders in egalitarianism.

Abir-Am (chapter 17) discusses three well-known couples who worked at the interface of the social sciences and social policy during the interwar and immediate post–World War II periods. Having chosen the Russells, Myrdals, and Mead and Bateson for her study, she is able to discuss not only collaborative couples in diverse social sciences but also the phenomenon of couples enjoying celebrity status mixed with some notoriety. Exploring details of the couples' relationships from their sexual radicalism through their innovative approaches to parenting, she probes above all the transdiciplinarity that was the hallmark of the couples' joint scientific work. She argues that, precisely as a heterosexual, creative couple with diverse disciplinary backgrounds, the Myrdals were able to draw on a wealth of personal and disciplinary interests in the writing of *Crisis in the Population Question*, the topics of which ranged from quantitative macroeconomics through childcare and fertility problems. Other fruits of cross-gender transdisciplinarity included the Russells' Beacon Hill School and Mead and Bateson's photographic approach to anthropological fieldwork.

Additional Comparisons: Creative Couples from Marginality and Radicalness to Academic Success

Scientific marriages tested the spouses' commitments to rethinking and remaking the institution of marriage. Openness to new forms of marriage was frequently associated with marginality in addition to that based on gender, or a general radicalness characteristic of one or both partners. Quite a few scientific spouses (Berkeleys, Coris, Marie Curie, Einsteins,

Helen Hogg, Helen Hughes, Abraham Jacobi, and Emily Nunn Whitman) were immigrants; some (Gerty Cori, the Einsteins, and Abraham Jacobi) were marginalized as the result of ethnic background; three wives (Huggins, Lonsdale, and Maunder) were born in Ireland but married and worked in England. Social class (Kathleen Lonsdale and Walter Maunder) helped to set still others apart.[30] Marriage between spouses from different countries, ethnic backgrounds, or social classes, by itself, implied a certain degree of social nonconformism. Some couples' experimentation with scientific marriage seems to have been linked to, or supported by, relatively egalitarian religious beliefs, such as Walter Maunder's Wesleyanism and the Quakerism that influenced Anna Comstock and the Lonsdales. The Curies, their daughter Irène, and the Jacobis seem to have been freethinkers; John Henry Comstock cared for no religion; and Emily Nunn Whitman was accused of agnosticism.

Many creative wives and husbands held liberal or even radical political and social views, and a few couples (Jacobis, Joliot-Curies, Lonsdales, and Russells) were political activists. Still, for some of the wives and husbands (Jacobis and Russells), political and social liberalism proved easier than the process of a couple's remaking marriage in mutually agreeable terms. Couples functioned in larger societies, whose gender-based mores were imposed upon them and, to varying degrees, internalized by them. Bertrand Russell's attitudes toward marriage, especially, were in the end more conventional than he would have owned.

At least three of the scientific wives (Hughes, Joliot-Curie, and Whitman) came from families that included mothers or grandmothers with college training or even advanced degrees. This early exposure to role models of accomplished and, in some ways, feminist women seemed to have two effects. It encouraged Joliot-Curie and the others to defy gender-linked restrictions on women in education and the professions, but it also challenged them to carve out identities and lifestyles of their own, in some ways distinct from their mothers'.[31]

It was education, above all, that brought most of the couples together and, in the most basic sense, made professional reciprocity possible. In many ways, the modern history of creative couples begins with the establishment of the women's colleges and the opening of major universities and advanced degrees to women in the late nineteenth and early twentieth centuries. All the scientific wives discussed in the present collection, with the exceptions of Elizabeth Gould and Margaret Huggins, were the beneficiaries of this major historical development. Already in 1913, H. J. Mozans predicted that new educational opportunities for women would result in increased numbers of (what we call) scientific marriages. "Women of the future," he wrote while discussing women's education, "will be more suitable companions for the rapidly increasing number of highly educated men of science. . . . they will not only share in the joys and the sorrows of their life-companions, but . . . they will also have a part in their thoughts, their studies, their labors, their achievements."[32]

Graduates of women's colleges are well represented among the creative wives. Three wives studied at Girton College, which had been established in 1869, and two at Newnham College, established in 1871. Both colleges offered the advantages of a women's college within a major university, Cambridge. Both recruited exceptionally talented women students and offered them superior academic training, encouraged their independence, and nurtured their achievement drive. As Girton College thus prepared students, like Grace Chisholm Young and Annie Maunder, for productive scientific partnerships, it perhaps left Dora Black Russell unprepared for the subordinate role in which Bertrand would eventually cast her. Grace Pickford was educated at Newnham, which Emily Nunn Whitman had also attended for a year and a half; Kathleen Lonsdale studied at Bedford College, the women's college of the University of London.[33] The American Helen Hogg graduated from Mount Holyoke, and Frieda Cobb Blanchard spent three years at Radcliffe before completing her B.A. at the University of Illinois.

The opening of some major universities and advanced degrees to women students in Europe and the United States helped to make possible quite a few other creative marriages. In coeducational institutions, wives and husbands met in the subordinate roles of student and instructor (see the essays of the second cluster) but, more and more, in the roles of student and student. Many of the creative wives had traveled to realize their still limited educational opportunities, and many were, in fact, double, triple, or quadruple pioneers, as the first woman or one of the first women to earn an advanced degree from their university, teach in a particular science department, publish in a particular journal, and so on.

A rarer "first" achieved by some of the scientific wives was a professorship—rare since even the early American women's colleges enforced employment discrimination based on a woman's marital status. These colleges refused to employ married women, and they expected women professors to resign upon marriage.[34] Most coeducational institutions seemed not even to think about hiring women faculty, and, when pressed to consider a scientific wife, found her a subordinate position or cited a more or less specific antinepotism rule that precluded her consideration (Hughes).

The case histories of the Nobelist couples, in particular, attest to the historical futility of using shared rewards, especially employment, as a necessary criterion for the success of a scientific couple. None of the creative wives of this collection, not even the Nobel laureates (with the perhaps qualified exception of Irène Joliot-Curie), totally escaped the "pattern of segregated employment and underrecognition," which Rossiter has documented so well for American women scientists through 1940.[35] This pattern is, in fact, highlighted by the dual biographical approach taken in many of the collection's essays, which permits some detailed comparison of the careers of spouses of comparable education and scientific accomplishment. Marie Curie rose to a professorship at the Sorbonne only upon Pierre's death. Following their Nobel Prize, Irène Joliot-Curie held the same chair that her father and mother had, but Frédéric was called to a chair at the Collège de France, from which he was able to build three laboratories. At least partially because of a gender-reinforced reluctance to lecture to large audiences (shared with Marie Curie and Frieda Cobb Blanchard) and hence to teach, Gerty Cori was underemployed from the beginning through the middle years of her career.

In Gerty Cori's case, and surely that of other creative wives (Hogg and Lonsdale), early underemployment was enabling—freeing the women of teaching and administrative duties. It was also constraining to both husbands and wives. Underemployment as well as unemployment of women kept the burden of breadwinning squarely on the shoulders of the husbands, some of whom (Einstein and Whitman) were themselves forced into early underemployment and at least one of whom (Lonsdale) seemed continually to modulate his scientific interests to changing market conditions. At the same time, underemployment kept the scientific wives economically and geographically dependent.[36] Alone of the wives, Mead followed her own career needs regardless of the features of her three marriages.

Still, seven wives (Comstock, Cori, Curie, Hogg, Joliot-Curie, Lonsdale, and Pickford) rose to the level of full professor, but only Joliot-Curie did it through a largely uninterrupted, progressive career path, comparable to that of the men scientists who were roughly her peers. Besides Nobel Prizes, historical events, national factors, and special professional and personal circumstances seem to have been required to assure the academic success of this select group as well as the more limited success of some of the other scientific wives. Major historical events, such as World Wars I and II and the second feminist movement, opened teaching and research opportunities for some (Berkeley, Hogg, Hughes, and Lonsdale). The French tradition of scientific dynasties helped to establish the Curies and Joliot-Curies academically. Perhaps, too, the colonial status of Canadian science, which lingered into the twentieth century, worked in favor of the Berkeleys. A few of the more successful wives (Berkeley, Britton, Comstock, Curie, and Lonsdale) found their niches in pioneering areas of scientific study, where at least initially competition with men scientists was less severe.[37] Not only Marie Curie but Kate Curran (Brandegee) and Helen Hogg benefited professionally from early widowhood.[38] Significantly, with the exception of the Nobel laureates, the wives obtained full professorships in specialties in which they did not formally collaborate with their present or ex-husbands (and thus in specialties in which they had escaped the Matilda Effect).

An especially telling case of an academically successful wife is Kathleen Lonsdale. She and Irène Joliot-Curie were the only of the collection's wives to train in a largely male-dominated field (physics) and to receive full professorships as married women. In addition to taking advantage of opportunities raised by the rebuilding of British science after World War I and by the new field of crystallography, Lonsdale had a powerful mentor, Sir William H. Bragg, and an unusually supportive husband ("the right husband," in her own words). She benefited from working in a field, crystallography, in which there was already in the 1930s a significant female presence. Moreover, Lonsdale possessed—as the very structure of Julian's essay so well captures—an amazing creativity in making a career out of unplanned moves, inadequate instruments, small grants, and even a haphazard collection of crystal specimens.

From the Creative Couple to Third Parties

Pursuing creative connectedness beyond scientific couples themselves, many of the essays of the collection suggest that third parties, most notably mentors like Bragg and relatives, played key roles in enabling or constraining the emotional and professional reciprocity to which all couples of the collection seem to have aspired. Pierre Curie's family took warmly to Marie Skłodowska, but Albert Einstein's parents actively opposed, and seem to have contributed to the deterioration of, his relationship with Mileva Marić.[39] If children seriously complicated the lives of some of the couples whose relationships deteriorated into estrangement or divorce, other couples seemed to adjust well to family additions at least partially because of considerable assistance from third parties. Seven of the creative couples (Brandegees, Brittons, Clementses, Comstocks, Hugginses, Maunders, and Pickford-Hutchinson) had no children, not always by choice. This circumstance kept the wives of these couples freer to engage in considerable scientific work, which all did. However, most of the creative couples, and this included the majority of the most professionally successful wives, had one child (Berkeleys, Coris, and Mead-Bateson) or two or three children (Blanchards, Campbells, Curies, Einsteins, Hoggs, Hugheses, Jacobis, Joliot-Curies, Lonsdales, Myrdals, Russells, and Whitmans). Mothers of some couples (Blanchards and Campbells) and even a father (Eugène Curie) helped couples to achieve emotional and professional reciprocity by providing childcare, at least on a temporary basis. Such support did more than alleviate practical problems: it evidenced family endorsement of what were largely alternative lifestyles, including the innovative dual-career family, of which many of the scientific couples were pioneers.

Both the Goulds and the Youngs had six surviving children, and here family support combined with a largely home-centered work environment to foster each couple's scientific reciprocity. The Youngs, for example, depended on Will's sisters as well as paid help for assistance with the care of their home and family. Eventually the Youngs' children, too, assumed household responsibilities.[40] Other couples (Brittons, Coris, Hoggs, Jacobis, and Lonsdales) seem to have been primarily or exclusively dependent on paid help, and Helen Hogg and Kathleen Lonsdale received scientific funding that was earmarked to cover the expenses of childcare.

Sisters encouraged some creative wives (Blanchard, Curie, and Hughes) in their pursuit of family and career, and a few brothers interacted with the couples or, more particularly, their married sisters (Gould and Putnam Jacobi) in supportive ways. None of the brothers mentioned in the collection, however, played a more central role in his sister's marriage than the prosperous Lucien Lucius ("L. L.") Nunn. In later years he, not Charles, emerged as Emily's confidant, and he eventually set up a trust fund to assure her a measure of economic independence.

Unlike L. L. Nunn, most of the "intrusive third parties" unearthed in the collection's essays—usually, third parties welcomed by one spouse but perceived as intrusive by the other[41]—were nonrelatives. When he renewed his earlier political collaboration with Abraham Jacobi, Carl Schurz wedged himself between Abraham and Mary. Similarly, there are suggestions that the "Olympia Academy"—an informal circle that gathered around Einstein in Bern to discuss science and philosophy—replaced Marić as his valued sounding board. Bertrand Russell left a key decision about coauthorship with Dora to a third party, his publisher. Albert Einstein, Bertrand Russell, and Dora Russell (the latter two of whom boasted of an "open" marriage that helped to blur distinctions between intrusive and welcomed third parties) found lovers who widened the emotional rifts in their marriages.

Other third parties were welcomed and enabling. Not only Kathleen Lonsdale but Marie Curie and Frieda Cobb Blanchard enjoyed the long-standing mentorship and active support of their major professors, Sir William H. Bragg, Gabriel Lippmann, and Harley Harris Bartlett, respectively.[42] Bragg and Bartlett guided their mentees through dissertations and into research careers and, in addition, helped them to work around the particular constraints imposed on women trying to combine family and science in the early twentieth century. From one perspective, Bartlett participated in, and benefited from, Frieda Cobb Blanchard's exploitation as a woman scientist confined to the secondary occupational role of assistant director of the Botanical Gardens at Michigan, which he directed. But he was also a close friend of the Cobb and Blanchard families, and one of Frieda's professional allies, who was willing to use his influence at crucial times to foster her career and the Blanchards' collaboration. When in her midthirties Frieda had her first child and fully expected to be fired from Michigan, Bartlett sent a crafty petition to the administration and preserved her position.

Other supportive third parties pass in and out of the career histories of the creative couples. Paul Schützenberger granted Marie Curie permission to use space at the Ecole municipale; John Stanley Plaskett allowed Helen

Hogg to share Frank's office and to use the telescope at the Dominion Observatory. Third parties preserved the memories of creative wives (Gould), secured pensions for them following their husbands' deaths (Huggins), and negotiated their successions to their deceased husbands' professorships (Curie). Such gestures attest to the professional advantages that followed the marriages of some women scientists to men who could connect them to the male-dominated scientific community. But associates of creative husbands did not automatically become supporters of their wives; the scene for inclusion had to be set carefully. By the time that Plaskett opened the Dominion Observatory to Helen Hogg, astronomical couples were no longer rarities, and Plaskett's son, a professor at the Harvard Observatory, had firsthand knowledge of some of the other successful couples as well as Helen's work. Helen's self-confidence was also a factor, and she and Frank—showing here a major characteristic of the more successful creative couples—were willing to press her case, when necessary.

The students and research associates of creative couples constitute another important group of third parties affecting and affected by spousal collaboration. A few of the collection's essays (most notably those written by Cohn, a research associate of Gerty and Carl Cori, and Julian, a postdoctoral student of Kathleen Lonsdale) begin to suggest the fruitfulness of analyzing the gender-nuanced relationships between women students or associates and their dual, scientific-couple mentors. Female mentees could benefit from professional and social relationships with creative couples without the gender-based risks entailed in similar relationships between women and unrelated men scientists. At the same time, female mentees and creative wives, like Cohn and Cori, enjoyed a special camaraderie within the largely male-dominated scientific preserves they came to share.

Contemporary Scientific Couples

Until recent decades, few were the women scientists who enjoyed the kind of mentor/mentee relationship of Gerty Cori and Mildred Cohn. Through the early 1970s, underrepresentation of women in science combined with the general historical invisibility of women scientists to rob many women science students and young women scientists of present and past role models. Through the same period women were being told not only that they could not become first-class scientists but also that they could not combine science and family. The invisibility of role models of married women scientists, in particular, left girls and women susceptible to internalization of the latter message.[43]

The present collection fits into the tradition of history of women in science, which has already done so much to correct the invisibility of women scientists as well as to identify some of the institutional barriers to women's participation in science.[44] The collection fits also into the growing movement to address the topic of science and family. Finally coming into their own, women scientists have begun to push discussions of women in science and scientific lifestyles to the forefront. Increasing numbers of senior women scientists have written memoirs or granted interviews that, unlike traditional accounts of men scientists, focus on their family arrangements as well as scientific work.[45] Women scientists have generously participated in sociological studies aimed at elucidating patterns of work and family,[46] and there has been unprecedented coverage of such issues in scientific journals and at scientific meetings.[47] On a larger scale, during the late twentieth century the second wave of feminism and changing socioeconomic conditions have combined in many Western countries to make the dual-career family more socially acceptable than ever before and, at the same time, a unit of scholarly and popular interest.[48]

The present collection adds a timely, historical dimension to these ongoing discussions. As the current evidence suggests, large precentages of contemporary married women scientists are married to men scientists,[49] and scientific marriage can be professionally advantageous in the present as it was in the past. Several studies indicate that scientific wives are well represented in the higher rungs of science. In the late 1980s, for example, Vitalina Koval found that, of the seven leading women scientists of the institute specializing in the social sciences of the USSR Academy of Sciences, one woman was single and never married, two were divorced, and four were members of scientific couples with children.[50] In the United States, scholars have presented data showing that marriage and children do not adversely affect a woman scientist's research productivity, and that women scientists married to men scientists publish more than women married to men in other occupations.[51]

The positive aspects of contemporary scientific marriage highlighted by the statistical studies fit with the anecdotal evidence and impressions offered by some contemporary women scientists. Thus Salome Waelsch— a contemporary geneticist, scientific wife, and collaborator with Carl Cori in his later years (see chapter 3)—has suggested that few husbands who are not scientists would really understand or tolerate a wife's "spend[ing] her nights in the laboratory." It is easier, she implied, for men scientists to realize compatible marriages to nonscientists even though "normal family lives" are difficult for both women and men scientists.[52] Here again, statistics fit general impressions, for only 11 percent of married academic men in social sciences have academic wives.[53]

The positive view of contemporary scientific marriage must be tempered by the realizations that the scientific memoir has traditionally been a genre for the successful and that interviews, statistical studies, and professional presentations have also largely focused on women who succeeded in doing science and, in many cases, combining science and family.[54] Longitudinal studies, such as those tracking whole classes of women science students from their college days through their life cycles, offer one method of addressing the existing imbalance in favor of the quantifiably successful.[55]

Too, scholars across the disciplines are continuing to analyze the socio-professional features of past and contemporary science and to determine how they have affected, and continue to affect, women and men, including those involved in scientific marriages.

While the conclusions on contemporary and, indeed, past scientific couples are thus preliminary, they are largely consistent with one another, with some significant twists resulting from changed historical circumstances. The historical, sociological, and auto/biographical literature suggests, for example, that the Curies' "anti-natural path,"[56] practiced to varying degrees, has been the lifestyle of necessity for many scientific couples from the late nineteenth century through the present. This is the austere lifestyle of which many contemporary married women scientists, particularly those with children, speak,[57] and it is, perhaps, this lifestyle that permits married women scientists to remain so professionally productive. In explaining how married women scientists can publish more than single women scientists, and women scientists with children more than those without children, Cole and Zuckerman have suggested that many contemporary women scientists "reduce their other obligations and activities to the bare minimum and concentrate almost entirely on work and family."[58] That women more than men continue to follow rigid work and family schedules is borne out by Arlie Hochschild's interviews of dual-career couples. Fathers reported having more leisure time for reading and other hobbies than mothers, even in families with considerable third-party childcare.[59]

Recommendations coming from the cross-national study of European women scientists, conducted in the late 1980s, suggest that contemporary women scientists may be paying an even higher professional price for following the "anti-natural path" than women of the past. In the late twentieth century, when science is "bigger" and more international than ever before, a rigidly organized way of life may deprive women scientists of informal occasions leading to "the acquisition of socially important information about opportunities for specialisation . . . , grants, vacancies, fundraising, etc."[60] Scientific marriage, especially that involving spouses within the same specialty, should however help to alleviate the socioprofessional drawbacks of a wife's or husband's rigid schedule, with one spouse's sharing of informal information with the other.

Women scientists, moreover, are increasingly voicing concern about the personal and family costs of a scientific career. If some scientific wives of the past boasted of their "anti-natural" way of life or their active temperaments (Curie and Lonsdale), which permitted them to do science even as they met (what they largely saw as) their family obligations, by the late twentieth century the "second shift" has become a well-recognized expression of gender inequity. Indeed, many contemporary women scientists and science students see the question of science and family from a new perspective. In unprecedented numbers, women are aiming at careers, including those in science,[61] and many are arguing for the option of combining career and family in a less austere, "natural" way. Thus a recent sociological study found a significant difference in attitudes toward science and family between senior and younger women scientists in the United States. As the researchers concluded, many senior women scientists have typically been compelled to compete with men scientists on the "traditional male model" of scientific practice, focusing narrowly on their research and career development and accepting pared-down personal and family lives. Some younger women scientists, women graduate students, and some younger men scientists and students, on the other hand, are calling for a more equitable balance between science and family. As a woman faculty member explained: "The women [graduate students] want to have another life; they want a family, [to] be able to socialize on the weekend. It doesn't have to be like that, but that's what they see; working on the weekends and every night."[62]

Like the creative wives discussed in this collection, contemporary women scientists continue to experiment with ways to accommodate their reproductive roles and biological clocks to the practice of science. The last decades of the twentieth century have brought significant changes in the availability of and social attitudes toward contraception and abortion, which have facilitated women's control over the number and timing of children, and hence increased their chances for adapting family to professional careers.[63] One modern trend, pioneered by a few of the scientific wives of this collection (Blanchard, Cori, and Joliot-Curie), is for a woman scientist to establish a career and then begin a family.[64] Other women scientists do both simultaneously. But still a third group choose underemployment or even unemployment during their childbearing years. These include young women scientists who do not seek, or in some cases decline or leave, full-time, postdoctoral or tenure-track appointments in order to allot more time to their families.[65] The persistence of underemployment, enforced on women scientists of the past and chosen by some in the present, strikingly highlights the century-old difficulty of combing science and family.

As more women entered and remained in the workforce during the late twentieth century, there was also a sharp increase in childcare. Although many of the scientific couples discussed in the present collection used some form of childcare, day care—especially that provided by nonrelatives or outside the home—had remained a questionable and relatively inaccessible option for scientific couples through the early 1970s. Now socially acceptable, American day care is however expensive, of varying quality, and unresponsive to the needs of professional couples whose work extends beyond normal working hours and usually includes travel.[66] Childcare arrangements prevailing in some European countries are more favorable. Contemporary Italian women physicists, for example, have attributed their striking success to many factors, including free day care for the children of all working women in Italy and the continuing Italian tradition of strong extended families, according to which relatives of working mothers (like the parents and sisters of quite a few of the collection's couples) willingly help with childcare, even beyond working hours.[67]

Informed by the pathbreaking gender scholarship of the late twentieth century, discussions of problems ranging from the underrepresentation of women in science through day care for scientists' children have led to the realization that gender equity in science is an important aspect of science policy,[68] and that equity entails not only eliminating harassment and discrimination in schools and the scientific workplace but also creating a " 'family-friendly' environment that does not penalize women and men who have other [nonscientific] responsibilities."[69] In short, the practice of science should be accommodated to the reproductive and other family roles of scientists.

There is modest institutional support for women scientists who experience interrupted careers.[70] Also, contemporary fathers have taken a more active interest in parenting than men of earlier generations, and this means that parental leave, flexible work schedules, job sharing, slowed tenure clocks, and other such institutional policies are now seen as concerns of both husbands and wives. But lest the current situation seem too rosy, feminist science policy analysts need only recall that, in the midst of her recommendations coming from the cross-national study of European women scientists of the late 1980s, Dorothea Gaudart stressed that, while there is an understanding that women, men, and society share responsibility for the upbringing of children, this understanding "has not yet been put into practice."[71]

Increasing opportunities for women in science have both improved and complicated scientific marriage. Greater representation of women at all levels of academia has meant greater availability of role models, including models of married women scientists with children.[72] The attitudes as well as sheer numbers of married women scientists have changed. Whereas Frieda Cobb Blanchard wrote of feeling that she had to apologize to the University of Michigan for having children, Corinne Manogue, a physicist at Oregon State University, recently said that the best thing she had ever done as a role model was walking around her department pregnant.[73]

Many contemporary scientific wives, unlike those of the past, enjoy economic independence, although the salaries of women scientists still lag behind those of their male counterparts.[74] Whereas the wives of the past suffered from employment discrimination and geographic mobility according to the dictates of their husbands' careers, contemporary married women scientists complain also of geographic immobility.[75] Into the 1950s and 1960s, scientific husbands competed for tenured positions at major research universities while many of their wives were limited to teaching positions at nearby colleges or research associate positions in the laboratories of their husbands or the latter's colleagues.[76] Since the 1970s, there

have been more employment opportunities for women scientists, especially for those willing to relocate themselves or their families. Although
geographic immobility and mobility are becoming shared problems for
scientific couples, wives continue to accommodate their careers to their
husbands' careers in greater numbers than vice versa.[77]

Still, the younger generation of contemporary couples has experimented
with various strategies for maximizing the professional opportunities of
both spouses. Egalitarian couples have tried alternating geographic moves
to pursue opportunities for the wife and then the husband, or vice versa.[78]
Others have embraced, or backed into, commuter marriages. Not an
option entertained by most of the creative couples discussed in the present
collection (with the qualified exceptions of the Myrdals, the Youngs, and,
for a brief period, the Coris), commuter marriage has added new obstacles
to the realization of shared intimacy and creativity and a new level of
practical complexity to the lives of scientific couples.[79]

Another solution to mutually satisfying employment for scientific couples rests with the larger society. Universities, colleges, and research institutes are increasingly being pressured to find suitable employment for
their staff's spouses. But double, tenure-track appointments at the same
institution raise once again the issues of nepotism. Research of the late
1970s showed that "previously formal anti-nepotism rules which are no
longer legal [in the United States] continue to operate informally." Department chairpersons admitted to worrying about the effects of a faculty
couple on departmental politics and on faculty evaluations of one another
as well as problems that a couple's marital strains might pose. In addition,
the Equal Employment Opportunity Act of 1972, according to which an
employer is supposed to select the best-qualified candidate for each job,
has complicated attempts in the United States to meet the employment
needs of scientific or other dual-career couples.[80]

Other general social trends, including greater access to and acceptance
of divorce, have changed, and are changing, the profile of scientific couples. As in Grace Pickford's case, divorce itself can be a professionally
liberating experience, especially for a wife-collaborator. Psychologists are
finding that divorced women and men can realize in second marriages
egalitarian relationships that eluded them in first marriages.[81] Changing
socioeconomic conditions have resulted in greater tolerance of couples in
which the wife is the primary breadwinner. This, combined with the
increased number of academic women, has made possible marriages where
the husband begins as a student or junior associate of his wife, or the wife
completes the Ph.D. before her husband, who then follows her during
the early years of marriage.[82] It is to be expected furthermore that the
study of contemporary scientific couples will soon broaden to embrace
unmarried heterosexual, lesbian, and gay couples enjoying intimate unions
that, in the past, were not legitimized by society and, for that reason,
remained largely unpublicized.

"Lows" and "Highs" of Creative Couples: A Synopsis

The many advantages that can come from creative partnership with a spouse notwithstanding, the life of the creative couple has not been for the fainthearted. Some of the scientific wives discussed in the present collection (Curie, Putnam Jacobi, Russell, and Young) are known to have hesitated before accepting proposals of marriage. Contemporary women scientists have been, and continue to be, warned against marriage, sometimes specifically marriage to men scientists. According to Fay Ajzenberg-Selove, Maria Goeppert-Mayer—herself a scientific wife and at that time a future Nobel laureate—told her that, although "it was hard to be a woman physicist, it was nearly impossible to be a married woman physicist."[83] More recently, Jean Langenheim, the past president of the Ecological Society of America, suggested that scientific marriage, with the employment compromises that it can entail, may be a reason for the low academic status of some contemporary women ecologists.[84]

Once married, wives, husbands, and couples as couples have been forced to make hard choices, and compromise professionally and personally. Wives compromised, and still seem to compromise, more: some have abandoned science; others have moved, or (in the contemporary cases) not moved, according to the dictates of their husbands' careers and pulled down second shifts as wives and mothers as well as scientists. The mothers discussed in the present collection (Blanchard, Campbell, Gould, and Myrdal), especially, worried about decisions to take or leave children behind as couples pursued fieldwork or other research opportunities; they and some contemporary scientific mothers have worried also about the very decision to leave so much of the raising of their children to third parties.[85] The more flexible scientific husbands of the past (Berkeley, Blanchard, Curie, Lonsdale, and Young), as well as increasing but perhaps still small numbers of contemporary scientific husbands, have modified research interests or traditional career tracks, at least partially in response to their wives' needs.

By the very fact of being married and professionally active, scientific wives have been constantly testing the limits of variance from the gender-dictated roles of their periods. Husbands as well as wives have sometimes paid a price for this variance. Early in his career Charles Otis Whitman was ruled out as a candidate for a professorship at Bryn Mawr because of Emily's "notoriety." When Frédéric Joliot's marriage to Irène Curie brought him firmly within the inner circle of French radioactivity, it also placed upon him the burden of proving that his scientific talents—and not his Curie connections—earned him a place in that circle. From the nineteenth century through the present, some couples have worried about the Matilda Effect on the creative wife who collaborated with her husband,[86] but Will Young expressed what has been perhaps a common (but usually unrecorded) fear of a husband's losing credit because of spousal collaboration. Scientific wives of the past and contemporary couples have continually suffered from the effects of formal and informal, legal and illegal, antinepotism policies.

Despite the difficulties of their lives, successful scientific couples of the past and present have reveled in the real joy of being intimately and creatively associated with equal or near-equal partners. As we have already seen, Frieda Cobb Blanchard wrote poignantly of the enjoyment of the unusually close companionship, intellectual and otherwise, granted to creative couples. Marie Curie remembered the years of the Curies' greatest collaboration, the difficult ones spent in the shed at the Ecole municipale, as the best years of her life. More recently, Ellen Williams—an American physicist married to astrophysicist Neil Gehrels—enthused: "Having an intellectual companion, somebody to talk to who can share your concerns, is wonderful. It's almost hard for me to imagine you could work out a successful marriage otherwise."[87] Still, no past or contemporary couple seems to have spoken quite so passionately or convincingly of the benefits of scientific marriage, combined with collaboration, as the Youngs. Sensitive to the material advantages that the Young family had lost as a result of Grace and Will's decision to pursue home-based, collaborative research, Will asked if couples who were three times as rich as they had "really as good a time" or remained as young. Grace, for her part, wrote: "We do collaborate in a delightful way." And Will, on another occasion, spoke most eloquently for all those creative couples who realized, and are realizing, the synergy of blended intimacy and creativity: "we are rising together to new heights."

PART I

Peaks of Collaborative Success: The Nobelist Couples

HELENA M. PYCIOR

Pierre Curie and "His Eminent Collaborator M*ᵐᵉ* Curie"

Complementary Partners

Unlike many other nineteenth- and early-twentieth-century women scientists who were married to men scientists, Marie Skłodowska Curie (1867–1934) was consistently recognized as her husband's collaborator. In 1901 the citation for the prix la Caze, which the French Academy of Sciences awarded to Pierre Curie (1859–1906), described her as Pierre's "eminent collaborator."[1] In the same vein, the "Presentation Speech" for the Nobel Prize in physics of 1903, which the Curies shared with Henri Becquerel, offered one of the first formal endorsements of spousal collaboration in science: "The great success of Professor and Madame Curie is the best illustration of the old proverb, . . . union is strength. This makes us look at God's word in an entirely new light: 'It is not good that the man should be alone; I will make him an help meet for him.' "[2]

Despite this early recognition of the importance of the Curies' collaboration, popular and scholarly accounts of the couple have largely failed to analyze them as a collaborative unit. Biographies of Marie Curie have generally, if vaguely, attributed the couple's scientific success to a complementarity of disciplines. Marie, it has been argued, was the chemist, and Pierre the physicist of the research team they constituted.[3] The present chapter goes beyond the latter thesis to rethink the Curies' fertile collaboration. The chapter explores, as background to their collaboration, Pierre's desire for a close partner in science following the loss of his brother, Jacques, as a scientific collaborator. Then the chapter argues that Marie and Pierre Curie's success as a scientific couple was due to a complex complementarity that included, but was not limited to, the partners' different commitments to chemistry and physics.[4]

The Curies' marriage brought together two extremely talented scientists with complementary minds, personalities, and scientific styles. Whereas

Pierre was a slow thinker who framed scientific conclusions soberly and cared little for priority and fame, Marie moved quickly from experiments to bold, published hypotheses. Whereas Pierre was intellectually restless, Marie was intellectually broad but at the same time persistent, capable of immersing herself in the study of radioactivity from 1897 through her death. Pierre's noncompetitiveness and disinterestedness, which may have inhibited his rise to scientific eminence, freed him to collaborate with Marie on equal terms, sharing both work and credit. Marie's decision to study radioactivity, her boldness, desire for recognition, and persistence—combined with the couple's scientific talent—assured not only that she attained scientific prominence but also that Pierre finally became, in the eyes of his contemporaries, a great man of science. Implicitly, the chapter argues for dual biography of creative couples: deeper understanding of Pierre is, in and of itself, deeper understanding of Marie, his complementary partner, and vice versa.

In the period prior to the Curies' marriage, Jacques Curie had been Pierre's primary collaborator, and, in many ways, Marie was Jacques's replacement. To Pierre, collaboration with a family member was a natural way of practicing science, since from his earliest years science had been a family affair.[5] As a youth he, alone or with family members, had gone into the French countryside to collect plant and animal specimens for Eugène Curie, his physician-father; and, according to Marie Curie, his father had first taught him the scientific method of gathering "facts" and interpreting them. In his late teens Pierre had started to collaborate with Jacques at the Sorbonne, where both were laboratory assistants. It was here that the brothers made the transition from domestic collaboration, done in the private sphere and more characteristic of the nineteenth century, to a collaboration that was familial and yet public. In the early 1880s the brothers jointly announced the discovery of piezoelectricity and the invention of the piezoelectric quartz balance. In 1883, however, they separated as Pierre became the director of laboratory work at the Ecole municipale de physique et de chimie industrielle of Paris (a newly established school with a program emphasizing science relevant to industry) and Jacques, the head lecturer in mineralogy at the University of Montpellier. Afterward their scientific collaboration was confined to vacation periods.[6]

Generally, the kind of scientific practice with which Pierre felt comfortable may be called "familial science." This was a scientific practice that, as far as possible, blended science and family. The laboratory was thought of as a home, and, ideally, coworkers were family members or others who came close to being regarded as such. Paul Langevin, one of Pierre's early students who became a distinguished physicist in his own right, wrote of Pierre's "need of living in the laboratory close to those he loved." Marie reported that for years Pierre had sought a laboratory of his own as a "quiet shelter."[7] Students as well as junior collaborators saw Pierre more as a "comrade" than a master; Charles Chéneveau remembered

Pierre's "immense kindness" and ability "to surround his simple collaborator with a great and tender affection."[8]

Behind Pierre's commitment to familial science there lay a faithful devotion to the study of nature. A positivist attachment to science seems to have taken the place of religion in his family. Eugène Curie had raised his sons as freethinkers and imparted to Pierre a strong social conscience. Pierre believed that science was the only sure route to a common human good, since social movements often failed and even successful social reformers could "never be sure of not doing more harm than good."[9] Not merely a part of Pierre's life, science was "the dominant preoccupation" of that life.[10] For him, there was no precedent, no time for a family disjoint from science.

Thus habituated to familial science from youth on, Pierre was ripe for spousal collaboration. Jacques's move to Montpellier left a void in Pierre's life that realistically only a scientific wife could fill. According to Langevin, Pierre foresaw that he would be able to lead a full life only in the company of a wife-collaborator.[11] As has often been pointed out, even before meeting Marie, he had bemoaned the fact that "women of genius" were rare, and worried about the detrimental effects on his science of an unsuitable marriage.[12] Moreover, during his engagement Pierre tried hard to help Marie to understand the relationship of familial collaboration that he and Jacques had shared. In love letters to her, he wrote of the days when the brothers had "lived entirely together." He emphasized that his bond with Jacques had been so close that they had "always arrived at the same opinions about all things, with the result that it was no longer necessary for us to speak in order to understand each other." The brothers had collaborated so well, he added, even though they "differed so entirely in character."[13]

Quite possibly, Pierre stressed how much Jacques was unlike himself because of his early appreciation of fundamental differences between himself and his future bride. It is likely that Marie, too, was already aware of the differences, and Pierre here sought to reassure her that a shared life would work for the couple as it had for the brothers. Indeed, a harmonious balance of similarities and differences seems to have attracted Marie and Pierre to one another. When initially introduced, each found in the other a kindred soul as they discussed science, first, and common social concerns, secondly.[14] Both were children of scholarly families of modest means, were shy, and were free of religious ties. Notwithstanding the importance of these and other similarities, their complementary minds, personalities, and talents certainly contributed to the couple's success as marital and scientific partners. There is strong evidence that emotionally and intellectually Marie was attracted to Pierre as "the dreamer absorbed in his reflections."[15] There is a suggestion that Pierre was drawn to Marie at least partially because of "the good little student" he saw in her.[16] Marie sometimes seemed to stand in awe of a husband who could lose himself in abstract thoughts for long periods. Perhaps Pierre similarly stood in awe of a wife who would set herself projects, from the mundane to the scientifically bold, and, despite obstacles, efficiently complete them.

"Grave and silent, he lived willingly with his thoughts and was unable to tolerate external disturbance." So Marie described Pierre in the preface she wrote for his posthumous *Oeuvres* of 1908.[17] In her later biography of Pierre and *Autobiographical Notes*, she highlighted his contemplative side as among the first characteristics she had noted in her future husband. She recalled observing "the grave and gentle expression of his face, as well as a certain abandon in his attitude, suggesting the dreamer absorbed in his reflections."[18] In later years her favorite portrait of Pierre was one in which from a dark background there protrudes his head, resting on his hand, his eyes gazing into space.[19]

This portrait is a paradigmatic visual depiction of the scientist as thinker, and yet, by all accounts, a telling depiction of Pierre Curie. At times, he simply lost himself in scientific reflection. He seemed compelled to think and rethink his scientific problems over and over again and in the most general contexts. This scientific style seems to have been at least partially a result of the nature of his mind. According to Langevin, Pierre was slow. Not intellectually ungifted, he was unsuited for quick study and quick conclusions. He did nearly everything slowly; he was almost never subject to outbursts of enthusiasm;[20] and so he was usually shielded even from the temptation to act quickly. Marie, too, emphasized Pierre's slowness, but explicitly associated it with his "dreamer's spirit." As she reported, he had received no formal elementary education (possibly because of his family's anticlericalism) and had not attended a lycée, since his parents had understood that his type of mind precluded the rapid mastery of a set curriculum; he "could not . . . [have] become a brilliant pupil in a lycée"; and Pierre himself said many times that he had a "slow mind." Declaring that the latter belief "was not entirely justified," Marie linked Pierre's slowness to his basic need "to concentrate his thought with great intensity upon a certain definite object, in order to obtain a precise result."[21] If Pierre sometimes worried about his "slow mind," Marie saw in his reflective, albeit slow, nature a positive and, yes, attractive difference from herself.

Pierre's "dreamer's spirit," high scientific standards, and positivist philosophy of science generally ruled out a quick and (what others might have seen as) efficient progression from identification of a scientific problem to publication of his results. Once he chose a problem for study, he tried to acquaint himself with all the literature surrounding it, in order to satisfy his compulsion for completeness and exactitude as well as to assure that he credited his predecessors appropriately.[22] Once engaged in a problem, he studied it from all possible angles. As Marie emphasized, he explored every detail of every problem with "the same conscientiousness."[23] This scrupulous scientific style fit his type of mind as well as his positivist philosophy, a synopsis of which he provided in an article of 1902 that he published with Marie. The article's relevant passage—which Langevin and Marie Curie acknowledged as an expression of Pierre's thoughts—explained that there were two basic ways of pursuing the study of phenomena. The scientist could "make very general hypotheses and then advance step by step with the help of experience," or the scientist could "make daring hypotheses in which he specifies the mechanism of phenomena." The latter was the potentially quicker route, since specification of a mechanism suggested which experiments ought to be performed. But a daring hypothesis usually included some error as well as truth, Pierre noted, and so it not only guided scientists but limited them as well. He therefore endorsed the former approach as the "sure but necessarily slow" route to scientific "progress."[24]

He published cautiously: he did not hurry into print; in his publications he shunned speculation; and he generally composed brief and elegant articles centering on experimental results. His caution followed from his high standards and positivism. According to Marie, he looked upon publication as "the logical consecration" of scientific results, and so published only when he was satisfied that he had "a group of facts or ideas clearly understood and bound together." Furthermore, his commitment to preserving an open mind in science led him to avoid "in his publications any assertion and even any presumption that was insufficiently founded."[25] Langevin, for his part, emphasized Pierre's "scrupulous conscientiousness" and overriding concern to avoid defiling science by incomplete comprehension.[26] Since scientific conduct is shaped by personal experience as well as philosophical beliefs, Pierre's caution in publishing perhaps stemmed also from an early experience that had taught him firsthand the pitfalls of publishing on a problem before examining all its major aspects. Indeed, in the early 1880s Pierre and Jacques Curie had seen a third scientist quickly draw a new theoretical conclusion from their pioneering work on piezoelectricity. In their papers of 1880, the brothers had established that the compression of certain crystals caused positive and negative charges on portions of their surfaces. They had thus discovered what came to be called piezoelectricity—or at least one of the two parts of that scientific concept. Within a year Gabriel Lippmann used the theory of thermodynamics to predict the reciprocal phenomenon, that an electric field would strain piezoelectric crystals. Lippmann quickly published his prediction, and Jacques and Pierre were left to do the delicate experiments necessary to verify it.[27]

A noncompetitiveness and disinterestedness, which his supporters believed were extraordinary, seem to have gone hand in hand with Pierre's scientific caution. Trained at home by his parents and then by a private tutor through 1875 (when he received his bachelor's degree), Pierre had been shielded from the schoolboy's conditioning into the competitive academic environment of late-nineteenth-century France. Possibly by nature and certainly by nurture, he shunned competition: in his research he gravitated toward scientific areas of low activity; the few times in his life when he was forced to compete, he did so with stress and poor effects. Langevin and others cited as a telling episode the series of visits that Pierre made in 1902 as part of his candidacy for membership in the French Academy of Sciences. Pierre's scientific friends, including Eleuthère Mascart, had pushed him to apply to the Academy in the hope that membership would improve his prospects for a university position commensurate with his talents. Custom required candidates to visit the academicians, and Pierre did so reluctantly and to no avail: placed in direct competition, he praised his rival, who eventually won the seat over him![28]

As this failed candidacy also shows, Pierre was, in many ways, a scientist who pursued "science for science's sake" rather than personal gain and fame. Not only Marie Curie but Langevin, M. D. Gernez, and others singled him out as a truly "disinterested" researcher.[29] Marie emphasized his comparative lack of concern for "career, . . . success, . . . even honor and glory." This facet of his character, according to her, helped to keep him immune to the temptation to publish prematurely to establish priority. As she reported, he had stated that the quality of a scientific work meant more than the name of its author.[30] Noncompetitiveness and disinterestedness also help to explain his reluctance to stand as a candidate for positions at the more established French institutions of higher education and to pursue honors, in addition to admission to the Academy, that supportive colleagues tried to arrange for him. When, for example, Paul Schützenberger, the director of the Ecole municipale, wanted to nominate him for the Palmes académiques, Pierre replied that he had decided to accept no decorations of any kind.[31]

Pierre himself, Langevin, and Pierre's detractors all commented on this unusual behavior, which—combined with Pierre's lack of formal ties to the Ecole polytechnique and the Ecole normale supérieure (which had trained so many of the other French scientists of the period)—disadvantaged him in the French scientific community. In a prenuptial love letter to Marie, he seemed to explain that the tension between French academic mores and his egalitarian principles wore him down. Referring to the necessity of earning a living, he wrote of "concessions" that everyone was "forced to make to the prejudices" of society. Through the mid-1880s he had tried to make no concessions: "I thought I had to exaggerate defects as well as qualities; I wore only blue shirts like workmen, etc., etc." Implying that by 1894 he had learned to compromise somewhat, he yet described concessions as "painful" and himself as "very old and feel[ing] greatly weakened."[32] His supposed concessions aside, years later Langevin wrote that "high mental probity and . . . sincerity hindered" Pierre from taking steps to court useful and powerful friends. Responding to unspecified misinterpretations of Pierre's conduct, Langevin emphasized Pierre's "detest[ation for] all that touching near or far on snobbism and vanity"[33]— a probable legacy of Eugène Curie, whose own views had been shaped by the revolutionaries of 1848. On the occasion of his election to the French Academy on his second try in 1905, when he had made only the minimal visits, Pierre confided in his close friend Georges Gouy that some believed that he was proud and would not "deign" to ask for votes.[34]

Like his reflective style, noncompetitiveness, and disinterestedness, Pierre's intellectual restlessness proved, to some degree, detrimental to his professional career. Restlessness helped to make him a scientist known for great breadth and, probably not coincidentally, a scientist who published no definitive memoir on any of his major research topics. Marie explained that "his natural curiosity and vivid imagination pushed him to undertakings in very varied directions."[35] Before marriage, his joint study of piezo-electricity with Jacques had been followed by independent research on two distinct physical subjects, symmetry and magnetism.

Even so, by the time of marriage Pierre had displayed exceptional scientific talent. He may have been, first and foremost, a dreamer, but he was a fertile dreamer whose slowly maturing thought sometimes led to fundamental theoretical insights. His research on symmetry, begun around 1883, resulted—albeit over a decade later—in the publication of (what are called today) Curie's laws of symmetry. He was also a dreamer who often converted his thoughts into ingenious and revealing experiments. Langevin praised him as "an admirable experimenter," with a special talent to see the experiment that was needed to elucidate a particular set of phenomena and then to improve on even that experiment.[36] During 1890–1895 Pierre conducted major research on the magnetic properties of substances at different temperatures. His publications on the subject offered "rich . . . experimental results," coupled with "sober" theoretical conclusions,[37] and came to "form the basis of all modern theories of magnetism."[38] He excelled, too, in designing scientific instruments, the only practical activity that he enjoyed. He constantly tinkered with instruments, designed new ones, and shared his designs with his coworkers (but rarely published on them).[39] If, by the time of his meeting Marie Skłodowska in 1894, he was then an accomplished physicist, he was not a well-rewarded or eminent scientist. A slow intellectual pace, noncompetitiveness, disinterestedness, the lack of ties to the elite écoles, and a restlessness that seemed to prevent him from pursuing definitive work in any one area of physics combined to deny Pierre but a handful of scientific patrons, including Schützenberger, Mascart, and Lord Kelvin. In 1894 he was still the director of laboratory work at the Ecole municipale, which could afford him only makeshift research facilities.

✦ ✦ ✦

Marie Curie was primarily a thinker-doer. At least that is how Irène Joliot-Curie, one of Marie's later scientific collaborators, sized up her mother in one short paragraph of her essay, "Marie Curie, ma mère." She stated that her parents were "very different" in character and "complemented one another marvelously." Her father was "an excellent experimenter . . . [and] also a thinker." "The thought of my mother," she wrote in contrast, was "more often directed toward immediate action, even in the scientific domain."[40] This characterization of Marie Curie as a thinker-doer (a person whose thought was "often directed toward immediate action") fits well with her life story and even Pierre's image of Marie as "the good school girl." Unlike Pierre's mind, Marie's was quick. Although Marie conjectured that Pierre could not have done well in a lycée, she had excelled in the gymnasium in Warsaw and again at the Sorbonne.

Marie appreciated Pierre's need for an active, complementary partner. Writing of the "beautiful and close collaboration" between Pierre and Jacques (but probably thinking also of her own collaboration with Pierre), she declared that "the vivacity and energy of Jacques were of precious aid to Pierre, always more easily absorbed by his thoughts." Analyzing a photograph of Pierre's family, she observed that Pierre's head was "resting on his hand in a pose of abstraction and reverie," whereas Jacques's "whole appearance" (including one hand on a hip) was one of "decision."[41] The contrast between Pierre the thinker-dreamer and, on the other hand, Jacques and Marie is of course not simply that of a thinker to a doer. It is rather that of a thinker-dreamer who ran the risk of losing himself in thought with limited action and that of thinker-doers who more readily converted thoughts into action. In the case of Pierre and Marie, the contrast is also that of a thinker-dreamer who reveled in broad reflections on nature and that of a thinker-doer whose "steadfast need for clarity" helped to bring such reflections to fruition.[42] Significantly, there are signs that by 1904 Marie—who was so attuned to visual messages and played such a large role in establishing the public image of Pierre the thinker (with whom some later scholars would contrast her as the plodder)[43]—appreciated herself as a thinker-doer. In a photograph of 1904, for example, she posed with her head resting on one hand, and the other hand on her hip.[44]

Marie and Pierre were complementary not only intellectually but also, to a certain extent, personally. Although she shared Pierre's disinterestedness with respect to material gain,[45] she seems to have been competitive and to have cared deeply about establishing a reputation for herself. As a Polish teenager in a gymnasium in Russian-dominated Warsaw and as a woman at the Sorbonne, she was an outsider and, as such, could not have afforded to remain aloof from the competitive system used to rank students at both institutions. These institutions, as well as her father and other Polish relatives and friends, helped to acculturate her to competition and other social mores of Western science—the very kind of conditioning that the young Pierre had escaped. If, during her trying service as a governess in the late 1880s, she had "lost the hope of ever becoming anybody" and transferred her "ambition" to her brother Joseph and her sister Bronya,[46] she regained her sense of self-worth when she moved to Paris and began her studies at the Sorbonne. Much later Joliot-Curie explained that her mother had shunned influential people not because she was modest but because "she had a very just sense of her worth and did not at all feel honored to meet the titled or ministers."[47] Marie Curie's desire to help Poland and the Polish people also moved her to pursue scientific recognition. She never forgot her native country and seemed to relish opportunities to stand as a symbol of the Polish intellectual spirit, which no foreign oppressor could extinguish.

As a person wanting to be somebody, as a Polish nationalist, as a beginning scientist, and perhaps as a pioneering woman scientist, she cared about publications and priority to an extent and in a way foreign to Pierre.[48] In the preface to Pierre's *Oeuvres*, she seemed alternately to respect and then to argue against her deceased husband's caution in publishing. She explained that "he never allowed himself to be carried away to premature publications intended to establish priority." Then she wrote: "When one spoke to him about questions of this kind he calmly replied: 'What does it matter that I have not published such a work, if another publishes it?' " The "one" just referred to was almost certainly Marie herself, who showed that she was very much aware of the adverse consequences of Pierre's strategy of publishing only mature works. The strategy, she implied, may have impeded the development of science. Pierre had spent a lot of time on certain scientific problems, had obtained interesting results, but had published nothing. Some of these studies had involved unique experiments, which—to the detriment of science—had then never been described in print. Pierre had hesitated even to publish descriptions of the scientific instruments that he had designed. If the latter had found application in laboratories around France, and Marie asserted that they had, it was in spite of his negligence in diffusing them.[49]

Pierre's caution was detrimental to him as a professional. In a rare moment of public candor, Marie stated that his publications, gathered in the *Oeuvres*, were: "very few in number; I will say even too few in number; it is then a result of his method of work." In the same paragraph she mentioned how Pierre had lost priority to other scientists. For example, he had contemplated a complete memoir on piezoelectricity, which he had studied in "a manner as complete as exact." But he had never published such a memoir. Rather Woldemar Voigt did, leaving Pierre to contemplate an even more general and complete memoir on crystallography—which, Marie noted, was still unfinished at the time of his death.[50] In her later biography of Pierre, she noted that his scientific "exposition . . . [was] limited to the strictly necessary" and that he "did not use his gifts as scientist and author in writing extended memoirs or books." He had planned such memoirs, she explained here, but "the difficulties with which he had to struggle during all his working life" had prevented their execution.[51]

Pierre Curie the thinker-dreamer who cared little for "career, . . . success, . . . even honor and glory" and Marie Skłodowska Curie the thinker-doer who wanted to be somebody were, then, different enough, and in just the right ways, to make a go of a scientific marriage that would eventually involve significant collaboration. In her preface to Pierre's *Oeuvres*, Marie (perhaps unconsciously) acknowledged the immediate, scientific importance of the marriage. In one sentence she recorded: "our marriage took place; I obtained authorization to work with him [Pierre] at the École."[52] The authorization came from Schützenberger, who thus proved himself a crucial, supportive third party for the newly married couple. Besides helping Marie to gain laboratory space and connections throughout France's scientific community, the marriage was of perhaps major symbolic value. It permitted Marie to establish herself as a wife-collaborator before she, as a widow, had to face the perhaps more formidable challenge of establishing herself as an independent woman scientist.[53]

Pierre, too, enjoyed profound personal and professional benefits from the marriage. He and Marie embraced what he had called the "anti-natural" existence, that is, a lifestyle in which there was time for science and family only.[54] Assisted by Eugène Curie, who, after his wife's death, moved in with the couple, Marie did her best to shield her husband from distractions (which he could not tolerate), even in the later years as the couple juggled research, at times three teaching positions, two daughters, and fame. She also realized an intimacy with him that seemingly none but a scientific wife could have. When he was called upon to teach a new course at the Ecole municipale, she assisted in preparing the lectures; when he retreated to the laboratory for hours on end, she conducted her research at his side. In the throes of their work on radioactivity, they enjoyed lunches and cups of tea amidst their scientific equipment as well as "happy moments" between experiments when they, lost in "quiet discussion," walked together around their research shed. As there were shared reflective episodes, they were sublime ones, especially the evening visits to the shed, where their luminous radium samples "stirred . . . [them] with ever new emotion and enchantment."[55] Indeed, according to Robert Reid, Marie saw radium as her third child,[56] an offspring of the Curies' interlocking creativity and intimacy. It was also a sibling rival of their two daughters, as witnessed by Eve Curie's (secondhand) account of the couple's late-night visits to their laboratory. Marie would watch over Irène until the toddler fell asleep and, in the very next hour, be in the laboratory, watching over radium with the same "attitude." Pierre would wait impatiently for Marie to come to him from Irène's bedside, but he stood by Marie's side and touched her hair gently as the two gazed on radium.[57]

The relationship with Marie improved Pierre's professional as well as personal life. Trying to capture Marie's special effect on Pierre, Langevin wrote: "she unleashed his power and completed the making of the great man."[58] Langevin, first, and Marie Curie, secondly, wrote of 1895 as a turning point in Pierre's life. It was the year in which he obtained his doctorate, became a professor of physics at the Ecole municipale, and married. Although neither Langevin nor Marie Curie said so, it is possible that Pierre's final steps toward the doctorate were hastened by his upcoming marriage and even taken at Marie's urging. Eugène Curie had earlier become anxious about Pierre's delay in finishing the doctorate, but his questions had been rebuffed by his son's declaration that "nothing pressed him" to complete the degree. At the age of thirty-five, seventeen years after the receipt of his *licence* in physics and four months short of his wedding, Pierre "finally consented" to present a thesis based on his work on magnetism.[59] Late in life, Marie still had vivid, romantic memories of attending the thesis defense in a "little room [which] that day sheltered the exaltation of human thought."[60]

Almost three years after their wedding, the Curies became partners in the study of radioactivity. Marie was fortunate to have Pierre—with his broad knowledge of contemporary physics, experimental talent, skill in designing scientific instruments, and noncompetitive personality—agree to put aside his current research on the growth of crystals in order to join her in the study of radioactivity. In many respects, Pierre, too, was fortunate to have the complementary Marie take him as her collaborator. Although the research on radioactivity would expose the couple to life-threatening radiation and eventually place Pierre in a more competitive situation than he liked, it, above all his other research, would assure his lasting fame.

Marie selected radioactivity as her dissertation topic in late 1897. She had already completed independent research on the magnetic properties of tempered steel, which, according to a committee of the Academy of Sciences, evidenced "great precision . . . [and] the best experimental methods" and which (published under the name "Madame Sklodowska Curie") helped her to begin establishing herself as a scientist.[61] Unlike magnetism, her dissertation topic was a daring choice, since radioactivity, discovered by Becquerel the prior year, was not an established field of study. From the beginning, perhaps, she was inspired by her incipient hypothesis that radioactivity was an atomic phenomenon. Becquerel's early work on the radioactive effects of uranium compounds had suggested as much, although there seems to be no evidence that he explicitly framed the general hypothesis.[62] As reported in her later doctoral thesis, she reasoned: "it was scarcely probable that radio-activity, considered as an atomic property, should belong to a certain kind of matter to the exclusion of all other."[63] Hence she hoped to find radioactive materials other than uranium among the "metals, salts, oxides and minerals" she quickly amassed and tested. Becquerel had detected the presence of "uranium rays" from their action on photographic plates and their ionizing effects on air (the rays make air a weak conductor of electricity). Whereas he had based his study of the intensity of uranium rays on such considerations as the time required to free an electroscope of a charge, Marie added new precision to the study of the radiation by measuring its ionizing effects with Jacques and Pierre's sensitive piezoelectric quartz.[64]

In April 1898 she independently reported to the Academy of Sciences that all compounds of uranium were active; a greater concentration of uranium meant greater activity; and compounds of another element, thorium, were radioactive. In addition: "Two minerals of uranium: pitchblende . . . and chalcolite . . . are much more active than uranium itself. This . . . leads one to believe that these minerals may contain an element much more active than uranium."[65] Thus, before Pierre formally joined her in the study of radioactivity, Marie had framed the atomic hypothesis, noted that there was a second radioactive element (thorium), and predicted the existence of at least one of the hitherto unknown elements that the Curies would eventually "discover." If Pierre's early experience with publication led him to become more cautious, Marie's may have emboldened her to continued speedy publication as she discovered that Gerhard Schmidt had announced the discovery of thorium's radioactivity in February 1898,[66] just two months ahead of her.

Nowhere in her independent paper did she state the atomic hypothesis, which nevertheless was basic to her conclusion that pitchblende and chalcolite contained a new element. In her final sentence, however, she wrote of elements and atomic weights in an early (incorrect) attempt to explain radioactivity.[67] At this time, Marie Curie—perhaps more than any other pioneer in the field—believed and was amassing proof that radioactivity was an atomic property. In later years she would persistently lay claim to formulating the early atomic hypothesis for radioactivity, and Ernest Rutherford would remember her "boldly relying on this hypothesis" in her search for the new elements of polonium and radium.[68] Well might Rutherford have assigned her credit for the hypothesis, since as late as January 1899 he had cautioned that "the apparently very powerful radiation obtained from pitchblende by Curie may be partly due to the very fine state of division of the substance rather than to the presence of a new and powerful radiating substance."[69]

In their first joint paper, sent to the Academy in July 1898, the Curies reported the discovery of polonium and laid claim to "a new method of investigation that we contributed to Becquerel rays."[70] The latter remark suggested concern about scientific priority. With the impetus probably coming from Marie, they now embraced a distinctive publication policy whereby, in papers published alone, together, or with others, they scrupulously claimed credit for and highlighted their individual as well as joint contributions to the study of radioactivity. Their paper on polonium, for example, not only asserted joint credit for the new, quantitative method that used the piezoelectric quartz, but also opened with a summary of Marie Curie's first, independent paper on radioactivity. "In an earlier work," the Curies began, "one of us has shown that . . . [the] activity [of pitchblende and chalcolite] is even greater than that of uranium and thorium, and has expressed the opinion that this effect was due to some other very active substance hidden in small quantity in these minerals."[71] In their paper announcing the discovery of radium, the Curies and Gustave Bémont expressed for the first time Marie's claim to the atomic hypothesis, based on her independent paper of April 1898 (a full citation of which they provided).[72]

This publication policy would prove to be an effective one that enhanced the Curies' reputation as a collaborative couple and also helped to solidify Marie Curie's reputation as an independent scientist.[73] Little noticed has been the symbolic value of the Curies' collaboration for Pierre: his work with Marie provided his first opportunity to serve as lead scientist in a research team. Perhaps because Jacques was a few years older or perhaps because of alphabetical ordering, Jacques's name had preceded Pierre's on joint papers. When Pierre collaborated with Marie, his name came first, probably because of patriarchal custom reinforced with some good reasons: until 1903 Marie did not hold a doctorate; she worked in Pierre's laboratory; and the Curies realistically sought academic advancement for Pierre. Still, the Curies seemed to want to prevent attribution of Marie's scientific contributions to Pierre. Here their publication policy—made possible by Pierre's noncompetitiveness, disinterestedness, and basic fairness—was critical. In highlighting their independent and joint work, the policy enabled Marie to overcome the double jeopardy of the Matthew and Matilda effects, according to which Pierre as the more established and man scientist would quite possibly have received the bulk of credit for their joint work.[74]

Although the papers in which the Curies introduced this publication policy contained enough evidence to make them and Bémont feel safe in arguing that there were two hitherto unknown elements, by 1898 they had isolated neither radium nor polonium. The isolation of radium and analysis of the rays associated with it and polonium became the foci of their continuing collaboration. Radium and polonium were present in pitchblende in small amounts. The Curies could not afford pure pitchblende, and only as a result of an exhausting extraction process did they eventually isolate radium from a residue containing pitchblende. Based on her character analyses of her parents, Joliot-Curie conjectured that Marie committed the couple to "the bold enterprise" of isolating radium before they had any assurances of personnel, money, or equipment. Similarly, George Jaffé, who spent 1904–1905 with the Curies, wrote that Pierre's "disposition made *him* stand aloof when *she* entered upon something like a romantic enterprise: the search for an unknown element."[75]

In 1899 and 1900 Marie and Pierre worked as full scientific partners on the chemical and physical aspects of radioactivity, as attested by Marie's later accounts and the Curies' publications of these years.[76] If during this period their partnership was not one of chemist and physicist, it was one of two scientists with somewhat different talents and styles. Independent, back-to-back papers that Pierre and Marie wrote at the beginning of 1900 provided glimpses of these differences. Here the Curies built on research by other investigators that had shown that the rays associated with radioactivity were deflected by a magnetic field. As Pierre explained in his paper, read to the Academy immediately before Marie's and published just ahead of hers in the *Comptes rendus*, the results of those earlier experiments were inconsistent for polonium. Friedrich Giesel had reported magnetic deflection of the rays coming from polonium that he had prepared, whereas Becquerel had observed no deflection for the Curies' polonium.[77]

In his paper, Pierre described an ingenious experimental setup, reported quantified results, and offered sparse general conclusions. His experiments showed, as the paper's title indicated, that some of the rays emitted by radium were deflected by a magnetic field and others were not. Furthermore: "the deflected rays are the more penetrating." He also "remark[ed] that, for all the samples [he had studied four radium sources], the penetrating, magnetically deflected rays are only a slight [*faible*] part of the total radiation." Based on a study of the activity of a sample of polonium, he concurred with Becquerel's findings: "The composites of polonium that I studied emit only undeflected rays." He then tied together the experiments on radium and polonium: "In the radiation of radium, the undeflected rays . . . appear entirely analogous to the rays of polonium."[78]

This was an important paper, representing a major step toward the eventual sorting of the radiation associated with radioactivity into alpha, beta, and gamma rays. Rutherford had already reported experiments "show[ing] that the uranium radiation is complex, and there are present at least two distinct types of radiation—one that is very readily absorbed, which will be termed for convenience the α [alpha] radiation, and the other of a more penetrative character, which will be termed the β [beta] radiation."[79] Pierre's experiments added a new dimension to the sorting of the rays: beta rays were magnetically deflected, but alpha rays seemed to be unaffected by a magnetic field. By limiting his presentation to the above-quoted general conclusions, Pierre however failed to highlight the significance of his work.

A synthesis, coming from Pierre's experiments and some of her own, was left to Marie. The forthright Marie began her paper by summarizing: "In the preceding note M. Curie showed that the radiation of radium was composed of two very distinct groups: the rays deflected by the magnetic field and the rays undeflected by the magnetic field." In the next paragraph she declared: "A more complete study of the penetration of the two kinds of rays . . . shows that their nature is entirely different and thus confirms the results obtained by the investigation of the effect of the magnetic field." Even before describing her own experimental setup, she reported her major finding: aluminum sheets were used to absorb the alpha rays, and the absorbability of the alpha rays increased with an increase in the thickness of the sheets traversed. Since this law of absorption was unlike that known for other types of radiation (including beta rays), she concluded that alpha rays—which were undeflected and easily absorbed—were "entirely different" from the deflected and penetrating beta rays. She further stated that the law of absorption of the alpha rays "recalls . . . the behavior of a projectile, which loses a part of its momentum in traversing obstacles." That is, she speculated (correctly) that the alpha rays were particulate.[80]

These back-to-back papers, then, testified to two different scientific styles and somewhat different philosophies. Although Pierre had shown that there were deflected and undeflected rays, he did not capitalize on the discovery. This was typical of Pierre, whose scientific style was perhaps best described by Marie when (as cited above) she analyzed his published work on magnetism: "So rich in experimental results," this work stated its theoretical conclusions "in as sober [*sobre*] a manner as possible."[81] Sobriety or moderation was not a characteristic of Marie's scientific style, at least not early in her career. As she had in her first paper on radioactivity predicted the existence of a new element, in her independent paper of January 1900 she pushed Pierre's and her experimental conclusions quite far in a theoretical direction. She synthesized and highlighted the conclusion that there were two "entirely different" kinds of rays emitted by radioactive bodies, which could now be sorted by magnetic deflectability as well as penetrability. She also proved willing to publish what the positivist Pierre probably thought of as a premature hypothesis—that alpha rays were particulate in nature. Possibly her insistence on publishing this hypothesis lay behind the couple's decision to report their work on deflected and undeflected rays independently. Perhaps significantly, in a slightly later, joint discussion of the effects of a magnetic field on the rays associated with radioactivity, they merged their independent papers, reproducing large sections of both papers word-for-word, with one major exception: they did not mention Marie's hypothesis.[82]

Despite this independently published, but related, work on the magnetic deflection of alpha and beta rays and their joint proof that the beta rays carried a negative charge, published later in 1900,[83] the Curies' formal collaboration soon abated. According to the standard bibliographies, they wrote but one joint paper from 1901 on.[84] In this paper, which appeared at the beginning of 1902, they chided Becquerel for hypothesizing that radioactivity was due to "atomic transformation." They urged scientists studying radioactivity to continue to use only "very general hypotheses."[85] With its eschewing of premature hypotheses, the paper seemed to reflect Pierre's philosophy more than Marie's, and, as noted above, she later attributed the paper's key positivist statement to Pierre. It is possible that Marie resented the philosophical concessions she made here. But there seems to be no direct evidence to that effect, and other factors help to explain the couple's diminished collaboration from 1901 on.

By 1901 the Curies had decided to split their research into two parts: the isolation of radium and the study of the rays associated with radioactivity. This was the point at which, to a certain degree, the couple divided into chemist and physicist. Marie "went ahead with the chemical experiments which had as their objective the preparation of pure radium salts," while Pierre "continued the investigations on the properties of radium." According to Marie, they believed that splitting their research was "necessary" for realizing their goal of isolating radium.[86] Complementary talents and personalities determined that Marie would focus on the latter goal. Pierre had never been taken by chemical research but had engaged in it when he thought it necessary.[87] Marie, on the other hand, seems to have embraced chemistry enthusiastically and with real talent. By 1901, moreover, the intellectually restless Pierre had apparently tired of the extraction process. A thinker-doer, a physicist-chemist, and a persistent genius, Marie completed radium's isolation.[88]

The couple's success, which resulted in uneven, gender-dictated rewards and opportunities for each, was also a factor in the abating of their collaboration. In 1900 Pierre was appointed an assistant professor with the Sorbonne's PCN program (established in 1893 to train medical students in basic science),[89] and Marie a physics instructor at the Ecole normale supérieure de jeunes filles at Sèvres, which trained young women for teaching. Although Pierre's new position carried no research laboratory and the Curies continued to study radioactivity at the Ecole municipale, the dynamics of their partnership changed. Pierre, having overcome the academic marginality linked to his position at the Ecole municipale and seeming to enjoy more self-confidence, took on a few new collaborators, including André Debierne and Georges Sagnac. Marie subsequently described the latter as collaborators with the couple,[90] but both men published with Pierre alone from 1900 through his death. A paper of 1900 with Sagnac was Pierre's first joint paper on radioactivity for which Marie was not a coauthor. In 1901 Pierre published a paper with Becquerel and four with Debierne. There thus began a pattern that lasted until Pierre's death (with the exception of the Curies' joint paper of 1902): he published with men-collaborators and sometimes alone; Marie continued to share his laboratory facilities and, for a while, worked closely with Debierne, but she published independently.

By July 1902 Marie, with Debierne's assistance, had finally prepared a radium salt from which she calculated the atomic weight of radium. Announcing the result in her own name,[91] she turned to writing her doctoral thesis. Defended in June 1903, published the same year, and subsequently revised and translated a number of times, the thesis was the only book-length statement on radioactivity that either Curie published while Pierre lived. The thesis (in different editions) was Marie's sole major publication during 1903–1905.[92] When in 1904 Pierre received a chair at the Sorbonne, she became the director of his promised research laboratory, and she more than Pierre threw her energies into establishing the laboratory. In addition, she endured two difficult pregnancies during the period, the first ending in a miscarriage in August 1903 and the second leading to the birth of Eve in December 1904. In her *Autobiographical Notes*, she recorded Eve's birth and added: "I had, of course, to interrupt my work in the laboratory for a while."[93]

From 1901 through 1904 Pierre, working regularly with multiple collaborators, published over twenty papers.[94] While continuing to report valuable experimental results on radioactivity, he and his formal collaborators, especially Debierne, debated the nature of radioactivity with Rutherford and Frederick Soddy. As is well known, the English scientists outlined and aggressively proved the transmutation theory during 1902–1903. Pierre continued to refuse to endorse a specific hypothesis—and so put himself and his collaborators at a disadvantage in what Rutherford saw as the "race" toward a general theory of radioactivity.[95] Yet, in at least two ways, Marie and Pierre Curie helped Rutherford and Soddy to establish the transmutation theory. In 1902, displaying noncompetitiveness and disinterestedness, they lent Rutherford a strong radium source, which enabled him to prove that, contrary to Pierre Curie's earlier results, alpha rays were deflected by a magnetic field and that, as Marie had hypothesized in 1900, they were particulate.[96] Secondly, in March 1903 Pierre and Albert Laborde proved that radium continuously released heat.[97] Despite this and other mounting evidence for transmutation, Pierre—and Marie, who touched on the topic in her thesis of June 1903—remained skeptical of the theory through the end of the year. Perhaps, as Marjorie Malley has suggested, Pierre now dominated Marie, and his obstinacy and "pride" obstructed the couple's acceptance of the theory.[98] Not a competitor by nature or nurture, Pierre took a while to accept defeat gracefully. For the rest of her life Marie said little about Rutherford and Soddy's victory. Rather she repeatedly detailed how she and Pierre had begun "a general scientific movement." When she broached the topic of transmutation, she highlighted the work of Pierre and Laborde as well as her own early suggestion of transmutation as one of a number of possible explanations of radioactivity.[99]

By 1904 Pierre was ill (probably from the effects of radiation), perhaps emotionally drained by the encounter with Rutherford and Soddy, and restless. If Marie was ready to continue to focus on radioactivity (as she did in an independent paper of 1906 and for the rest of her career), he was ambivalent. In 1898 radioactivity—with its limited bibliography and few researchers—had seemed an ideal field for him, one in which he was able to work "in peace and without hurry."[100] Within a few years, however, it had become popular and competitive. The linguistically talented Marie helped to keep Pierre current with the foreign literature,[101] but, still, he "did not adjust well to the feverish output . . . of publications" and "was often tempted. . . . to take refuge in regions of science that were calmer and more favorable to mature reflection."[102] When appointment to a chair at the Sorbonne in 1904 seemed to rejuvenate him, he devoted his energy to the preparation of a series of lectures on crystallography and planned a memoir on the subject, encompassing piezoelectricity and symmetry.[103] In 1905 he published nothing, and in 1906 one, posthumous article, in which he and Laborde reported on the radioactive gases coming from mineral waters.[104]

Pierre died in an accident in April 1906. At that time his memoir on crystallography was still unfinished. For some years Marie—who did not want to let go of the couple's collaboration and perhaps appreciated that its symbolic value extended even beyond Pierre's death—spoke of completing the memoir as well as carrying on the couple's joint work on radioactivity.[105] She never did the former, but she did take other steps to conserve Pierre's memory,[106] including writing the preface to his *Oeuvres* and a biography.

At the time of his death, Pierre was a Nobel laureate, a titled professor at the Sorbonne, and a member of the Academy of Sciences. It is certainly possible to argue, following Langevin, that marriage and radioactivity made Pierre a great scientist. Prior to marriage his substantial contributions to physics, including his laws of symmetry and fundamental work on magnetism, had been insufficient to raise the noncompetitive, disinterested, and restless Pierre to scientific eminence. According to Langevin, Marie consciously desired that Pierre be great.[107] Not the only creative woman to embrace such a wifely aspiration, she had the scientific talent, persistence, bold style, and—as importantly—gifted but noncompetitive husband, with a special attachment to familial science, to permit her to march hand in hand with him to scientific distinction. Both Curies also enjoyed considerable support from key third parties, including Eugène Curie, Schützenberger, and Debierne, who dared to work with the woman-scientist Marie as well as Pierre. In 1903 the complementary Curies shared the Nobel Prize in physics, and in 1904 with Pierre's chair at the Sorbonne came Marie's paid appointment as the director of his laboratory. They were as near-equals as French academia would permit in the early twentieth century, but not equal enough for the couple to codirect the French effort on radioactivity. During his final years Pierre led a circle of formal collaborators; "his eminent collaborator," Marie, bore the couple's second daughter, established his laboratory, taught at Sèvres, and published independently. Despite her Nobel Prize, Marie's professorship at the Sorbonne came only upon Pierre's death, and even then required the special intervention of the couple's colleagues.

◆ 2 ◆

BERNADETTE BENSAUDE-VINCENT

Star Scientists in a
Nobelist Family
Irène and Frédéric Joliot-Curie

For most people, the names Irène and Frédéric Joliot-Curie are forever intertwined. The image of the couple has been fixed in our memory by a famous picture of Irène and Frédéric, in white chemists' coats, at work in their laboratory in the Radium Institute in the 1930s. This picture, signed by a well-known professional photographer, André Kertész, was clearly intended to reproduce another famous picture, that of Pierre and Marie Curie in their laboratory at the Ecole municipale de physique et de chimie industrielle in the 1890s. In fact, the photograph of the Joliot-Curies was taken in 1932, in the earlier years of their collaborative work.[1] Thus, even before they made any significant discovery, Irène and Frédéric were already recognized as a couple of potential scientist stars.

Irène Curie (1897–1956) was her mother's research associate at the Radium Institute in Paris. There she met her husband, Frédéric Joliot (1900–1958), when he came to conduct Ph.D. research under the guidance of Marie Curie in 1925. Frédéric and Irène Joliot-Curie started a fruitful collaboration in the early 1930s, culminating in the joint discovery of artificial radioactivity in 1934, for which they were awarded the Nobel Prize for chemistry in 1935.

However, the common coupling of both partners' names with both family names as Joliot-Curie obscures the fact that Irène and Frédéric coauthored only thirty-three papers of their total publications. The number of papers each published alone shows a remarkable quantitative parity: Irène alone or with other collaborators published forty-two articles, and Frédéric alone or with other collaborators published forty-four.[2] Moreover their collaboration did not last more than four or five years (1930–1935). Following the award of the Nobel Prize late in 1935, their work took divergent paths: Irène Joliot-Curie, like Marie Curie before her, was ap-

57

pointed Professor at the Sorbonne. She retained her laboratory at the Radium Institute, where she pursued the research program initiated by her parents. Frédéric Joliot was appointed Professor at the Collège de France. He began a second research program on nuclear reactions that eventually led to his becoming the first high commissioner of the Atomic Energy Commission (Commissariat à l'énergie atomique, hereafter CEA), upon its foundation in October 1945. His brilliant career, energetically conducted amidst the tensions and competition surrounding World War II, pioneered the general trend within the French scientific community from traditional individualistic research to large research teams in laboratories.

Therefore, it seems important to revise the stereotype of the Joliot-Curies as star scientists. Considering Irène Curie and Frédéric Joliot as distinct figures, this paper will stress the contrast of their characters. It will focus on Irène's career because hers is a unique case in the history of science: the daughter of a famous Nobelist scientific couple, Marie and Pierre Curie, she was herself a partner in a Nobelist couple. Her life and innovative work are usually described as a brilliant anomaly in the mostly male, conservative scientific community of French physical scientists, a community clinging long to a tradition of experimental physics, while ignoring some of the more recent developments in atomic physics in the interwar period.[3]

Without minimizing Irène Joliot-Curie's outstanding achievements, I would like to suggest that, when placed within the milieu of her family and the broader historical and social context of French science, the life and work of Irène Curie seem not only less anomalous but even, to a certain extent, conformist.

The first section of this essay (Childhood and Family Context), covering Irène Curie's childhood, emphasizes the importance of her family as a milieu that prompted her choice of career, especially the importance of her widowed mother as a role model. The second part (The Curie Scientific Style), devoted to her scientific achievements, considers the kind of collaboration she had with her husband and their position within the international physics community. The third part (Public and Private Life), which is concerned with Irène Curie's public life, highlights her success in designing her own style as a woman scientist.

Childhood and Family Context

Throughout the nineteenth century, when France went through many political upheavals and changes of government, science emerged as one element of stability. Under the supreme control of the French Academy of Sciences, scientific activity revealed a remarkable continuity, exemplified by scientific dynasties such as the Jussieus, a family of natural scientists—Antoine, Bernard, Antoine-Laurent, and Adrien—whose destiny was for nearly a century closely intertwined with the history of an institution, the Muséum d'histoire naturelle, formerly known as the Jardin du roi.[4] The physics section of the Academy of Sciences continuously had a Becquerel among its members from its creation in 1828 until 1908. The Becquerel dynasty covered three successive generations: Antoine (1788–1878), who did important work on electric batteries; his son Edmond (1820–1891), who joined the Academy in 1863 after doing research on magnetism, electricity, and luminescence; and the grandson Henri, discoverer of uranium rays in 1896 and cowinner of the Nobel Prize in 1903 with Pierre and Marie Curie.

Thus, the end of one scientific dynasty that influenced nineteenth-century science was the beginning of another illustrious scientific dynasty that deeply influenced twentieth-century science. The Curie family, in turn, had three successive generations who joined the Academy of Sciences: Pierre Curie in 1905, Frédéric Joliot-Curie in 1942, and Pierre Joliot in 1982. Possibly a fourth generation is on its way, since Alain Joliot, who recently finished his Ph.D., seems to have made a promising start in his career in the life sciences.

Yet, in contrast to former scientific dynasties, the Curie-Joliot family was a dynasty of couples: over three generations, the women did not content themselves with the traditional female role of a scientist's wife, or with taking charge of the social activities related to the operation of scientific patronage.[5] Marie Curie (1867–1934), Irène Joliot-Curie, and Hélène Langevin-Joliot (born 1927) were at least as active as their husbands in scientific research. It must be stressed, however, that none of the female spouses, two of whom were Nobelists, was elected to the male-dominated Academy of Sciences.

Both Irène and Hélène chose to work in the same field as their parents: radioactivity for Irène, who, significantly, took polonium, an element discovered by her mother, as the topic of her early research. (Her doctorate was on particles emitted by polonium.) Moreover, Irène stayed at the Radium Institute, founded by her mother, during her entire career. Her daughter, Hélène Langevin, now works in nuclear physics at the University of Orsay, a department founded by Frédéric Joliot-Curie in the 1950s. (Her brother, Pierre Joliot, chose a related but distinct field, biophysics.)

Both Irène and Hélène married within the "tribe" of the scientific elite, thus consolidating the French tradition of "scientific endogamy," since intermarriage between scientists' families was already frequent in early-nineteenth-century Europe.[6] Irène met her husband at the Radium Institute among her mother's students. He had the same background as her father, having been trained at the Ecole municipale de physique et de chimie industrielle, where Pierre taught. Hélène met her husband inside the inner circle of scientists surrounding the Curie family; she married Michel Langevin, grandson of Paul Langevin (1872–1946), the French physicist who had a love affair with her grandmother Marie Curie in 1911.[7] Like her father and grandfather, Hélène's husband was a physicist.

The Curie family lived in a scientifically stimulating milieu. Pierre and Marie Curie were close friends of Jean Perrin (1870–1944) and Paul Langevin, two leading figures of French physics; Emile Borel (1871–1957), a leading mathematician; and Charles Seignobos, a famous historian.[8] All these families used to spend their summer vacations swimming, sailing, and biking at l'Arcouest, a little fishing port on the Brittany coast, which they jokingly nicknamed "Sorbonne plage." The children not only shared their summer activities year after year but also studied together in Paris. Instead of attending the state primary school, they were taught at home by their parents. Marie Curie was in charge of teaching physics, Paul Langevin of mathematics, Jean Perrin of chemistry, and Henriette Perrin of history and geography. When she was about twelve years old, Irène went to a private school, the Collège Sévigné. She passed the *baccalauréat* exams in July 1914, at the age of seventeen, and became a student of chemistry at the Sorbonne. Brought up apart from public schools, Irène never had the apprenticeship of sociability normally provided by the schoolyard. This special education may have determined a striking, dominant feature in her adult life. Irène Joliot-Curie was rather shy and not very talkative. Some collaborators said she often behaved in a clumsy and tactless way and was absolutely unfit for public relations.

Irène never left the niche of the Curie family to venture into an unknown world. It was so obvious to her that she had to follow the same path as her mother that she never contemplated the possibility of choosing a different way of life. Looking back to her childhood, it would not be too difficult for a psychoanalyst to provide some explanation in terms of, say, a "Jocasta complex."[9]

Irène Curie was born in Paris on September 12, 1897, the same year her mother started her doctoral research on elements emitting uranium-like radiations. As a child she was initially cared for by a nurse and later by her grandfather, Dr. Eugène Curie, who came to live in the house of his son Pierre when he became a widower. Pierre and Marie's scientific activity reached its peak during Irène's early childhood. In 1898 they discovered radium, and the following years were devoted to painstaking and exhausting experimental work for testing the elementary nature of radium and polonium.[10] In 1903 the Nobel Prize brought a flood of public attention, with many journalists and visitors at their home. Though tenderly loved by her grandfather, the young Irène could not help but feel lonely, often abandoned by her parents, who spent most of their time with their other child: radium. Jealous of radium, then jealous of her younger and prettier sister, Eve (born in 1904), Irène came to demand exclusive love from her mother, apparently trying to capture her mother's attention through what is now termed anorexia.[11]

However, after Pierre Curie's accidental death in 1906, Marie Curie's depression led to her seeking increasing affection from her eldest daughter, Irène, eventually establishing a mother and daughter collaborative team during World War I for the mobile X-ray units, and later in the laboratory. Following Dr. Eugène Curie's death in 1910, the Curie family consisted of three women. The two sisters chose contrasting roles. Whereas Irène was inclined to emulate her mother—she was fond of mathematics, swimming, sailing, and trekking, and did not care for pretty clothing—Eve, who was seven years younger than Irène and who felt excluded from the learned conversations between Marie and Irène, chose an alternative model. She became a charming, attractive, talkative child, fond of the piano, who developed a different relationship with her mother, one of mutual protection. When Marie became increasingly blind and fragile in the 1920s, it was Eve who took care of her, in a motherly way.

In contrast, as Irène matured, she gradually took the place of the missing Pierre Curie in her mother's life as a scientific partner and companion. Though she was only seventeen years old when World War I broke out, she assisted Marie Curie on the northern front with the mobile X-ray units and served as a military nurse until the war's end.

The Curie Scientific Style

In 1918, shortly after the demobilization, Irène Curie entered the Radium Institute, founded and managed by her mother, as a *préparateur*. In France, this was a regular position for junior scientists preparing a doctorate with a senior scientist as a patron. Frédéric Joliot occupied the same position when he came to the Radium Institute in 1925, on the recommendation of Paul Langevin. Though she started in the modest position of a beginner in the laboratory, Irène soon became Marie Curie's favorite assistant. She devoted herself to investigating the fluctuations of alpha rays emitted by polonium and gamma rays emitted by radium, two elements discovered by her parents. Though Irène would later discover new radioactive isotopes as well as new radioactive phenomena, the feeling of continuity in the family circle was so strong that when Eve Curie published an article about Irène in 1936, she described the discovery that prompted her sister's Nobel Prize in 1935 as the "discovery of artificial radium."[12]

While reproducing Marie Curie's model of collaborating with one's husband, Irène's career was worlds removed from the difficult conditions encountered by her mother. She never had to work in a poor shelter or live in a miserable student room. Moreover, Irène started her career with a salary of Fr 20,000 given by the government to her powerful patron, her mother. In 1921 she accompanied her mother on her triumphal visit to the United States, traveling in luxurious conditions and being introduced to influential scientific circles. When she married Frédéric Joliot, one of her mother's top students, in 1926, Irène also received an apartment from her mother.

Moreover, Irène worked in a well-funded, prestigious institution that attracted scientists from all over the world.[13] She entered her career with an already famous last name. As early as 1925, when Irène defended her thesis, a crowd of journalists gathered in front of the Sorbonne.[14] She was perceived to be a future star scientist and a potential Nobel Prize winner, even before she had done any work on her own. Irène, unlike Marie, never had to fight for recognition.

Her comfortable position in her early career, together with the special education she received as a child, made her independent of public opinion. Her self-confidence was so deeply rooted in the family niche that she is said to have been absolutely indifferent to her public image and to all marks of social prestige.

In her "predetermined" career, the main source of surprise came from the side of Frédéric Joliot. Though he belonged to the same little world of the Radium Institute and was also fascinated by Marie Curie, who supervised his early research, his cheerful character contrasted with Irène's. He was as lively, talkative, and charming as she was shy and lofty. He had a remarkable capacity for fantasy and was always seeking to captivate people around him. At first glance, Irène and Frédéric made as odd a couple as Molière's Alceste and Célimène, with a reversal of the male and female roles. Irène could evoke the misanthropic Alceste and Frédéric the charming Célimène. Frédéric also showed great political concern, stimulated by his first "patron" at the Ecole municipale, Paul Langevin. Having broken the excessively close relationship between mother and daughter, Frédéric provided Irène with an equilibrium she could not reach in her own family. There are many signs that she was happier as a wife. When Irène became a mother, she felt much more attached to her children and presumably more dependent on them than Marie Curie had been with regard to her own daughters, until later in life. Irène's closer relationship to her children can be viewed as a feature of her own, one she did not draw from her mother's model. Conceivably, the only domain in which Irène might have pretended to improve upon her brilliant scientist-mother was in her own everyday life as a mother.

The close collaboration between Irène and her husband did not start before 1929, or three years after their marriage, and was aimed at investigating the alpha rays emitted by polonium. To a certain extent, as Noelle Loriot rightly emphasized,[15] the Joliot-Curie couple reproduced the complementarity between Pierre and Marie Curie: Frédéric, much like Pierre Curie, who had built a highly sensitive electrometer that allowed Marie to test very weak radiations, proved an expert in the design of new scientific instruments.[16] In 1931 Frédéric Joliot-Curie built an improved Wilson chamber, an apparatus for visualizing trajectories of ionized particles through a gas saturated with water vapor.

In the early years of the Joliot-Curie collaboration, Irène, scientifically somewhat older and more expert, was undoubtedly the leader: she had already accumulated a great deal of Radium Institute experimental skill. In their collaborative work, Frédéric's clever and hard work on the electrolysis of polonium conducted for his Ph.D research enabled him to define precisely the chemical properties of polonium and to assign it a place after tellurium in the periodic table of chemical elements. He thus became an expert in preparing very thin metallic sheets, which proved very useful in further research on the interaction between matter and radiation.

Radiochemistry required not only the painstaking manual tasks of concentrating, purifying, and crystallizing the active substance, exemplified in Marie Curie's early research on radium, but also a specific skill for making fast and accurate physical measurements in experiments on the interaction between rays and particles. To grow up in the Radium Institute was undeniably a great advantage for Irène, because it provided her with a great deal of tacit knowledge. Over these early years, the Joliot-Curie couple appeared to be a cross-gender mirror of the earlier couple of Pierre and Marie. With her seniority and expertise, Irène acted like Pierre, while the latecomer Frédéric, ambitious as he was, played the role of Marie.

Like Marie and Pierre Curie, Irène and Frédéric Joliot-Curie never acted as a single person. Nevertheless, whatever their contrasting psychological dispositions, they shared a common formative scientific culture, the local subculture of the Radium Institute. In contrast with other leading laboratories studying radioactivity, such as the Cavendish Laboratory in Cambridge, Enrico Fermi's laboratory in Rome, and the laboratory of Otto Hahn and Lise Meitner at the Kaiser Wilhelm Institute in Berlin, the Radium Institute of the 1920s offered a subculture that can be characterized as chemistry oriented. In all these laboratories, scientists performed similar experiments, bombarding various elements with alpha particles, then detecting and identifying the rays or particles emitted. They used different methods, including the scintillation technique, used by Ernest Rutherford after 1908, or the droplet technique, initially designed by Robert Millikan. The latter was displaced by the Wilson chamber and the Geiger-Müeller counter in the 1930s.

However, at the Cavendish, the main purpose of Ernest Rutherford's and James Chadwick's experiments was to elaborate a theory of the atomic nucleus. In Berlin, at the Kaiser Wilhelm Institute of Chemistry, headed by Otto Hahn, Lise Meitner's work was aimed at clarifying relationships between alpha and beta rays in order to shape a nuclear theory. Working in close relationship with Max Planck and quantum physicists, Meitner did her best to apply quantum theory to nuclei, and in some cases— especially the spectrum of primary electrons—she disagreed with Chadwick's conclusions. Fermi's laboratory in Rome and the Radium Institute in Paris were more concerned with the purification and identification of radioelements than with the theoretical explanation of radioactive phenomena in terms of atomic structure.[17] Physical measurements were one important aspect of the daily work but were used as means for discovering new chemical elements. The significance of the local culture of the Radium Institute can be seen in both the successes and missed opportunities of the Joliot-Curies.

Though abstention from whiggism is the prime commandment of the historian of science, taking a retrospective view of a period is sometimes convenient to emphasize contrasts in scientific styles. Among the above-mentioned leading teams working on radioactivity in the 1930s, the Joliot-Curie couple probably had the highest potential for experimental tacit knowledge. Their ability culminated in the difficult and hazardous art of preparing strong sources of polonium through electrolysis or thermal volatilization. But they lacked the theoretical background that proved to be indispensable for the identification of particles or rays emitted during experiments. Neither Irène nor Frédéric was ready to grasp the theoretical interpretation of the experimental data they had described. Thus the Joliot-Curies paved the way for but missed the discovery of the neutron.

In 1930 two German physicists, Walter Bothe and Herbert Becker, described a new penetrating radiation resulting from the bombardment of a light element such as lithium with alpha rays emitted by polonium, which they identified as gamma or electromagnetic rays. Irène and Frédéric studied the effects of the Bothe and Becker radiation penetrating through matter with a Wilson chamber, the window of which had been covered by a very thin sheet of aluminum. Irène and Frédéric observed a striking phenomenon: when paraffin wax, a substance rich in hydrogen, was exposed to the penetrating radiation, protons (i.e., hydrogen nuclei) were emitted with a high velocity. They concluded that, unlike gamma rays, the penetrating radiation was able to eject nuclei of light atoms. Chadwick (1891–1974), who reproduced the Joliot-Curies' experiments in the Cavendish Laboratory, identified a new particle, the neutron, in the penetrating radiation, with a mass nearly equal to the mass of the proton but which was not visible in the Wilson chamber because it was electrically uncharged.

Far from being dispirited, the Joliot-Curies turned to experimental studies of neutrons and were able to determine their mass and their velocity at emission. They proved that the penetrating radiation contained not only neutrons but also gamma rays capable of creating light particles with nearly the same mass as electrons but running back toward the emitting source in a magnetic field. They concluded that they had factual evidence for the positive electrons, or positrons, recently predicted by Paul Dirac on a theoretical basis and just discovered in cosmic rays. In October 1933 the Joliot-Curies reported their results at the Seventh Solvay Conference in Brussels and ventured a bold explanation: the emission of a positive electron, Irène declared, was the result of an induced transmutation related to the creation of a neutrino. Their paper was immediately criticized by Meitner, who raised serious objections about the Joliot-Curies' experimental results, objections that left them disarmed.

Back from Brussels, Irène and Frédéric resumed their experiments, and on January 11, 1934, as they were bombarding a thin aluminum sheet placed over the window of a Wilson chamber with alpha particles, they were surprised to observe that the emission of positive electrons did not stop when the removal of the source stopped the neutrons' flow. The emission of positive electrons lasted a few minutes and decreased in a way suggesting a radioactive phenomenon. Their experimental skill in radiochemistry was manifest in the fact that they were able to identify an isotope of phosphorus, with a half-life of only three minutes and fifteen seconds, generated by the transmutation of aluminum atoms.

It was "the first atomic nucleus created by man," as Frédéric Joliot announced proudly before relatives and friends, whom they had immediately called to see their experiment. In 1935 they were awarded the Nobel Prize in chemistry, the third Nobel Prize given to the Curie family. Whereas at the Nobel reception in Stockholm in 1905, only Pierre Curie delivered the address, with Marie sitting silently at his side, in 1935 the Nobel lecture was delivered by both laureates: Irène summarized the process of discovery while Frédéric described the chemical identification of the artificial isotope.

With the Nobel Prize came academic positions that not only put an end to their spousal collaboration but came to emphasize contrasting paths for Irène and Frédéric. Gradually, Irène adopted the more conventional female role of a brilliant "second," while Frédéric launched a new career that was to make him one of the most influential and powerful scientists in France.[18]

In 1937 Frédéric Joliot was offered a chair at the Collège de France and was given the facilities for equipping three laboratories with all the modern and heavy instruments required for the study of nuclear physics. While Irène continued her investigations with alpha rays emitted by small quantities of polonium, Frédéric had a cyclotron in his laboratory at the Collège de France that could produce as many alpha rays as one-hundred kilograms of radium. His laboratory at Arcueil had inherited the 1.2-million-volt Van de Graaff generator displayed at the Palais de la Découverte on the occasion of the International Exhibition held in Paris in 1937. For his third laboratory of "atomic synthesis" at Ivry-sur-Seine, near Paris, Frédéric ordered a cyclotron with a 2-million-volt generator. With two young collaborators from among the émigré scientists in Paris, Hans Halban from Austria and Lev Kowarski from Russia, Frédéric Joliot set up a new research program on the effects of neutrons bombarding uranium nuclei.

Whereas Frédéric Joliot escaped the Curies' niche, Irène stayed at the Radium Institute. Like her mother, she got a chair at the Sorbonne in 1937. Also like her mother, who published a treatise on radioactivity in the early 1920s, she published a textbook entitled *Les radioéléments naturels: Propriétés chimiques, préparation, dosage* (Paris, 1946). But unlike her mother, she did not become the director of the Radium Institute. Until 1946 it was André Debierne, Marie Curie's former collaborator and the discoverer of actinium in 1899, who managed the Institute founded by Marie Curie.

Though Irène was the scientific leader of the Radium Institute, which she habitually represented at scientific meetings, she was not in charge of the daily management of the Institute.[19]

Irène Joliot-Curie continued the Radium Institute's tradition of radio-chemistry with new collaborators. Once again she displayed the efficiency and limitation of the Radium Institute's local culture. In 1938, when she bombarded uranium with neutrons (working with Pierre Savitch, a collab-orator from Yugoslavia), she noticed the production of a radioelement with a half-life of 3.4 hours and identified its chemical properties as very similar to those of actinium and lanthanum. But once again she missed the theoretical explanation of her observation, which was suggested by her emerging rival in science, Meitner.

Born in Vienna, Lise Meitner (1878–1968) was nineteen years older than Irène and a pioneering woman scientist.[20] Like Marie Curie, she had to work in a poor shelter, a carpenter's workshop, for some time, in her earlier career, because the Chemical Institute headed by Emil Fischer did not allow women in its laboratories. Like Marie Curie and Irène, she volunteered as a radiologist-nurse during World War I, but in the army of her native Austria. Unlike Marie and Irène Curie, she was more of a physicist than a chemist. In the Hahn-Meitner collaboration, Hahn acted as the chemist, primarily concerned with the discovery of new elements and the identification of their properties, whereas Meitner, who was head of the physics division of the Chemical Institute headed by Hahn at the Kaiser Wilhelm Institute, was more interested in the properties of radiation in cosmic physics and in the investigation of the nuclear structure of the atom. Though Meitner was able to master rapidly any new experimental techniques, she was also a theoretician.

Meitner's theoretical work culminated in 1938, when, while in Stock-holm after having escaped from Berlin in haste because of her Jewish origin, she heard that the bombardment of uranium with neutrons pro-duced isotopes of radium. Because heavier decay products were expected, she asked Hahn and Fritz Strassmann to double-check the identity of the decay products. After reproducing the experiments, Hahn wrote her that a radioisotope of barium was also produced. Meitner and Otto Frisch, her nephew, then working with Niels Bohr in Copenhagen, met and discussed this even more puzzling result because uranium nuclei were twice as heavy as barium nuclei. They soon realized that Bohr's model of the nucleus could provide a clue and interpreted the phenomenon as a partition of the uranium nucleus into two nuclei with nearly the same mass. Using the equivalence between mass and energy, Meitner was able to calculate the large amount of energy that would be released by the partition from the difference between the mass of the uranium nucleus plus the bombardment neutron and the total mass of the two fragments produced. Meitner and Frisch described the nuclear fission in a joint paper, and a few months later they established that there were no transuranic elements in the products.

Thus Irène, like Fermi, who also searched for transuranic elements, was prepared for but missed the discovery of nuclear fission. Bertrand Goldschmidt, who was then Irène's young assistant at the Radium Institute, remembered Frédéric's bitter remark that "had he collaborated with Irène, they would have discovered nuclear fission before the German team." In fact, the German team was dismantled in 1938 by Meitner's exile, but she was still able to grasp the theoretical meaning of the experiments performed by her former collaborators. Nevertheless, Meitner also missed one discovery. After identifying the isotope U-239, formed by the capture of a neutron in U-238, she noticed a very weak radiation that was later identified as neptunium by Edwin Macmillan. If we recall that a few years earlier the Joliot-Curie couple had missed the discovery of the neutron, it becomes clear that missing or not missing a discovery is less dependent on the abilities of a creative couple than on the whole research program and local culture of the institution where they are working. Thus, Frédéric Joliot's bitter remark mainly betrays his increasing self-confidence in his own role as Irène's indispensable collaborator.

Following his own path, Frédéric Joliot became the French leader in nuclear physics when he set up a nuclear chain reaction with his collaborators in 1939. After World War II, he was appointed scientific director of the Atomic Energy Commission (CEA), while Irène contented herself with the subordinate position of head of CEA's chemistry section. Increasingly, she came to play the more traditional part of a female spouse working in the shadow of her brilliant and powerful husband.

How are we to explain such a reversal of power in the Joliot-Curie couple? In order to understand Irène Joliot-Curie's transformation from a senior partner in a Nobelist couple into a modest wife, one needs to consider now the balance between her public and private life.

Public and Private Life

Unlike Marie Curie, Irène accepted public political offices on several occasions. After her mother's death in 1934, she joined the Comité de vigilance des intellectuels antifascistes and wrote a number of papers on the social role of women in fighting for equal status in professional activities. Under the Front populaire, the socialist government elected in 1936, she became the first minister of research and, indeed, the first woman minister in France. However, she resigned after two months, and the position was given to Jean Perrin, a 1926 Nobelist physicist. Why did she accept this official position, and then resign so quickly? Conceivably she accepted this offer in the heyday of great enthusiasm among the socialist scientists who had helped to campaign for the Front populaire.[21] Above all, she seemed motivated by the hope of convincing women that they could reach the highest public responsibilities. But the gap between the feminist symbol and the daily duties of the ministry proved too large for her. She soon realized that she was bored by, or not prepared for, administrative and political work. Some witnesses recalled frequent quarrels between Irène and Frédéric during the time she was a minister, because Frédéric complained that she did not take her new job seriously.

It has also been suggested that Irène Joliot-Curie's political commitments were instigated by Frédéric Joliot, who came from a republican family and had been very much influenced by his former mentor Paul Langevin, a French leader of the World Peace Council and of the Human Rights League.[22] Though Frédéric Joliot played a key role in her commitments, one need not overestimate his influence or assume that Irène lived first under her mother's domination and then under her husband's influence.

In reality, Irène's political concern can be dated back to the influence of her grandfather, Dr. Eugène Curie, who was a liberal and a freethinker. If she became involved in politics in 1934, it was not because her mother died that very year but rather because the political circumstances prompted many French intellectuals to fight against fascism. The Comité de vigilance des intellectuels antifascistes was founded by Paul Langevin, Paul Rivet, a natural scientist, and Emile Auguste Chartier (better known as Alain, his usual signature), a famous French philosopher, in March 1934. Three months later it counted ten thousand members with 250 provincial sections. It would have been impossible for Irène, a leading figure of science known for her liberal tendencies, to remain outside the tide of the committed intellectuals.

Moreover, within her liberal milieu, Irène's feminism could be perceived as a rather conventional attitude. Compared with the activities of politically more active women, such as Eugénie Cotton,[23] Irène's public statements in favor of women's suffrage sound rather timid. She wrote in the magazine *Femmes françaises*:

> I think that the decision of giving women the vote is a measure of justice that has been too long delayed. Indeed one can expect that a great majority of the female electorate is not prepared for its social role. The problem is unavoidable in the beginning; it is because women have been banished from political life that many of them are not interested in it. The only argument that catches my attention against women's suffrage is the following: in several countries, women's votes contributed to the election of reactionary deputies who promptly reduced the civil and economic rights of their electresses. Hence the paradox: in several countries where women were allowed the vote, they became economically more dependent than in France, where they did not have the vote. Nothing like that will happen in France, because of the changes that occurred in political life during the war. However we have to be vigilant, for usually antifeminists, whether men or women, are also against progress in general.[24]

Presumably, Irène Joliot-Curie's feminist commitment never went beyond these moderate words, which placed women's rights within a broader concern for social progress.

Similarly, whereas Frédéric Joliot joined the Communist Party during World War II and later became one of its most active propagandists, Irène never became a member.[25] Furthermore, unlike her husband, she was not sympathetic to the Communist model. By 1939, however, Irène and Frédéric too had signed a protest against the Soviet-German Alliance, or the Molotov-Ribbentrop Pact. In a letter of 1940 to Marie (Missy) Mattingly Meloney, the famous American journalist who had campaigned for Marie Curie's gram of radium, Irène wrote: "The events of this past year clearly demonstrated that fascism and communism are both international. Fascists from different countries help each other and the communists do exactly the same. If democracies fail to raise a movement of international solidarity, they will certainly be destroyed."[26]

Irène did not radically change her opinion about communism as did other scientists—most notably, Frédéric Joliot and Paul Langevin—during the Resistance. Unlike many women of her milieu who were deeply involved in the Resistance movement and were deported to concentration camps, Irène Joliot-Curie was not active during World War II. It was over the war years that her way of life most profoundly parted from Frédéric's. Her passivity after her resignation from the government contrasted with Frédéric's ever increasing political activism. The busier he was in the public arena—first in 1940 with regard to the smuggling of heavy water from the Netherlands to the United Kingdom and France, then from 1942 on as the leader of a Resistance network in Paris—the more she retired from public affairs.

How can we explain such a contrast? In fact, at the beginning of the war, she did her best to preserve her scientific activity, while the couple's two children spent a full year with a nurse in Brittany without seeing their parents. Later, from 1942 to 1944, Irène spent much of her time in a sanitarium in the French Alps because she (like her mother and her grandmother, Bronisława Skłodowska) suffered from severe tuberculosis. However, her poor health was not the only reason for her attitude. From her letters, it is obvious that she missed her children: she felt that Hélène, born in 1927, and Pierre, born in 1932, needed their mother's affection and care. Like every ordinary wife in this period, she was worried about her husband—and Frédéric was in more danger than most. Her family life seems to have captured all her attention, just at the moment that her husband started to assert his authority in public life. In 1944, Frédéric was warned by a German officer, his former assistant Gentner, that he and his family were in imminent danger of arrest and execution. Thus on the eve of the Allies' landing in Normandy, he sent Irène and the children to Switzerland, where they met Paul Langevin, and he himself fled to a clandestine life near Paris.

During the Cold War, Irène was a victim of her husband's political commitment as a Communist, rather than being herself an actor on the public stage. For example, the U.S. Immigration Service denied her entrance in 1952. Her contract as *chef de la section chimie* with the CEA was not renewed after her husband had been dismissed as CEA director in April 1950. In March 1953 Irène Joliot-Curie's application for membership in the American Chemical Society was rejected. The hot debates raised in American scientific journals by the rejection of a Nobel Prize winner reveal the impact of McCarthy politics on the scientific community as well as the difficulty of keeping international scientific life outside political events.[27]

A stark contrast emerged between the young girl who, with her mother, was at the forefront of the war effort during World War I and the worrying mother and wife during World War II. Hence a paradox: while her mother, a widow and a pioneering woman scientist who had to struggle for acceptance in French society, was above all a scientist, public celebrity, and manager, Irène Joliot-Curie, whose career was so much easier, seemed to hesitate more between her various roles as a scientist, public figure, wife, and mother. She was an elusive character. In an informal interview, her son Pierre Joliot remembered that, while he frequently opposed his father on political matters when Frédéric became a dedicated Communist leader, he never had to fight against his mother. He had a close immediate understanding of her, but realized that she lived in a different world. She seemed to belong to another planet, the Curie planet. The "Curie planet" presumably referred to that specific system of ethical values generated by Pierre and Marie Curie's own dedication to scientific research, mixed with the political ideals transmitted by Dr. Eugène Curie and with private education by a team of outstanding teachers, which impregnated Irène Curie during her early childhood.

Conclusions

One of the most distinctive features in the life of the Joliot-Curie couple is that the initial contrast between the two partners' characters, far from decreasing, rather increased over the years of their common life. While Frédéric more and more acted on the public stage, Irène withdrew to her own planet. The most remarkable point is that the continuity in their psychological behavior led to a reversal of their social status and public image, which shows a shift in the relative prestige of the two partners. In the early 1930s, when they acted as a collaborative team, the husband enjoyed the high renown of his in-laws; significantly he, like his wife, used the hyphenated dual last name Joliot-Curie. In the late 1940s, as a result of their divergent celebrity in the aftermath of the Nobel Prize as well as during and after World War II, Irène in turn could have enjoyed the prestige of being the wife of the French leader in nuclear physics. In fact, she appeared to care neither about social prestige nor about the humiliations that her husband's political fate brought her.

The striking evolution of the Joliot-Curie couple can certainly be understood if one considers the historical circumstances of interactions between the war and the culturally hegemonic model of couples in French society. However, in this case, the sociostructural dynamics seem less decisive than the familial tradition. Irène's education in the Curies' planet overshadowed all external influences.

Does this mean that, throughout her life, Irène Joliot-Curie complied with the role of a "second"—second to her mother, who acted as a pioneer, then second to her husband, who became a leader in the public avenues? In stark contrast to her sister Eve, who tried to make her own way through life, Irène—and to a certain extent her daughter Hélène, who became a physical scientist—deliberately chose to follow Marie Curie's path. However, Irène did not simply reproduce her mother's "anti-natural" pattern. She designed a role of her own, out of multiple constraints as a scientific heiress, spouse, and mother. Brought up in a scientific and feminine milieu, she did her best to spend all her life "at home"—not as a housewife but as a woman scientist working at the Radium Institute—while perpetuating a brilliant scientific dynasty.

◆ 3 ◆

MILDRED COHN

Carl and Gerty Cori
A Personal Recollection

"**O**ur efforts have been largely complementary, and one without the other would not have gone as far as in combination."[1] Thus spoke Carl Cori on the occasion of the Nobel banquet in 1947, when he and his wife Gerty were awarded the Nobel Prize in physiology and medicine. And those who knew them could not agree more. They had a very special relationship, as though they had a nonverbal method of communicating with one another. She might start a sentence and he would finish it. When one expressed an idea, the other would elaborate, and then they would alternate expanding on it. Theirs was a rare intellectual symbiosis, perhaps the result of their coming together so early in their lives when they were both eighteen-year-old medical students. Their first research paper, in the field of immunology, was published when they were medical students. The Nobel award culminated careers that from the beginning were a joint research enterprise.

Carl Cori and Gerty Radnitz discovered each other in their first year of medical school at the German University (Carl Ferdinand University) of Prague, which was at that time part of the Austro-Hungarian Empire. They found that they had much in common: both were born in 1896; both came from families that were Austrian in origin but had been citizens of Prague for at least six generations; both were enthusiastic about outdoor activities, particularly mountain climbing; and both soon shared a deep interest in research.

Yet their backgrounds were dissimilar. Carl came from a Catholic family with an academic tradition on both sides,[2] and Gerty came from a Jewish family. Carl's father, Carl I. Cori, was one of the leading zoologists and marine biologists of central Europe.[3] When Carl was two years old, his father left his position as a professor at the Carl Ferdinand University to head the Marine Biological Station in Trieste, also a part of the Austro-

Hungarian Empire. It was there that Carl, together with an older and younger sister, grew up. He was exposed not only to the broad scientific and archaeological erudition of his father during expeditions on the Biological Station's boat but also to the many prominent scientists who visited his father. His summers were spent in the Tyrolean Alps with his extended family, including his maternal grandfather, Ferdinand Lippich, a professor of mathematical physics at the German University of Prague, and his uncle, Friedrich Lippich, a professor of chemistry at the same university. Carl's great grandfather, Wilhelm Lippich, was an anatomist at the University of Padua and later a professor in Vienna. With this family background, it is hardly surprising that Carl eventually followed an academic career.

Less is known about Gerty's family background,[4] since, unlike Carl, who wrote an autobiographical memoir, Gerty never did so, and she rarely spoke of her family. Gerty's father, Otto Radnitz, was a successful businessman. Trained as a chemist, he devised a method of refining sugar and became a manager of sugar refineries. Her mother was a cultured woman, a friend of Kafka. It may have been a maternal uncle, Robert Neustadt, a professor of pediatrics, who sparked her interest in pursuing a medical career.

Carl attended a conventional eight-year gymnasium in Trieste with emphasis on Latin and Greek and upon graduation decided to go back to Prague for his medical studies. Gerty, the oldest of three sisters, was tutored at home until she was ten, not an uncommon educational path for daughters of well-to-do parents at that time. She then attended a private school for girls at Teschen and graduated in 1912. Having decided to enter the university medical program, in her characteristic strong-minded way, she made up many years of Latin and other requirements for entrance in record time, two years at a gymnasium.

Carl and Gerty were immediately drawn to each other when they met at medical school; they made an attractive couple. Carl described her at that time as a young woman "who had charm, vitality, intelligence, a sense of humor, and love of the outdoors."[5] She was slender, brown-eyed with auburn hair, and he was handsome—tall, blond, blue-eyed (in a photograph taken when he was twenty-one, which Gerty kept on her office desk, he resembled the movie actor Alan Ladd).

The first two years at medical school passed pleasantly for Carl and Gerty; they studied together and spent leisure time together on excursions to the country and skiing expeditions. World War I seemed remote until Carl was drafted into the Austrian army in 1916. His experiences in the war were harrowing. The war shook his faith in human behavior. He also became disenchanted with the ability of medical practitioners to cure human disease. He was particularly shocked and frustrated by his experience at the end of the war, when he was attached to a hospital for infectious diseases near the front lines on the Piave River in Italy, where a flu epidemic decimated the poorly nourished army and civilian population. During this period, Gerty was an assistant in the medical school, giving patients physical examinations and doing routine laboratory work.

With the war over, the young pair were reunited, resumed their medical studies, and published their first joint research project in 1920 on the immune bodies in blood in various diseases. This work was carried out at the University Medical Clinic in Prague under the mentorship of Professor R. v. Jaksch. Both Carl and Gerty were awarded their medical degrees in 1920, and in August of that year they were married in Vienna, where they had gone to do postdoctoral work. They worked separately, but not by choice. Carl did laboratory work in the Internal Medicine Clinic at the University of Vienna in the morning and in the university's Pharmacological Institute in the afternoon. At the latter institution, Carl Cori was the only one who was able to engage in research, because his father, once again at the German University of Prague, had been able to send him some frogs needed to study the mechanism of seasonal variation of vagus action on the heart. Materials and facilities were scarce in defeated Austria; the empire had collapsed as a consequence of World War I. Another institution, the Karolinen Kinderspital, gave Gerty a position where she did clinical work and also, in spite of limited facilities, research on temperature regulation before and after thyroid therapy in a case of severe congenital hyperthyroidism. She published several clinical research papers on disorders of the blood in this period.

The young couple decided that they preferred to spend their lives doing biomedical research rather than practicing medicine. The decision was not easy to implement, since conditions in postwar Vienna were chaotic, with almost the whole population near starvation. The probability of obtaining paid research positions was very slim. At the clinic where Carl worked, a free meal a day was given in lieu of pay. In fact Gerty Cori developed xerophthalmia due to vitamin A deficiency because the physicians in charge of her hospital would not accept a dietary supplement offered by a United States relief organization. She was cured by returning home temporarily to Prague, where a better diet was available. Nevertheless not all aspects of life in Vienna were negative; the young couple enjoyed the free cultural riches available, the art treasures in the many museums and the excellent exhibits in the Museum of Natural History.

In the early summer of 1921, Dr. Gaylord, the director of the State Institute for the Study of Malignant Disease (now Roswell Park Memorial Institute) in Buffalo, New York, was seeking to lure a European research worker to his institute. Upon the recommendation of H. H. Meyer, who had been impressed with Carl Cori's work on the frogs, Gaylord interviewed him. Since Carl thought nothing would materialize from this episode, he accepted a position with Otto Loewi in the Pharmacology Department of the University of Graz while Gerty remained in Vienna at the Children's Hospital. Loewi, who had discovered that acetylcholine was the substance which produced vagus stimulation of the heart, proved to be a most exciting teacher and an invaluable mentor. It was during this intellectually stimulating period that Carl first conceived of how to investigate intestinal absorption and the fate of sugar in the animal body.[6] An anecdote ascribes Gerty's interest in sugar metabolism to her father, who had developed diabetes and then challenged her with, "You're a doctor now. Why don't you find a cure for diabetes?"[7]

In spite of his enthusiasm for his work with Loewi, Carl found other conditions at Graz decidedly unsettling. It was necessary to prove his Aryan descent to be eligible for employment at the university. Gerty visited him from Vienna and was appalled by the poor living conditions. Since conditions conducive to the pursuit of research careers seemed highly unlikely to develop, they were determined to leave Europe if they had the opportunity. Their desire was so strong that they applied to the Dutch government to serve as physicians for five years among the natives in Java. Fortunately, after six months at Graz, an offer arrived for Carl from Gaylord in Buffalo. Carl and Gerty welcomed the opportunity to emigrate to the United States, and they never regretted that decision.

Carl Cori came to the Institute as a biochemist in 1922, and Gerty joined him six months later when a position as assistant pathologist became available at the Institute. (Undesired separations during the early years of marriage are more typical of dual-career couples today than in the early 1920s.) Once Gerty arrived in the United States, they were never separated again. The obstacles to their establishment of a joint research program, however, were not entirely removed. Initially Gerty was told that she would lose her job unless she stayed in her laboratory in the Institute's Pathology Department and stopped working with Carl. She was restricted to microscopic examinations required by her superior. Undaunted, in her spare time she did research that required only a microscope and that led to her paper in 1923 on the effect of thyroid extract and thyroxine on the multiplication rate of paramecia.

After their first year in Buffalo, opposition to the Coris' research collaboration evaporated. Some thirty years later Gerty told me of another occasion when she burst into tears upon being reprimanded for collaboration with her husband; it was labeled "un-American" by the professor who was offering Carl a faculty position at a nearby university. Carl related this incident in his autobiographical memoir, stating that he found two requirements for the position unacceptable, namely, that he stop working on insulin and that his collaboration with Gerty cease. Gerty was told that she was standing in the way of her husband's career and that it was un-American for a man to work with his wife. Carl assured her that it was merely unusual, and he commented in 1969 that "in some respects it still is." He added: "It is a delicate operation which requires much give and take on both sides and occasionally leads to friction, if both are equal partners and not willing to yield on a given point."[8]

Although some of their Buffalo papers were coauthored with other investigators, the bulk of their eighty publications in the period 1922–1931 resulted from their joint research. Thirteen papers dealt with questions about the regulation of blood glucose (sugar) levels and the role of hormones, epinephrine and insulin, in the process. Experimentally, it was their meticulous attention to the development of quantitative analytical methods for determining glucose, glycogen, lactic acid, and inorganic and organic phosphate, wherein their strength lay. Studies of glucose metabolism in rats traced the breakdown of muscle glycogen (the macromolecular storage form of glucose) as influenced by epinephrine and insulin, leading to lactic acid in the blood. When lactic acid reaches the liver, glycogen is formed. The liver in turn releases glucose to the blood, which transports it to the muscle, where it forms glycogen. These experiments and their deep knowledge of physiology led the Coris to formulate what came to be known as the "Cori cycle" of glycogen metabolism.

In spite of their joint work, Carl's status was quite different from Gerty's. In 1924, Carl Cori's name appeared on the heading of the Institute's stationery as "Biological Chemist," together with the names of five specialists in other areas and the director. Gerty Cori's position did not warrant such recognition, although she was sole author of four publications in addition to the joint ones with Carl. Another reflection of her status may be gleaned from an acceptance letter to her in 1924 from F. Peyton Rous, editor of the *Journal of Experimental Medicine*, for a paper that was not coauthored by Carl, which was addressed "Dear Mrs. Cori"; all other authors were addressed as Dr.[9]

It was during the years in Buffalo that the Coris adapted to their new environment and familiarized themselves with the customs, attitudes, and culture of their adopted country, of which they became citizens in 1928. They came to prefer it to Europe, although they never forgot their European heritage. They read widely, learning about the history, politics, and literature of the United States. Being theater buffs, they attended plays in New York City whenever feasible. To satisfy their love of the outdoors, they explored the east, particularly upstate New York, the Adirondacks, the White Mountains, and Cape Cod. Through meetings of the Society for Experimental Biology and Medicine held at various universities in the Buffalo area, they met and made friends with some outstanding scientists. Their circle of friends then and later was not limited to scientists but included musicians and other nonacademics.

In 1931 Carl Cori wrote a masterly review on carbohydrate metabolism, which established him as an authority in the field. In the same year he was rather surprisingly offered, and accepted, the chairmanship of the Pharmacology Department at Washington University School of Medicine in St. Louis. His previous academic experience was limited to a one-year adjunct assistant professorship of physiology, which he held at the University of Buffalo during his last year in New York. Gerty was given a research position in the Pharmacology Department at Washington University. Some twenty years later, when she was often hospitalized because of her illness, she told me: "I don't feel guilty about accepting my full salary in spite of my frequent absence, because when Carl and I first came here, they paid me 10 percent of what they paid him."

The years in St. Louis were most rewarding for the Coris in both their personal and professional lives. There were no impediments to their joint research, and there ensued a most productive collaboration, which ended only with Gerty's death in 1957. They made breakthroughs in the path and mechanism of glycogen utilization, and in 1936 their son Thomas was born. Gerty continued to work full-time and always retained a full-time housekeeper who took care of Tom.

When they first arrived in St. Louis, Carl had to organize a department, equip a research laboratory, and spend considerable time teaching medical students. Devoting all her time to research, Gerty never taught medical students in either pharmacology or biochemistry. Although Gerty was an inspiring teacher and mentor in a one-on-one relationship, she was uncomfortable lecturing to large audiences and generally avoided doing so. Although a first-rate experimentalist, Carl Cori ceased working at the laboratory bench when, after 1946, his other duties became too demanding to allow it.

Because of their careful analytical methods, in particular measuring hexose phosphate by two different methods, namely, phosphate content and reducing power, and finding a discrepancy between the two, the Coris discovered a new intermediate of glycogen breakdown. In addition to the well-established glucose-6-phosphate, which has equal phosphate content and reducing power, they described an earlier intermediate, glucose-1-phosphate, which has the same phosphate content but no reducing power. This compound became known as the Cori ester, and its structure was established in collaboration with Sidney Colowick, their first joint graduate student. The enzyme that catalyzed its formation was named phosphorylase. The Coris showed that the reversal of the phosphorylase-catalyzed reaction led to the formation of glycogen, the first demonstration of the synthesis of a biological macromolecule in a test tube.

In collaborative work with Arda Green, a research associate, the enzyme glycogen phosphorylase was crystallized from muscle and its physical and chemical properties thoroughly investigated. The Coris also described two forms of phosphorylase, designated a and b, and demonstrated the conversion of a to b and the need for adenylic acid to yield an active b form, foreshadowing the regulatory mechanism of glycogen breakdown by protein phosphorylation and allosteric effectors. The totality of their fundamental discoveries involving the enzyme-catalyzed chemical reactions of carbohydrate metabolism led to their sharing the Nobel Prize in physiology and medicine with Bernardo Houssay from Argentina in 1947. The Coris were cited "for their discovery of the course of the catalytic conversion of glycogen," and Houssay was cited "for the discovery of the importance of the anterior pituitary hormone for the metabolism of sugar." Following the work on phosphorylase, the Coris plunged into enzymology, purifying and characterizing the enzymes involved in the individual steps of the glycolytic pathway in the conversion of glycogen and glucose to lactic acid. Their research became more and more biochemically oriented, but they always returned to the physiological implications of their enzyme studies.

Concurrently with the trailblazing work on phosphorylase during the years of World War II, Carl spent a good deal of 1942–1945 directing a secret war research project whose goal was to develop protective measures against the effects of mustard gases. As project director, he was burdened with writing frequent reports and attending monthly meetings in Washington, D.C. The latter were time-consuming because he traveled between St. Louis and Washington by train.

Although Carl Cori began his career in St. Louis in the top academic rank, professor and chairman of a department, Gerty's climb up the academic ladder was, as might be expected, exceedingly slow. From 1931 to 1937, her rank was research fellow in the Pharmacology Department, and in 1937 she became a research associate. It was the policy of the Pharmacology Department to give faculty appointments only to those who taught medical students. Helen Graham, a female colleague who was very active in teaching, was promoted to the rank of associate professor of pharmacology in 1937. The Department of Biological Chemistry did not have this restriction. For example, in 1942 Sidney Colowick was listed as research assistant in pharmacology and instructor in biological chemistry, and Arda Green was listed as research associate in pharmacology and assistant professor of biological chemistry. In that year Gerty was still listed as research associate of pharmacology, but in 1943, in recognition of her research productivity, she was appointed an associate professor of pharmacology and of biological chemistry, although she had no teaching responsibilities.[10] In 1947, a few months before being awarded the Nobel Prize, Gerty was promoted to a full professorship.

In the early 1950s, Washington University formally announced its antinepotism rules, which stated that a husband and wife could not have appointments in the same department. However, the chancellor, Arthur Compton, wrote Gerty a personal letter assuring her that the rule did not apply to her. Two years after I came to the university, I received a letter from The Johns Hopkins Medical School offering me a position. The last sentence in that letter read, "We understand that your husband is a physicist, we shall try to find him a position too." I showed this unusual letter to Gerty, who said, "Mildred, frame it; I have never received such a letter."

To induce the Coris to remain at Washington University, the university offered Carl Cori a shift from the chairmanship of pharmacology to biological chemistry as well as additional space. Their joint research was well recognized, and in 1946, at the end of World War II, they were both offered professorships at many prestigious institutions such as Harvard, Berkeley, and the Rockefeller Institute. However, the Coris really liked St. Louis, where they had a wide circle of friends and a well-designed modern house and garden, to which they were much attached; she nurtured the flowers and he grew the vegetables. In addition, they approved of the philosophy of the School of Medicine, where research was the primary priority and top-quality research people were appointed regardless of professional or personal background. Dr. P. A. Shaeffer, who was responsible for bringing the Coris to Washington University, was dean of the school and chairman of the Biological Chemistry Department, although no longer very active in the latter capacity. In 1946 he voluntarily stepped aside three years early so that Carl Cori could assume the chairmanship of the department, and as a result the Coris remained in St. Louis.

The Coris had already hosted some outstanding scientists in the Pharmacology Department, among them three who were destined to become Nobel laureates: Severo Ochoa came from Spain via England in 1939; Earl Sutherland began as a medical student assistant in 1939 and joined Cori's department as a faculty member in 1946; and Luis Leloir arrived from Argentina in 1943. Among other outstanding scientists, Herman Kalckar was welcomed when he could not return to Copenhagen because of the outbreak of war in Europe. Arda Green, Gerhard Schmidt, and Sidney Colowick participated in the work on phosphorylase that led to the Coris' Nobel award. All of these colleagues remained lifelong friends of the Coris.

In the fall of 1946, my husband having accepted an assistant professorship of physics at Washington University, I became a research associate in Cori's Biological Chemistry Department. The previous spring I had the opportunity at a luncheon in New York hosted by my mentor, Vincent du Vigneaud, to inform Carl Cori that I was coming to St. Louis and would like to work in his department. He spent about half an hour in my laboratory discussing my research plans with me and promised to let me know his decision. After considerable delay engendered (as I later learned) by his indecision about remaining at Washington University, he informed me that I could join his department

I did not meet Gerty until I came to St. Louis, and I shall never forget her first remark: "I understand you are more fortunate than I. You have both a daughter and a son; I have only a son." What a warm welcome from a great woman scientist. The Coris treasured their son, Thomas (named after the Czech leader Tomáš Masaryk), but did have some problems while he was growing up, partly because of the cultural gap between their European heritage and the reality of an American boyhood. He seemed not to be as studious or intellectually inclined as Gerty would have wished. Nevertheless he was sufficiently influenced by the scientific ambience at home to obtain a Ph.D. in organic chemistry. Thomas, who seems to have inherited his maternal grandfather's entrepreneurial talents, became the successful president of one of the largest biochemical–chemical companies.

From the outset of my stay in St. Louis, Gerty was continuously supportive of my career. Her laboratory was across the hall from the room where I was assembling a mass spectrometer. Whenever she had a waiting period in her experiment, she would come into my lab and essentially give me an enthusiastic tutorial in enzymology. What a treat for me. Upon discovering that I was not a member of the American Society of Biological Chemists, she nominated me, and I was elected. The first time I was invited to present my work at another university, she was almost as excited as I was. When I presented her with a first draft of our joint paper, she was lavish in her praise of my writing.

Carl Cori's support was essential for the development of my career. Not only did he permit me to work independently from the time of my arrival, when I held the rank of research associate, he also supplied the financial resources to implement my research including placing an electronics engineer at my disposal to aid in the construction of a mass spectrometer. When I sought his advice on research problems, he was always available and helpful. Toward the end of my stay in his department, he groomed me for a position on the editorial board of the prestigious *Journal of Biological Chemistry*. In addition to instructing me on reading a paper critically, he said, "Even if the manuscript isn't very good, if any worthwhile science is in it, try to save it; it is the duty of the Society's journal to the scientific community." I was the first woman to be a member of that editorial board.

The Coris were supportive and encouraging to the many scientists, young or mature, who flocked to their center of enzymology from all over the world once World War II was over. In the period 1946–1950 the research fellows included, among others, Helen Porter from England (later elected to the Royal Society), Shlomo Hestrin from Israel, Otto and Eva Walaas from Norway, Jean Wiame, Christian de Duve, and Claude Liebecq from Belgium, Olav Lindberg from Sweden, Tadeusz Baranowski from Poland, Theodore Pasternak from Switzerland, and Arthur Kornberg, Edward Krebs, Victor Najjar, and Charles Park from the United States. Most of us participated in the daily departmental brown-bag lunches in the library, which were enlivened by Gerty's animated conversation and Carl's wit. We were awed by the breadth and depth of their knowledge and interests. They were avid readers; Gerty specialized in history, biography, and modern novels, whereas Carl favored archaeology, poetry, and art. Carl also had remarkable linguistic ability, being equally comfortable conversing in German, French, Italian, and English.

With the help of an excellent housekeeper, the Coris entertained frequently in their home with its book-lined living room. They were particularly hospitable to the foreign visiting scientists and the unattached members of the department. One of Gerty's graduate students, Patricia Keller, recalls with gratitude that they not only invited her to their home but also took her to the opera and art exhibits. An evening in their home was always stimulating, and the guests would invariably include nonscientific friends—sociologists, artists, or musicians.

The announcement of the Nobel award engendered great excitement and joy within their laboratory group in St. Louis and among their friends elsewhere. Their pleasure in receiving the Nobel Prize was marred by the knowledge that Gerty had developed an incurable illness, myelosclerosis, which led to anemia. They had discovered her problem at high altitudes while indulging in one of their favorite vacation activities, climbing in the Colorado Rockies, in the summer of 1947. She suffered with the disease for ten years, gradually weakening. As Carl described it many years later when he was terminally ill, "Gerty was heroic." And indeed she was. Just observing her in the laboratory, one would never have suspected that she was ill. Devoted to her throughout her illness, Carl took an active role in monitoring her anemia and even devised a treatment that maintained her for about a year, just as other treatments had, until she was overwhelmed by side effects. In order to help Gerty rid herself of her smoking addiction, essential because of her low hemoglobin levels, Carl stopped smoking too. Only twice during the ten years did she comment discouragingly to me, the first time about halfway through the period: "If something like this happens to you, it's better that a ton of bricks should fall on you." Toward the end of her life she said one day: "I gave a party to squelch the rumor that I was dead." Otherwise she was upbeat.

Her scientific work went on unabated and with the same dedication. It was during this period that, in collaboration with a graduate student, Joseph Larner, she discovered the debranching enzyme that was responsible for degradation of the 1,6 linkages at the branch points in the glycogen structure. Her most elegant work, which was the focal point of her research in her last years, was the analysis of the individual enzymatic defects in inherited glycogen-storage disease in children. In collaboration with a postdoctoral fellow, Barbara Illingworth, she established that there were four different types of the disease corresponding to different enzymatic defects, the first demonstration that an inherited human disease can be traced to a defect in an enzyme. This outstanding work was recognized by an invitation to present the prestigious Harvey lecture in New York in 1952.

Public recognition came to both of them, but for Gerty the honors were fewer and came later, most after the 1947 Nobel Prize. For example, Carl was elected to the National Academy of Sciences in 1940, but Gerty only in 1948. They received two joint awards, the Midwest Award from the American Chemical Society in 1946 and the Squibb Award from the American Society of Endocrinology in 1947. In 1946 the prestigious Lasker Award from the American Public Health Association was presented to

Carl alone, as were the Sugar Foundation Prizes in 1947 and 1950 and the Willard Gibbs Medal from the American Chemical Society in 1948. Gerty was awarded the Garvan Medal, an American Chemical Society award reserved for women chemists in 1948; the Sugar Research Prize from the National Academy of Sciences in 1950; and the Borden Award from the Association of American Medical Colleges in 1951. Carl held memberships in several elite foreign societies, the Royal Society (London), the Royal Danish Academy of Science, and the French Academy. Both were members of the American Academy of Arts and Sciences and of the American Philosophical Society. Many more universities awarded honorary degrees to Carl than to Gerty, and he was a more public figure in the scientific community. He was elected president of the American Society of Biological Chemists in 1949 and was also chosen president of the International Biochemistry Congress in Vienna in 1958. He served for many years on the editorial boards of two biochemical journals. Because of his remarkable diplomatic skills and his uncanny judgment, Carl Cori's counsel was widely sought within the university and by many outside foundations and institutions. Gerty Cori's sole public service role came when she was appointed by President Truman to the National Science Foundation Board.

For their sixtieth birthdays in 1956, the Coris' collaborators honored them with a Festschrift of original research papers published as a special issue of *Biochimica et Biophysica Acta*, an international journal. Among their collaborators who contributed were six future Nobel laureates: Ochoa, Sutherland, Leloir, Kornberg, de Duve, and E. G. Krebs. Houssay, who had shared the Nobel award with the Coris, wrote an introductory appreciative essay.

Although Carl and Gerty were coauthors of many papers in the years after 1946, there were also a significant number of publications in which only one of their names appeared among the authors. Gerty's research focused more on the structure of glycogen determined enzymatically and on glycogen-storage disease,[11] whereas Carl's was more oriented toward hormone effects on glycogen metabolism, the mechanism of enzyme action in the pathway of glycogen and glucose metabolism, and tissue permeability to sugars. He made unique and incisive contributions in all these areas.[12] A few graduate students and postdoctoral fellows were joint, but most worked exclusively with one or the other.

In such a close and complementary collaboration as existed between the Coris, it becomes almost impossible to separate and evaluate the contributions of each individual. Nevertheless some members of the scientific community attempted to do so, voicing opinions ranging from "she was the leader and he was the follower" to the opposite extreme, "he had the ideas and she carried them out." I think the consensus among those who knew them well would be that they were equally creative, both were gifted laboratory workers, and neither ever deviated from a direct path to their scientific goals. But temperamentally they differed considerably.

Gerty was vivacious and outgoing; her mind was quick and sharp. Her enthusiasm for and dedication to science were infectious, and she supplied enough motivation for both of them. She was a tireless worker in the laboratory, arriving with Carl promptly at nine in the morning, beginning her experiments immediately, and continuing until six in the evening, totally involved in her research every moment. It was obvious that the experiments had been planned jointly with Carl the previous evening at home.

On one occasion when Gerty and I were working together, the analytical results on the samples she had given me were not self-consistent. When I said, "I think the samples were mislabeled; the experiment will have to be repeated," her almost tragic response was, "A whole day wasted!" Although she, like Carl, did not suffer fools gladly and could be harshly critical, she was usually kind and compassionate—finding suitable housing for a visiting scientist, arranging proper medical attention for an ailing secretary.

Carl generally appeared somewhat aloof and austere but never pompous and frequently displayed an engaging gaiety and wit. His intellectual and personal impact was so compelling that he rarely failed to influence those who interacted with him. The economy and rigor of his writing style are quite in character. When one reads their joint Nobel address, part written and delivered by each of them, there is no difficulty in discerning where Carl ended and Gerty began. His section reads like a proposition in Euclidean geometry.

Their own words best express their attitudes and personalities. In an interview with Edward R. Murrow in the radio series *This I Believe*, Gerty ends her statement with: "Honesty, which stands mostly for intellectual integrity, courage and kindness are still the virtues I admire, though with advancing years the emphasis has been slightly shifted and kindness seems more important to me than in my youth. The love for and dedication to one's work seems to me to be the basis for happiness. For a research worker the unforgotten moments of his life are those rare ones, which come after years of plodding work, when the veil over nature's secret seems suddenly to lift and when what was dark and chaotic appears in a beautiful light and pattern."[13]

Carl Cori ends his autobiographical memoir with: "The frontiers of physics, astronomy, and biology and the instrument of their study, the human mind, fill one with wonder as do the great creations of art and architecture, past and present. From these and from contact with nature, love, and friends springs the joy of living, but the poet's creative urge most often springs from unhappiness and sorrow. Humanism may be as important to mankind as competence in a particular field of science."[14]

Gerty died at home in 1957; only Carl was with her. Thus ended one of the most exemplary, productive, and enduring collaborations in biomedical science. To have had both of them as friends and mentors was a privilege; for me, it was a unique experience, the inspiration survives.

Epilogue

In 1960 Carl Cori married Anne Fitzgerald-Jones, with whom he happily shared interests in archaeology, art, and literature until his death in 1984. Upon his retirement from Washington University in 1966, he became a visiting professor of biological chemistry at Harvard Medical School and remained actively engaged in research in his laboratory at Massachusetts General Hospital until the last year of his life. During his years in Boston, he pursued a productive collaborative research program with a distinguished woman scientist, Salome Gluecksohn-Waelsch, a professor in the Genetics Department of Albert Einstein Medical College in New York. Their studies involved the discovery and elucidation of a hereditary deficiency of glucose-6-phosphatase in mice. Gerty and Carl Cori had discovered a hereditary defect of this enzyme in man fifteen years earlier. The first joint paper of Cori and Gluecksohn-Waelsch was published in 1968 and the last in 1983.

◆ PART II ◆

Couples Beginning in Student-Instructor Relationships

✦ 4 ✦

JANET BELL GARBER

John and Elizabeth Gould
Ornithologists and Scientific Illustrators,
1829–1841

$$T$$he romantic but tragic lives of two people who lived and loved and worked for science in London in the decade of the 1830s are here recounted. John Gould, taxidermist to the king and to the Zoological Society of London, and Elizabeth Coxen, governess, both twenty-four years of age, were married on January 5, 1829. Before the decade was out, John was the most knowledgeable ornithologist of both England and Australia and Elizabeth a highly regarded scientific illustrator, whose paintings and lithographs of birds were praised for their beauty as well as for their accuracy. And then suddenly it was over—not John's career, which carried him to greater successes and even fame, but Elizabeth's life, and thus their marriage and scientific enterprise together.

They had met, it is speculated, either at the Aviary of the Zoological Society, where John was employed and where Miss Coxen may have brought her young charge, or through mutual acquaintances. She was charming and intelligent; he brimming with youthful self-confidence and enthusiasm.[1]

Elizabeth felt that her life as a governess was sometimes "miserably, wretchedly dull," as she had no one with whom she could share her thoughts or feelings.[2] John held rare and coveted paid employment in science; in 1827, at age twenty-three, he had won a competition for the post of curator and preserver for the fledgling Zoological Society of London.[3]

Less than two years after their marriage, with the support of the Zoological Society and fifty prepaid subscribers, John and Elizabeth Gould produced their first folio of illustrations: four plates of birds from the Himalaya Mountains recently acquired by the Society.[4] N. A. Vigors (1785–1840) presented papers on them at meetings of the Zoological Society in November 1830.[5] Gould, who attended the meetings and was responsible for arranging the display of the beautiful birds, correctly gauged the market for illustrations of exotic birds.[6] The lithographic plates were executed by Elizabeth from her own detailed watercolors, for which she referred to John's sketches and the stuffed birds posed in front of her.[7] Each print was colored by hand by one of a staff of colorists. The folio, the first of an eventual twenty issues of *A Century of Birds*, was an instant success. The work was subscribed to by over three hundred naturalists and nobility of England, plus museums in England and on the Continent.[8]

The Goulds' productivity was extraordinary. Before the last folio of Himalayan birds came out in March 1832, Elizabeth and John were at work on their second publishing venture, *Birds of Europe*, closely followed by monographs on toucans and trogons and a synopsis for the *Birds of Australia*. Elizabeth drew and lithographed most of the 160 plates for those volumes, assisted by the artist Edward Lear (1812–1888) of nonsense rhyme fame.[9]

John Gould made three trips to the Continent to observe living birds and to obtain fresh specimens, including one with Elizabeth in the summer of 1835, before the twenty-second and last part of *The Birds of Europe* was published in July 1837, with 380 plates drawn by Elizabeth.[10] Then in 1837–1838 John Gould wrote the text and Elizabeth prepared the 50 plates for the volume *Birds*, for Darwin's *Zoology of the Voyage of H.M.S. Beagle*.

Production of the folios was accompanied by production of Gould babies, eventually numbering eight, six of whom survived. The Goulds' marriage was one of genuine affection, and their scientific relationship was unique, their talents complementing one another to an extraordinary degree. John Gould would certainly have been a success without his "Eliza," as he fondly called her—he had a remarkable aptitude for business combined with an unerring eye for detail and abilities to sketch and to place stuffed birds in lifelike poses—but Elizabeth's artistic talent gave him an advantage he could not otherwise have enjoyed.

John Gould was born on September 14, 1804, at Lyme Regis, Dorset, and began collecting flowers, insects, and bird eggs while still a child. He said that he had been charmed by nature as an infant, when his father held him up "to see the verditer [greenish] blue eggs in a hedge-sparrow's nest."[11] When John was fourteen, his father was appointed foreman gardener at Windsor Castle. John collected birds and made preparations of their skins, reportedly to sell to the boys at nearby Eton.[12] At age eighteen he was apprenticed to a gardener at Ripley Castle, but he had another career in mind; two years later (c. 1824) Gould was in London, apparently having set himself up as a taxidermist.[13] Employment by the Zoological Society opened up new opportunities in science for Gould. In November 1828 a new British warbler (*Sylvia tithys*), found by Gould, was exhibited to the Society; John Gould's first published paper (of over three-hundred published during his career) concerned this discovery.[14]

John Gould always maintained that he had had no formal education. Albert Günther at the British Museum once remarked that he had helped Gould with the invention of scientific names, because Gould was deficient in the "classics."[15] Gould was, however, very well educated in those matters that concerned him. The Linnaean system was easy to master; Gould not only did so but became one of the world's expert taxonomists. He learned to draw, developing "an extraordinary facility" for capturing the characteristics of any bird on paper. He wrote many letters, and his composition was deemed entirely satisfactory. He became an outstanding success at book publishing, producing, with the aid of his artists, forty-one volumes of large folio works with 2,999 hand-colored plates, "an average of more than one plate for each week of his fifty years of working life."[16] And Günther had to admit that no one of his age contributed as much to ornithology as Gould.

It is difficult to say whether Gould had any thoughts on theoretical issues that were so much discussed in midcentury. His letters and writings reveal no interest in politics, religion, or anything except birds, and occasionally fishing or shooting. One of his daughters later recollected a father who was constantly busy with his science. Gould's biographer has said, "That he loved his family there is no question; but he lived his birds."[17]

Although Gould seemed uninterested in Darwin's theory, he contributed materially to it, through his keen observations and his interest in the migrations and geographical distribution of birds. Darwin has been described as being "frankly stunned" by Gould's telling him that the Galapagos finches were a peculiar group of thirteen species, all closely related to one South American finch, that Galapagos mockingbirds belonged to three distinct species from different islands, and that twenty-four of twenty-six land birds were from separate species found nowhere else in the world. Four months after his meeting with Gould, Darwin began his first notebook on the "Transmutation of Species."[18] In the *Origin of Species* Darwin stated that it was Gould who suggested to him "that in those genera of birds which range over the world, many of the species have very wide ranges"; Darwin used it as an argument against special creations.[19] Moreover, Gould was keenly aware of variability within species. When criticized for too brightly coloring some of his illustrations, he is said to have replied, "Well, there are sure to be *some* specimens brighter than we do them."[20]

Elizabeth Gould, born on July 18, 1804, in Ramsgate, Kent, was one of four (out of nine) children of Nicholas and Elizabeth Coxen who survived to adulthood.[21] Nothing is known of Elizabeth's education except that it enabled her to gain a position as a governess. The one surviving letter written before her marriage reveals only that she taught her nine-year-old pupil French, Latin, and music; she did not mention art. Her remaining extant letters, most of which were written from Australia (by then she was an established artist), reveal a lively interest in and knowledge of her surroundings, and a good command of English usage.[22] John Gould appreciated his wife's talents from the first. In 1870 he told his friend R. Bowdler Sharpe that when he broached the subject of producing a volume of Himalayan birds to Elizabeth, she asked, "But who will do the plates on stone?" His reply was, "Why, *you* of course!"[23]

There were few precedents to Elizabeth's work. The method of lithography, invented in 1798 in Germany, was first used for natural history illustration in England by William Swainson (1787–1855) in 1820. Some of Elizabeth's drawings of Himalayan birds were displayed at the first British Association for the Advancement of Science meeting at York in September 1831, alongside Audubon's etchings of American birds.[24] Although Elizabeth experienced the almost annual arrival of a child, she continued with her drawing, made possible by servants and by her mother's ability to care for her children when Elizabeth was abroad with her husband.

It may be safely said that external perceptions of the Goulds and their scientific relationship mirrored internal reality, until the memory of Elizabeth Gould was lost from public consciousness. Early praise of Elizabeth's work came from every quarter. When plates for *A Century of Birds* were exhibited at a meeting of the Zoological Society in November 1830, they were hailed by members as the most beautiful since those of Thomas Bewick and Swainson.[25] Sir William Jardine (1800–1874) and Prideaux John Selby (1788–1867), who had published illustrated bird books of their own, observed in 1831 that Elizabeth's first attempts at stone drawing were "very tolerable,"[26] and with each publication raised their estimate of "her drawing and attitudes," until in 1836 Jardine pronounced her plates to be "beautifully executed."[27] The ornithologist Alfred Newton is said to have praised Mrs. Gould's early work,[28] and Darwin, in his introduction to *Birds*, in the *Zoology of the Beagle*, stated, "The accompanying illustrations, which are fifty in number, were taken from sketches made by Mr. Gould himself, and executed on stone by Mrs. Gould, with that admirable success, which has attended all her works."[29] Edward Lear, who worked with Elizabeth Gould for six years, wrote that in the early years John Gould "owed everything to his excellent wife, & to myself, without whose help in drawing he had done nothing."[30] Botanist Charles Pickering (1805–1878), who visited Elizabeth in Australia in 1839 while on the United States Exploring Expedition, reveals that she was then well known in America; he wrote of meeting "Mrs. G., to whose talent and industry the world is indebted for the celebrated Ornithological Illustrations. I had the pleasure of seeing the lady at her pencil, and was surprised at the rapidity of her execution."[31] Finally, zoologist Hugh Strickland (1811–1853) in an 1844 talk before the British Association for the Advancement of Science, spoke highly of John Gould's books and particularly praised Elizabeth: "Nor should it be forgotten that the talents of Mr. Gould were most ably seconded by his amiable partner, who, up to the time of her decease, executed the lithographic department of his various works." Strickland added that *A Century of Birds of the Himalaya Mountains* "at once established the fame of this admirable artist."[32] A recent critic has written that when Elizabeth was enabled to view living birds, her "illustrations, full of character and life, make one feel that those responsible for them knew their subjects intimately in the wild."[33]

In nine years Elizabeth had made over 650 lithographic plates for seven books, but she had more to do. John had found a project that would satisfy both his desires and the public's interest. Australia was a sparsely colonized, largely unexplored part of the empire, and its birds were almost unknown. John Gould has been called "the Father of Australian Ornithology" for good reason: the two years he and Elizabeth spent studying, collecting, and drawing the birds of Australia amounted almost to a second discovery of the subcontinent. In 1826 Vigors and Thomas Horsfield (1773–1859) had published a preliminary account of the Australian birds in the Linnaean Society's collections.[34] Gould studied the birds closely, and in 1833, before the Zoological Society of London, he named his first new Australian bird, the Blue Wren, *Malurus pectoralis*. By 1836 he had exhibited and named some twenty-seven species of Australian birds, making him the recognized authority on the birds of the continent.[35]

When he married, John Gould gained not only a wife and partner in his work but an opportunity to acquire new specimens. Elizabeth's brothers, Stephen and Charles Coxen, who had migrated to Australia, collected birds and shipped them to John.[36] Gould obtained enough specimens, when combined with those at the Linnaean Society, to produce the first folio volume of his proposed series.[37] The Coxens could not, however, provide the numbers of birds he required. Moreover, without knowing the "habits, manners, and general economy"[38] of birds—knowledge obtainable only by seeing them alive—he could neither describe nor sketch them properly. He decided to go to Australia, to see the birds, to hear their calls, and to do his own collecting. Elizabeth would accompany him, to draw the birds in a fresh state and as much as possible in their natural surroundings. He departed with Elizabeth, their oldest son, John Henry (then seven), their nephew Henry William Coxen (age fifteen), collector John Gilbert, and two servants, on May 16, 1838, for a two-year sojourn in the antipodes, leaving the three younger children with Mrs. Coxen.

A trip to Australia in 1838 was a bold venture, as the colony was only fifty years old and the interior of the continent largely unknown.[39] The Goulds were well aware of the novelty of their undertaking. When John wrote to Jardine about their plans to "circumnavigate the globe!!!" he added, "What an age." But Gould trusted his own abilities. He wrote in the same letter to Jardine: "The voyage will certainly be a very expensive one but I doubt not my work will have a great sale in consequence and if so amply repay me for my exertion."[40]

The journey took four months, as did every letter sent home and each reply. Life was a precarious matter in the early nineteenth century, and no one took survival for even another day for granted. Elizabeth wrote to her mother on October 8, 1838, three weeks after arriving, "Could I be sure of meeting you all again in health I could be content, but there is the anxiety. . . ."[41] Nearly every one of Elizabeth's letters echoes that concern and her heartache at being absent from her young children: "Oh, my dear Mother, how happy shall I be if permitted to see you once more and my dear children."[42]

But her letters were increasingly optimistic about the task the Goulds had set for themselves. Although Elizabeth was not, before meeting John, an enthusiast in the "ornithological department," she appreciated her husband's—and her own—places in science. She wrote her mother again not long afterward:

> Persons to whom we have been introduced are exceedingly kind, and John is acquiring a vast fund of information in the ornithological department, which must, I think, prove interesting to the lovers of that science.
>
> We got here in just the right season [early spring], and I assure you he has already shown himself a great enemy to the feathered tribe, having shot a great many beautiful birds and robbed others of their nests and eggs. Indeed John is so enthusiastic that one cannot be with him without catching some of his zeal in the cause, and I cannot regret our coming though looking anxiously forward to our return.[43]

Elizabeth realized that they were making history.

Gould chose to begin his visit with Tasmania, partly because many of the known species of Tasmanian birds were found only on that island.[44] He set out almost immediately to explore the country near Hobart Town and along the River Derwent.[45] Soon after arriving in Hobart Town, the Goulds met the famed Arctic explorer Sir John Franklin, then governor of Tasmania, and Lady Jane Franklin. The Goulds were invited to stay at the Franklins' cottage, and when John left in December with Lady Franklin at her invitation for an expedition to the south coast of Van Diemen's Land on the government schooner, Lady Franklin prevailed upon Elizabeth to stay at Government House during their absence.[46] In January 1839 Elizabeth wrote her mother of Lady Franklin's kindness, of her own activities, and of her homesickness:

> We are very comfortable at Government House, and Lady Franklin will not hear of our going from it as long as we remain in town. By *we* I mean Henry and I. . . . during John's absence, I find amusement and employment in drawing some of the plants of the colony, which will help to render the work on Birds of Australia more interesting. All our sketches are much approved of and highly complimented by our friends. I wish you could hear some of the speeches that are frequently made us, because I know you like dearly to hear your daughter praised. . . . Are the children yet all at home? Give them 20 kisses for me.[47]

Elizabeth Gould made seventy-four drawings of plants, mostly watercolored, for incorporation into the backgrounds of the plates to *The Birds of Australia*.[48] Lady Franklin became one of Elizabeth's admirers; she wrote Elizabeth the year after the Goulds' return to England: "I will rejoice if this interesting country is still further illustrated by you. Its *Flora*, you know, I have long destined for yr. clever pencil. As long as we are here, your home is with *us*."[49]

On their expedition in the government schooner, Lady Franklin intended that Gould should visit both the south coast of Tasmania and Port Davey on the west coast, but stormy weather prevented the party from proceeding farther than Recherche Bay, where the boat was windbound for a fortnight. Lady Franklin remarked in her journal about John Gould: "His good humour under his prolonged disappointment had never failed, but his time was precious to him."[50] In spite of their reverses, Gould observed thousands of giant petrels and collected new parrots, albatrosses, ducks, and gulls, as well as petrels, and the nests and eggs of penguins, oyster catchers, and emus.

In the winter of 1839, while Elizabeth drew plants in Hobart, Gould explored northern Tasmania, the islands of the Bass Straits, and the area of Australia near Sydney. In Sydney he visited the explorer Charles Sturt, and with Sturt explored southern Australia. He went on to the homestead of his brothers-in-law, Stephen and Charles Coxen, in the upper Hunter district, where he collected lyrebirds to send to Richard Owen, the comparative anatomist. By this time he had collected "about 800 specimens of birds, 70 of quadrupeds . . . and the nests and eggs of above 70 species of birds, together with skeletons of all the principal forms."[51]

Elizabeth's letters to her mother that winter echoed that account of John's accomplishments and expressed a growing interest in her husband's efforts:

> John is . . . so persevering and indefatigable in his exertions, that. . . . He has already done more in obtaining nests and eggs, making skeletons of the various forms of birds, etc., and in getting information of their habits, than has been hitherto done. . . . Many of the birds possess very curious habits, which have not been ever publicly noticed. I think the great mass of information John has obtained cannot fail to render our work highly interesting to the scientific world.[52]

Elizabeth remained in Hobart Town, carrying on her share of their work, until after Franklin Tasman Gould was born on May 6, 1839, and was old enough to travel. John Gould returned to Tasmania for his son's birth,[53] but by the end of May he was in Adelaide, and in June explored the Murray River district in southeastern Australia.[54] He returned to Hobart again for his wife and two sons, and together they departed for Sydney on August 21, arriving September 3.

Elizabeth kept a diary for five weeks in 1839, from August 21 through

September 30. It provides one of the few clues we have to Elizabeth's attitudes toward her role in science. Her diary entries suggest that she had developed an interest in and knowledge of plants; she wrote about her first visit to the Sydney botanic garden: "quite a novel scene to me. A great many beautiful native and Cape of Good Hope plants and shrubs."[55]

In September Elizabeth and her sons accompanied John on a trip to Newcastle, to explore the islands near the mouth of the Hunter River north of Sydney.[56] Her diary presents a picture of the rain forest, since destroyed, on Mosquito Island, where John and his servant had set up camp:

> 17th . . . had a pleasant row to the island. found the tent pitched in a cleard [sic] spot in the midst of the bush where nature appeared in her wild luxuriance. Immense parasites twining around the trees taking root some of them at the tops of the trees and hanging down to the ground, others surrounding the trees like a crown—heard the bell bird [Manorina melanophrys] &c.—[57]

That such an experience was rare for a woman is revealed in a letter written to Elizabeth by Lady Franklin, who was herself unusually well traveled: "I almost envied you to hear of your living in tents on the Hunter, and what do you do next?"[58] Elizabeth did have other glimpses of the wild, although she did not go camping again; her diary reports:

> 23rd [September] Crossed the river with John went into the bush much pleased with its wild scenery—of a different character to that at Newcastle.

John wanted his artist to experience the Australian wilderness and its flora and fauna firsthand.

After leaving Newcastle, the Goulds continued on to the homestead of Elizabeth's brothers. Several times while on the way,[59] Elizabeth recorded in her diary that she was "employed all day making drawings" of both birds and plants but "walked out in evening" or early in the morning. Her entries, such as the following recorded after one of her walks, reveal to us her increasing knowledge of natural productions:

> the air balmy and delightful, great numbers of the blue mountain parrots [rainbow lorikeet, Trichoglossus moluccanus] were making their morning meal on a large kind of the Eucalypti—two of the beautiful Nankeen night herons [Nycticorax caledonicus] passed over our heads and we heard the curious note of the coul [cowl] bird or bald-headed friar [Philemon corniculatus]. . . . in the evening John . . . brought me some beautiful specimens of a climbing plant bearing thick clusters of cream colour blossoms. . . . [September] 29. Saw some beautiful Eucalypti blossoms—the native pear [Xylomelum pyriforme] in blossom and also in fruit now . . . great quantities also of the Kangaroo apple [Solanum aviculare] in the bush.[60]

The birds she saw were no longer just beautiful creatures being killed by John. During her sixteen months in Australia Elizabeth had developed a real interest in them for their own sake, as well as for her task of painting them.

While Elizabeth stayed at the Coxens', John, with a party including aborigines, explored the upper Hunter–Liverpool Range–Namoi area.[61] Elizabeth wrote her mother on December 6, 1839: "John has gone to the Liverpool Plains and will be absent 7 or eight weeks, after which we shall leave Stephen and go to Sydney. . . . It is particularly unpleasant to us to be so frequently separated, but of course my going with him is out of the question; he will sleep under a tent all the time."[62] It was John Gould's "most ambitious and successful expedition."[63] Not only did he sleep under a tent all the time, he was constantly on the move, attempting to collect the maximum number of species in the time available. It would have been folly indeed to bring a baby into the bush, where campers experienced both extreme drought and torrential rain; they had to carry their own food and water, and sanitary facilities were nonexistent. There were not just hardships; there were real hazards. Gilbert, who collected birds all over the continent, was killed there in 1845, and he was but one of a number of persons who perished while collecting for Gould in Australia.[64]

In February 1840 the Goulds returned to Sydney; John explored the Illawarra district, then left for home with his family on April 9, 1840. They put the shipboard time to profit, as they had on the voyage out, John in observing and collecting oceanic birds, and Elizabeth in drawing them. By the time they returned to England on August 18, 1840, they had, as they planned, circumnavigated the globe.[65]

The Goulds brought with them several live birds, including two budgerigars or grass parakeets (raised by Charles Coxen), a breed they introduced to England. Elizabeth's portrait shows her with a pet cockatiel (*Nymphicus hollandicus*) brought back from Australia.[66] On August 25 John Gould read his notes on bowerbirds before the Zoological Society of London. The habits of these birds[67] are said to have become the parlor talk of England. Within four months the Goulds published the first of thirty-six parts of *Birds of Australia*, covering six hundred species of Australian birds. For an edition of 250 copies, colorists hand-colored 1,500 plates a month for eight years; the last volume came out in 1848. The Goulds' book on Australian birds, for which Elizabeth painted most of the original watercolors and executed 84 of the 681 plates,[68] has not been surpassed. During the same period John Gould also published a volume on kangaroos, *A Monograph of the Macropodidae or Family of Kangaroos*, and began work on his three-volume *Mammals of Australia*. For this notable achievement, he was elected a full fellow of the Zoological Society of London in 1840 and a fellow of The Royal Society in 1843.[69]

In May 1841 John Gould wrote to Jardine: "I have just taken a cottage at Egham [near Windsor] on the banks of the Thames from this time untill [*sic*] September in order to give Mrs. G. and our little folks the benefit of country air. I shall also get some fishing although I must be in London nearly half my time."[70] On August 10 Sarah Gould was born, and five days later, less than a year after their return to England, Elizabeth Gould died of puerperal fever. She was thirty-seven.

Because she died so young, and because her husband went on to receive greater and greater acclaim for his books, the memory of Elizabeth Gould faded among those outside her immediate acquaintance. After John Gould died in 1881, and until some of Elizabeth's letters were brought to light in 1938, she was almost forgotten. Except for Henry, her own children could hardly remember her. There was only one grandchild, the offspring of a girl who never really knew her mother.[71]

She was not entirely forgotten, however. In the 1890s, as reported by R. Bowdler Sharpe, who completed John Gould's publications after his death in 1881, there were still those alive who remembered Elizabeth at work in the rooms of the Zoological Society on Bruton Street.[72] Edward Lear, who had worked alongside her doing some of the illustrations between 1831 and 1837, wrote to John Gould after Elizabeth's death, "I should like some little sketch—nothing of value to you—done by Mrs. Gould, as a memorial of a person I esteemed and respected so greatly."[73] Neither was she forgotten by the Australians, who have published the most literature on her.

John Gould did not forget his Eliza. He wrote in December 1841 to Hugh Strickland: "this part [part 5 of *Birds of Australia*] was published almost a month earlier than scheduled. . . . I shall be glad of a line saying how you like the present part; almost the last of the work of my Dear and never to be forgotten partner."[74] Gould named a finch, *Amadina gouldiae* (now *Chloebia gouldiae*), found by Gilbert about 1840, described as having all the colors of the rainbow, after his wife.[75] Finally, John Gould memorialized Elizabeth in print, in his introduction to *The Birds of Australia*:

> At the conclusion of my *Birds of Europe*, I had the pleasing duty of stating that nearly the whole of the plates had been lithographed by my amiable wife. Would that I had the happiness of recording a similar statement with regard to the present work; but such, alas, is not the case, it having pleased the All-wise Disposer of Events to remove her from this sublunary world within one short year after our return from Australia, during her sojourn in which country an immense mass of drawings, both ornithological and botanical, were made by her inimitable hand and pencil, and which has enabled Mr. H. C. Richter, to whom, after her lamented death, the execution of the plates was entrusted, to perform his task in a manner highly satisfactory to myself, and I trust equally so to the subscribers.[76]

John Gould was usually "all business," but he truly missed his Eliza. He never married again, and it is reported that in his old age he clung to his daughters for emotional support.[77]

• • •

Systematists are invaluable to the science of biology; in order to communicate experimental results, one must accurately identify organisms, a task measurably aided by illustration. Scientific illustration was a burgeoning field in the nineteenth century, and the Goulds were at the forefront of it. For John Gould, art was a part of his science, and for Elizabeth, science was a part of her art. Their success depended on John's eye for details of both structure and habit, on his skill at shooting, his taxonomic knowledge, his business acumen, and on the accuracy and attractiveness of Elizabeth's lithographs. Because of their work, the name Gould became synonymous with Australian ornithology.[78] Outside Australia and apart from her descendants in England, Elizabeth Gould, unlike her husband, is still relatively unknown, but with Hugh Strickland (1844) we can say that Elizabeth Gould should not be forgotten.

BARBARA J. BECKER

Dispelling the Myth of the Able Assistant

Margaret and William Huggins at Work in the Tulse Hill Observatory

In October 1910, the recently widowed Margaret Lindsay Huggins wrote a long and mournful letter to Joseph Larmor. Her husband, William Huggins, who was recognized by his contemporaries as one of the founding fathers of stellar spectroscopy,[1] had died earlier that year at the age of eighty-six. Larmor had served as secretary of the Royal Society during William Huggins's tenure as its president. Together, the two men administered the Society's affairs and became close friends in the process. Margaret Huggins knew that Larmor shared her grief. She promised him one of William's favorite signet rings as a keepsake, and described the agony of her first pilgrimage to the memorial where her husband's ashes were kept. Finally, she thanked Larmor for his role in obtaining a pension for her years of work as her husband's collaborator:

> No doubt you know about my Pension. £100 a year has been granted me, "for my services to Science by collaborating with" my Dearest. This I *could* accept without *any* reflection on the memory of my Dearest—& with *honour* to myself as well as to *him*. I do regard the Pension as an honour *to him* though it is honourable also to me, & I humbly hope,—*really earned* for the 35 years of *very hard work*. None of you know *how* hard *we* worked here just our two unaided selves.[2]

Accounts of the work of William Huggins always mention that he was assisted in his research by his wife.[3] In Charles Mills and C. F. Brooke's *Sketch of the Life of Sir William Huggins* (1936), for example, we are told that "William Huggins did not allow his marriage to interfere with his work"; rather, he "derived great benefit from his wife's able assistance." They cite Margaret's own enumeration of her contributions to the work of the Tulse Hill Observatory which, in addition to observation, included such things as arranging instruments, monitoring batteries, mixing chemicals, dusting and washing up the laboratory, doing "small things," and being "generally handy."[4] From the description given, the reader is led to understand that while these tasks provided essential support to the work at hand, they were nonetheless subordinate to the research agenda designed and directed by William Huggins.

This chapter will present a different interpretation of the collaborative work of William and Margaret Huggins. This new view has emerged from an examination both of their notebooks, which are held at the Whitin Observatory at Wellesley College,[5] and of their extensive correspondence.[6] These documents provide a vivid description of ongoing daily activity in the Hugginses' laboratory and observatory, and make possible, for the first time, a more definitive assessment of Margaret Huggins's role in the work at Tulse Hill. An analysis of these documents has revealed the complexity of Margaret Huggins and the influence of her observational, interpretive, and supportive contributions. Margaret Huggins was more than an able assistant, amanuensis, and illustrator, whose work conformed to her husband's research interests: Her very presence and expertise not only strengthened but also shaped the research agenda of the Tulse Hill Observatory.

It has been difficult until now to assess the nature and value of Margaret's contributions to the research done at Tulse Hill. First, the historical figure of William Huggins—that of the solitary stellar explorer seated alongside his star spectroscope in his private observatory, an image crafted in part by Margaret Huggins herself—has loomed large in retrospectives on the origins of what came to be known as astrophysics.[7] Second, there is an alluring internal consistency in the information gleaned from the Hugginses' published scientific papers and reminiscent accounts of their collaborative work that has enhanced their authority over the years and blinded earlier researchers to the need to delve beyond the public facade.

Finally, traditional gender roles mimic those which we use today to distinguish principal investigators from support personnel on research teams: Principal investigators are evaluated on the basis of such things as originality and insight into the theoretical and practical problems encountered in ongoing research; support personnel's success is measured on the basis of how well they tend instruments, follow instructions, and work cooperatively as part of a team.[8] There is a risk that when investigative partners are also husband and wife, a hierarchical evaluative structure may appear applicable to their joint scientific work. This risk is enhanced when the body of published papers contains what appear to be clear signatures of something like today's hierarchical teamwork structure: husband as observer, wife as instrument tender; husband as principal interpreter of data, wife as recorder; husband as analyzer of measurement error, wife as corroborator, and so on. Thus, it may seem fitting to evaluate the husband's work in terms of its originality and theoretical insight, and the wife's contributions in terms of the peripheral support it supplies the research effort.

Marilyn Bailey Ogilvie has compared the collaborative dynamics of three scientific couples, including William and Margaret Huggins. She pointed out that because the Hugginses' joint published papers are written principally in William's voice, reliance on these documents alone to determine how they divided the work, particularly observations, yields a limited picture of Margaret's contributions. Ogilvie suggested that a more accurate assessment of the scientific merit of Margaret Huggins's contributions to the work done at Tulse Hill would require an examination of relevant primary material. It was her hope that such an examination would make it possible to contrast Margaret's scientific "originality" with her abilities as data collector and supportive team member.[9]

Ogilvie was right to suggest an historical analysis of the primary sources that underlie the Hugginses' published papers. The notebooks reveal that Margaret and William did not work together as principal and secondary researchers; rather, they worked as collaborative investigators. The constant give-and-take on which such a work relationship is based blurs the usual markers that distinguish the originator of an idea or plan from its executor. Thus, I would argue, "originality" fails to function as a useful indicator of either William's or Margaret's worth as a scientific investigator. There is, however, an alternative dimension of their collaborative work that presents itself as an insightful evaluative tool, namely that of "individual initiative." A careful reading of the notebooks has uncovered evidence of Margaret's initiative in such diverse activities as problem selection, instrument design, methodological approach, and data interpretation, thus providing a clearer sense of the nature, extent, and value of Margaret Huggins's scientific contributions to the work done at Tulse Hill.

Margaret Huggins was born in Dublin in 1848, the second child and elder daughter of John Murray, a solicitor, and his wife, the former Helen Lindsay.[10] She lived with her family in a Georgian town house overlooking the sea at 23 Longford Terrace in Kingstown (now Dun Laoghaire), Ireland, and was about eight years old when her mother died.[11] The information in obituaries and memorial essays can be assembled into a plausible, but apocryphal, story of her childhood.

Margaret's enthusiasm for astronomical research marveled her acquaintances. After her death, her friends drew on childhood anecdotes, many no doubt related to them personally by Margaret, to speculate on the extraordinary circumstances that may have predisposed her to engage in what they viewed as a unique vocation for a Victorian lady. Some attributed her early astronomical interest to her grandfather, others to her reading an article on spectroscopy in a young people's magazine. A number even suggested that the author of this magazine article was none other than William Huggins. "Before [Margaret Huggins] reached her teens," wrote Sarah Frances Whiting after her friend's death, "she worked with a little telescope. . . . Later, inspired by anonymous articles in the magazine, *Good Words*, she became interested in the spectrum, and made a little spectroscope for herself by which she detected the F[r]aunhofer lines. It was the romance of her life that she afterwards became the wife of the astronomer who wrote the papers, and with him made many discoveries with the magic instrument."[12] Some version of this story is repeated with authority in a variety of widely read sources. However, a search through all the volumes of *Good Words* published before 1875, the year of Margaret's marriage, has uncovered only one series of anonymous articles on astronomy. Given the style and content of the articles and William Huggins's eclectic interests at the time, it is unlikely that he was their author.[13]

Articles on a variety of astronomical subjects written by prominent astronomers of the day appeared in the magazine during Margaret's youth. Her awareness of and interest in the work of contemporary astronomers undoubtedly resulted from a combined influence of several of these articles read over an extended period. However, there is one article by Charles Pritchard, then president of the Royal Astronomical Society, that is worth noting. This article, "A True Story of the Atmosphere of a World on Fire," appeared in the April 1867 issue. The "world on fire" that Pritchard described was a recently discovered nova. After recounting the flurry of interest this object generated throughout the international community of astronomers, Pritchard introduced W. A. Miller and "Mr. Huggins" of the Tulse Hill Observatory. According to Pritchard, these gentlemen were experts in spectroscopy, and thus able to analyze the nature of such an unusual star. Pritchard's use of such kinetic phrases as "sudden bound," "strong impulse in a new direction," and "no longer . . . restricted" conveyed the advantages he felt spectroscopy brought to astronomical research. He provided simple directions for constructing and using a spectroscope complete with suggestions to guide the reader's expectations and insure successful observations.[14] If Margaret's interest in astronomical spectroscopy was indeed piqued by reading one particular article in *Good Words*, Pritchard's is a likely candidate.

◆ ◆ ◆

Margaret and William Huggins were married in September 1875 at the Monkstown Parish Church near Margaret's family home.[15] She was about twenty-seven years old, and he was fifty-one. In one account of their first meeting, the pair were introduced by Howard Grubb, William's Dublin-based instrument maker. A more romantic version tells how "the two star-gazers stopped their investigations long enough to 'exchange eyes.' "[16] Unfortunately for the study of the Hugginses as a collaborative couple, there are no known documents substantiating either story.

It would be helpful to be able to trace the sequence of events that sparked William's willingness to work full-time with a collaborator. The few things known with certainty about the early life of William Huggins portend a solitary lifestyle. Born in 1824, William was the sole surviving child of a London silk mercer and his wife. While accounts of his boyhood vary, they all state that he received little formal education aside from some private tutoring at home. When he reached university age, William remained at home and worked as an assistant in his father's shop.[17] Huggins's scientific activities during this period, while not specifically identified, have nevertheless been described as self-motivated and directed—activities pursued independently in his spare time providing him with opportunities for "thinking out problems for himself."[18]

The mid-1850s marked a series of major changes in William Huggins's life. His parents sold the family business in Cornhill and moved to 90 Upper Tulse Hill Road in Lambeth, a growing suburb just south of London.[19] At about this same time, William formalized his interest in astronomy—becoming a fellow of the Royal Astronomical Society, purchasing an improved telescope, and acquiring a bound notebook in which to record his observations. His family's new home provided him with a garden in which to construct a small observatory to house his astronomical instruments.[20]

The notebooks reveal that although William invited others to his observatory at his convenience, and collaborated on select projects with his neighbor, the chemist William Allen Miller, he was principally a lone observer until his marriage in 1875. Margaret's presence changed both the kind of work done at Tulse Hill and its organization. William's terse notebook jottings were replaced by Margaret's lengthy and detailed entries. More importantly, photography suddenly appeared as a new method of recording what had previously been purely visual spectroscopic observations. Evidence gleaned from the notebooks points to Margaret Huggins as a strong impetus behind the establishment of Huggins's successful program of photographic research.

Photographic work at that time was a complex and often frustrating activity even for those with experience and considerable skill. Until the 1880s, when ready-made photographic plates became widely available, few were prepared to invest the time, money, and energy required to make photography an avocation. William Huggins's solo photographic experience seems to have been limited to producing some lunar photographs in the late 1850s.[21] His spectroscopic work on nebulae and the motion of stars in the line of sight was done at a time when rapid improvements were being made in photographic processes.[22] Nevertheless, he pursued this taxing research using only visual observations.

In 1863, William Huggins briefly became involved in an effort to photograph a star's spectrum. This was part of the collaborative research that he did with his friend and neighbor, William Allen Miller. Miller, who taught chemistry at King's College in London, was an old hand at spectroscopy and a skilled photographer.[23] During the course of their 1863 investigations, Miller and Huggins reported they had twice captured an image (albeit disappointing ones) of Sirius's spectrum on a wet collodion plate. Undaunted, they announced at the time, "we have not abandoned our intention of pursuing [our photographic experiments]."[24] Their apparent enthusiasm aside, neither Miller nor Huggins immediately followed up on this plan.

In fact, it was not until December 1876, over a decade later, when the first paper devoted solely to spectral photography done at Tulse Hill appeared in the *Proceedings of the Royal Society* with William Huggins as its author. Huggins wished readers of this paper to view his newer photographic work on star spectra as a resumption of his earlier work with Miller. To establish that connection clearly, he prefaced the 1876 paper with the statement, "In the year 1863 Dr. Miller and myself obtained the photograph of the spectrum of Sirius." Then, following a reminder of the intention to continue experimenting with spectral photography that he and Miller had expressed, Huggins deftly compressed the intervening years into a moment's hesitation by announcing, "I have recently resumed these experiments."[25]

Many accounts credit Margaret with having learned the basic principles of photography in childhood or adolescence. A close friend went so far as to say that Margaret's skills in photography were self-taught and that she mastered them before she made her spectroscope.[26] The number of women who achieved some degree of renown for their work during the early history of photography was small, but not inconsequential. Queen Victoria was an early patron of the Photographic Society of London. This royal enthusiasm for photography may have encouraged women with both leisure and financial means to experiment with the emerging art form.[27] Margaret's interest in photography identifies her as one of an adventurous group of young women of her day.

While the facts behind Margaret's training in photography remain unclear, there is sufficient evidence available in the laboratory notebooks to demonstrate that her practical and technical photographic expertise was considerable by early 1876, when she began, with her husband, a rigorous program of photographic experiments, and assumed the task of making the notebook entries. March 31, 1876, marks Margaret's first entry in the notebooks:

Photographed Sirius. Wet Plate, 9 minutes exposure. Photograph on the edge of the plate in consequence of want of adjustment. 3 lines across refrangible end of spectrum.[28]

The next entry, on April 3, states:

Took a photograph of Venus with a wet plate and 8 m. exposure. . . . Afterwards tried to photograph Betelgeuse with a Dry plate and exposure of 30 m. No image[,] which may be accounted for by the sky being overspread with thin white haze.[29]

Nearly every entry thereafter contains some mention of photographic work. The process employed at the start was wet collodion, although the entry for May 7 notes a comparison of the wet and dry process. Margaret reported:

> The dry plate gave best results. . . . These results were so good that [I] thought I might endeavour to photograph the spectrum of Venus using the same narrow slit I had from the Solar Spectrum.[30]

Here we see her first use of the first person in what was probably a reference to herself. It is, of course, possible but unlikely that she was merely transcribing her husband's personal notes about the work that was done, and that the "I" refers to him.[31] By December 1876, however, Margaret was using the first person plural and sentences in the first person singular. In July 1879, she began to use the initial "W" to single out William's contribution to the work at hand, while she referred clearly to her own work in the first person singular.[32]

Margaret's entries soon revealed her interest in experimental design. On May 9, 1876, for example, she wrote that she "took one or two photographs of Solar spectrum with a view to determining how wide I might open the slit and still obtain lines."[33] By June, she was demonstrating her expertise in improving and adapting instruments as well as methods to the new photographic tasks at hand:

> I had a new and much smaller camera made to use in connection with the above described apparatus. . . . I was occupied upon all favourable days in testing and adjusting this photographic apparatus upon the solar spectrum: at the same time testing different photographic methods with a view to finding, relatively to different parts of the spectrum the most sensitive, and relatively to the whole spectrum the quickest method for star spectra. I found that although otherwise desirable wet collodion processes are open to serious objection on account of oblique reflection—a second spectrum in greater or less degree being invariably present. . . . After this I used in turn Emulsion, Gelatine, and Captain Abney's Beer plates and obtained some excellent photographs of the solar spectrum both by direct sunlight reflected by a Heliostat and by diffused daylight.[34]

As the years went by, Margaret continued to take the initiative whenever photography was employed at Tulse Hill. Her interest in problem selection, for example, came out in a note she added to a letter from her husband to David Gill in 1879. "If only a few nights sufficiently clear come," she wrote, "I want to try photographing a nebula. The difficulty would be to keep it on the slit:—but difficult as it would be I am most eager that we should try & get some result. It would be valuable."[35] By 1887 she was sufficiently confident in her own interpretive skills to note, "I cannot feel sure there is anything on the nebula plate but William fancies there is. Well if there be anything it's practically useless it's so faint."[36] She developed her own ideas about what counted as a quality photograph and what was required to obtain one. "I was . . . unable to be in the Observatory," she wrote in November 1893, "but W[illiam] insisted on working alone. Again tried [the globular cluster] Messier 15. . . . Developed next day and delighted to find a spectrum good enough to tell us something. It is not however as strong as I should have liked & I regret much that W[illiam] would not take my counsel & have left the plate in so that it might have had continued exposure the next fine night."[37]

While published articles about the early photographic accomplishments at Tulse Hill make no mention of Margaret Huggins, it is clear from the notebooks that her photographic skill made possible an important shift in the observatory's research agenda. William Huggins's notebook entries during the five years preceding his marriage show that, in addition to his visual spectroscopic work, he participated in the ongoing astrometric program guided by Greenwich and the Royal Astronomical Society. Thus, he timed occultations and eclipses, observed visual changes in comets, and drew pictures of planet surfaces. But photography played no role in this enterprise until the year after his marriage.

✦ ✦ ✦

Throughout the 1880s, the Hugginses were engaged in two principal efforts. The first involved repeated attempts to photograph the solar corona without an eclipse. The second centered on examining different nebulae to resolve the nature of what came to be known as the "chief nebular line," a green emission line that William had noted some years earlier in the spectra of several nebulae.[38] This line is located tantalizingly close to, but not precisely coincident with, spectral lines associated with several terrestrial elements. Margaret contributed actively to both of these research projects. Both projects embroiled the Hugginses in controversy over methods, instruments, and interpretation of received data.

It was the second effort, determining the nature of the principal nebular line, that became the subject of the first paper on which Margaret Huggins appeared as coauthor with William. As such, it serves as a benchmark in the Hugginses' collaborative relationship.[39] It is not clear why this particular paper was their first coauthored paper. Given William's awareness of its potential for controversy, he might have introduced Margaret on a gentler slope.[40] In the paper's introduction, he explained, "I have added the name of Mrs. Huggins to the title of the paper, because she has not only assisted generally in the work, but has repeated independently the delicate observations made by eye."[41] These "delicate observations" required making repeated direct visual comparisons of the spectrum generated by burning magnesium in the laboratory against that produced by a nebula. The brilliance of the burning magnesium was blinding, while the faint light of the nebulae tested the limits of human visual sensitivity. Nevertheless, this "labourious & anxious task," Huggins wrote in a letter to George Stokes, "I and Mrs. Huggins, who is now a very trained observer of such things, have done to the utmost of our ability, and with the greatest possible care."[42]

What circumstances prompted William to publicly acknowledge the value of Margaret's contributions to their collaborative research at this time? After all, by 1889, she had been serving as his invisible research partner for more than a decade. The notebook records bring to light Margaret's important role in this particular research effort, and provide clues as to the extraordinary factors that came together to push William over the conventional brink to coauthorship with his wife.

The Hugginses' work on this project began in the fall of 1888. On October 12, Margaret described their plans to begin comparing the spectrum of magnesium with that of a nebula:

> We are very anxious to try and determine whether the Mg line which Lockyer asserts is coincident with the 1st Nebula line, and the Mg line which he also asserts to be coincident with our new Nebula line,—really are so coincident. To try and throw light on this very important point—for much may turn on it—we wish to examine by eye various nebulae for the 1st line, with the 15″ and compare directly with the nebula line the spectrum of Mg."[43]

J. Norman Lockyer—editor of *Nature*, professor of astronomy at the Normal School of Science (later the Royal College of Science) at South Kensington, and member of the Committee on Solar Physics—was by this time an archrival of William Huggins. Their clash of personal style coupled with the similarity of their research interests had long since placed them on a collision course. The issue of the identity of the chief nebular line provided one more opportunity for them to meet head-on.

Lockyer contended that all celestial bodies were comprised of swarms of meteorites in various stages of evolutionary development. In his view, the heat generated by collisions of large numbers of meteorites in space made them incandescent. Varying numbers and intensities of these collisions were responsible for the individual differences observed in the population of known nebulae. Lockyer was encouraged in this view by his observation that when magnesium, an element common to meteorites, was brought to a sufficiently high temperature, a line appeared in its spectral signature that was virtually coincident with that of the chief nebular line.[44]

William Huggins, on the other hand, considered the nebulae to be gaseous. When Huggins first observed the chief nebular line in 1864, he suggested that its proximity to a known line of nitrogen indicated the possibility that nebulae contained some exotic form of that element.[45] Subsequent observation dissuaded him of this view, however, and by 1889 he and Margaret were of the opinion that nebulae might be composed of some new and as yet undiscovered material.[46] Hence, they were inclined to argue that the nebular line, while very close to, was distinct from any associated with magnesium.

In order to gather conclusive evidence for their view, the Hugginses set out to make a direct comparison of several nebular spectra with that of burning magnesium. To do this satisfactorily required the spectrum observed through their telescope to be perfectly aligned with that of the comparison apparatus. This alignment was achieved by making use of the magnesium *b* band, a closely spaced series of Fraunhofer lines in the green part of the solar spectrum. Thus, the bright emission lines in the *b* band of the laboratory spectrum were matched up with the dark absorption lines of the daytime sky. This calibration was critical, and much care was taken in its execution. "It has taken much trouble to get everything satisfactorily arranged," Margaret explained. "First, we directed the telescope & spectroscope to the sky. . . . Then Mg was flashed in as required. We did not leave the apparatus until we both felt satisfied that coincidence between the dark *b* [band] of daylight & the bright *b* [band] of the burning Mg was perfect."[47]

Throughout the fall and winter, the Hugginses directed their attention to the problem of the nebular line comparison. In February 1889, William began keeping a separate record of his own observations in an old notebook.[48] The occasional overlap in their notebook entries during this period provides sparse but valuable insight into their individual research interests, methods, and concerns. On March 6, after a number of visual observations and one attempt to secure a photograph of the nebular spectrum in direct comparison with that of burning magnesium, William drew a sketch of what he had observed. With the crosswire centered on the magnesium line, the chief nebular line appeared to him to be just a bit to the left, or the more refrangible side of the magnesium line.[49] In Margaret's entry of March 6, she mentioned that they were thinking of sending a paper to the Royal Society about their work. To that end, she recorded that on March 9, they rechecked the calibration of their apparatus.

> W[illiam] then put the spectroscope on the Moon bringing in the *b* group so that I might observe whether the Mg lines coincided exactly with those of the *b* group. I thought *not* decidedly.[50]

William's entry of March 9 makes no mention of the recalibration, but rather concentrates again on confirming his previous observation of the placement of the nebular line in comparison with that of magnesium.[51] Two days later, Margaret wrote that the calibration was checked once again:

> W[illiam] thought the bright line did fall coincident. Then I observed. I found a difficulty in getting good observations. A number, did rather give me the impression that it was after all coincident: but one thoroughly good observation showed me as distinctly as I saw it on Saturday night that the bright line was not truly coincident but on one side. I left off feeling certain on the point.[52]

In spite of the care they had taken earlier in calibrating their instruments, the telescopic spectrum was now observed to be no longer coincident with that of the comparison apparatus. Margaret observed this lack of coincidence, and, although William was initially unable to confirm her observation, she felt "certain" of it. William apparently became convinced of the alignment problem on the night of March 11, and remarked casually in his notebook entry that "this state of adjustment is satisfactory for comparison of nebulae, and can be allowed for."[53]

While William may not have been aware of it at the time, Margaret's discovery of the lack of alignment in the comparison apparatus averted what would have been for him a professional embarrassment.[54] William's constant references in his published papers to the care and accuracy of his observations were underscored by his use of four significant digits in reporting his results. If the reliability of his data was ever brought into question, it not only would have meant the loss of the debate with Lockyer on the question of the chief nebular line but also would have damaged his credibility in the wider community of astronomers.

By the time the misalignment was recognized, the season for observing the Orion Nebula was coming to an end. There was no time to recalibrate the instruments and take chances on the weather's providing enough clear nights to make a second set of observations. Besides, their only successful photograph had been taken just a week before discovering the misalignment. Difficulties in getting good photographs of the nebular spectrum had already persuaded them to abandon further photographic attempts for the moment and concentrate their energies on visual comparisons.[55] Still, the importance of having some photographic evidence to support their visual observations was uppermost in the Hugginses' minds.[56] Thus, because of Margaret's confidence that their spectroscope was "not *shifty*" and hence that the observed disparity was constant over all their observations, the paper came to be based on data acquired when the instruments were ever so slightly out of alignment.

The separation between the nebular line and that attributed to magnesium was clearly small. In spite of the slight displacement of their equipment, which reduced the separation of these lines even more, both Margaret and William were confident they each had observed it. This, Margaret believed, added even greater strength to their argument:

> Now this observation is important for it showed that our arrangements really *displaced the Mg line slightly to the left towards the neb. line*, thus making it more difficult to observe the doubleness by reducing the true separation between the neb. line & the Mg oxide one. . . . This observation that our Mg lines are displaced slightly towards the *left*, surely gives great force to our observations showing duplicity.[57]

In their paper on the spectrum of the Great Nebula in Orion, the Hugginses artfully converted this potentially disastrous turn of events into a forceful argument in favor of their view on the nebular line. They perfunctorily declared the serendipitous misalignment to be a purposeful one—an experimental design chosen to give their opponents every conceivable advantage:

> Indeed, to prevent any possible error in the observation of apparent want of coincidence of the nebular line . . . the arrangement was purposely made that the lines of magnesium were seen to fall . . . a very little on the more refrangible side of the middle of those lines . . . if under such circumstances, the nebular line was seen on the more refrangible side of that of magnesium the observation would be much more trustworthy.[58]

Was it Margaret's suggestion to transform this accident into a conscious instrumental adjustment? Did this maneuver win her coauthorship? Undoubtedly it played an important role. But it should be pointed out that the observations being made were exhausting ones. Margaret frequently complained that the dazzling light of the burning magnesium tired her eyes. She felt this made her observations less reliable. Margaret would have been about forty at this time while William was sixty-five. Were his eyes so resilient that these observations could have been managed without a collaborator? This kind of work made teamwork essential. While William may have been able to overlook Margaret's earlier contributions, these efforts would have been harder to ignore. Today, it is tempting, but probably mistaken, to suggest that her coauthorship was the consequence of her husband's pioneering support of women engaged in scientific work.[59]

There is another factor that needs to be considered, and that is the Hugginses' interest in preserving and perpetuating William's image as an observer whose extreme care and caution urged him to reject interpretational speculation. In spite of the Hugginses' precautions, the work on the chief nebular line had presented both Margaret and William with tremendous difficulty. It was conceivable that their observations might not be confirmed by others in future trials. Some unforeseen circumstance could force them to modify or even retract their claims. Such a concern would have been much on William Huggins's mind as he contemplated sending the Orion Nebula paper to the Royal Society in March 1889. In fact, just a few months earlier, Norman Lockyer, as part of his effort to find additional observational support for his meteoritic hypothesis, reviewed the existing literature on cometary spectra.[60] This led him to question the accuracy of a diagram William Huggins had included in a paper written a number of years earlier on Comet b, 1881.[61] It turned out that Margaret had drawn the diagram.

Feeling the need to respond to Lockyer's 1889 criticisms and defend her illustration, if only to herself, Margaret added the following comment to her original notebook entry made in June 1881:

> In making the diagram [of the comet spectrum] I was in consultation with my husband. The measures upon which the diagram was laid down were made for the most part by both of us[,] one checking the other[,] and were many times repeated. . . . But no matter how much care is taken in a matter of this sort, accuracy can only be approximate. The characters and intensities of the lines and groups are I think fairly truthfully represented—but the number of the lines given in the faint group between H and h is only a guess. . . . But while acknowledging these guesses, I wish to state that they are not guesses at haphazard, but are guesses founded upon careful examination of our comet plates compared with others concerning the interpretation of which there can be no doubt.[62]

Meanwhile, William wrote to George Stokes:

> The diagram was drawn by Mrs. Huggins & there is a slight error in the relative strength of H & K, but the diagram was not intended to be a *picture* of the solar spectrum, but simply to show the relative positions of the solar lines & the new lines.[63]

It would have been useful to have Margaret in a more visible position this time in 1889 to help shoulder the burden of proof should Lockyer eventually turn up some damning evidence to counter their nebular-line work.

✦ ✦ ✦

William and Margaret Huggins lived and worked together for thirty-five years as collaborative investigators. This interpretation of their working relationship differs from that drawn from the published record and reminiscent accounts. Given the rich store of extant primary source material providing insight into their lives and work, why has the full extent of their collaboration only recently come to light? Why has the image of William as the principal investigator and Margaret as his able, but subordinate, assistant persisted for so many years? It may be argued that the correspondence is too widely scattered, or that the notebooks are not readily accessible to the historian. These are serious obstacles. But there is a more formidable barrier that must be overcome, and that is the power of the Hugginses' historical image.

The traditional and romanticized image of William as the principal investigator and Margaret as his able assistant is largely their own creation. It has endured because it has been verified and amplified by the published accounts, and because it has fit the needs and expectations of those who have retold the tale. William and Margaret worked hard to present themselves as classic representations of Ruskin's ideal Victorian couple: "[The man] is eminently the doer, the creator, the discoverer, the defender. His intellect is for speculation and invention. . . . But the woman's . . . intellect is not for invention or creation, but for sweet ordering, arrangement, and decisions."[64] The strength of this legendary image is captured in the photograph of William seated alone beside his star spectroscope. The absence of Margaret is telling. If we now turn back for a moment to Margaret's impassioned statement to Larmor, that "none of you know *how* hard *we* worked here just our two unaided selves," we catch her in a rare moment of candor that offers us a brief glimpse of the truth.

PAMELA M. HENSON

The Comstocks of Cornell
A Marriage of Interests

$\mathbf{B}$oth individually and as a couple, Anna Botsford Comstock (1854–1930) and John Henry Comstock (1849–1931) forged new paths in nineteenth-century natural history. Their story does not fit a neat pattern of male professional/female assistant or shared professional careers. John Henry was chair of the Department of Entomology at Cornell University, noted for his research on the evolution of insects and his innovative teaching methods. Anna dropped out of college before she married and entered science only as an assistant to her husband. She finished her degree in science and pursued training as a scientific illustrator to help John Henry with his teaching and research. Combining these new skills with her inherent love of nature and the aesthetic, she became a leading figure in the nature-study movement of the 1890s. Both Comstocks retired from Cornell as full professors, with Anna better known nationally than her more technically oriented husband. The road to Anna's remarkable success was not without obstacles, but the couple derived strength from a close partnership in which each influenced and enhanced the work of the other.

Anna Botsford was born in 1854 in a log cabin in western New York State. The only child of older parents, Anna grew up in a loving home on a prosperous farm. Her mother, Phoebe Irish Botsford, a quiet Hicksite Quaker, shared her love of the natural world with Anna on walks through nearby fields and woods and taught her the names of wildflowers and the constellations in the night sky. After one particularly beautiful sunset, Phoebe commented, "Heaven may be a happier place than the earth, but it cannot be more beautiful." This delight in the natural world remained central to Anna's life and career. Young Anna also learned the importance of home. When she was three, the Botsfords moved to a comfortable frame house, which, in the Quaker tradition, was always open to those in need of shelter and comfort. Throughout her life, Anna retained the belief that a home was a gift to be shared.[1]

Anna's parents were widely read and encouraged their daughter's education. She gained her first experience as an educator at the age of fourteen, when a teacher became ill and Anna was asked to teach for six weeks. In 1871 she entered the Chamberlain Institute and Female College in nearby Randolph, New York, to prepare for a college education. She enjoyed her studies but resented efforts to make her "experience" religion. Like her Methodist father, Marvin Botsford, Anna was of an independent mind when it came to faith, and Chamberlain left her with a distaste for dogma. She changed her plans to attend the University of Michigan at Ann Arbor when a Chamberlain alumnus told her about Cornell. Not only was it nonsectarian, but her friend warned her, "It is a great place for an education; but if *you* go there, you won't have such a gay time as you have had here, for the boys there won't pay any attention to the college girls." Anna concluded Cornell would be an ideal place for her education, since it had "the advantages of a university and a convent combined."[2]

The college's founder, Ezra Cornell, and its first president, Andrew D. White, established a nonsectarian campus, devoted to liberal and practical education, where students of limited means could work their way through school. Cornell admitted its first class, of men only, in 1868; women entered Cornell for the first time in 1872. Anna came in the second term of 1874, boarding at the new Sage College. Other early Cornell women included Julia Thomas, later president of Wellesley College; Martha Carey Thomas, who assumed the presidency of Bryn Mawr College in 1894; Harriet May Mill, who devoted her life to the suffrage movement; and Susanna Phelps, who became a noted embryologist and comparative anatomist. The lively young Anna, however, had little of the seriousness of purpose of these future leaders of women's education and suffrage. She found the Thomas sisters admirable but cool and aloof, and she did not join them in the field of women's rights. She did, however, form a lifelong friendship with Phelps, with whom she shared an interest in natural science.[3]

Although early Cornell men did generally ignore their female counterparts, Anna found university life both challenging intellectually and exciting socially. Dances, musicals, sleigh rides, and steamboat excursions on the lakes leavened the hours spent on studies. But Anna's life took its most decisive turn in 1875, when she enrolled in the introduction to zoology course taught by Professor John Henry Comstock, a recent Cornell graduate. "Harry" and Anna soon formed a close friendship. The young professor boarded at Sage College where Anna lived and sat at table next to her. On long walks through the neighboring woods, they collected plants, mosses, and insects. However, Anna left school and returned home, leaving no explanation for why she ended her education after such a promising start. Harry kept in touch and visited her at home frequently, sending her drawing tools so she could learn to draw insects for him. Her well-known piece *The Brook* was engraved at this time. They were betrothed in 1877 and married in September 1878 at her home in Otto, New York.[4]

The young couple was a study in contrasts, in physical appearance and personality. Anna was tall, dark, calm, and outgoing. He was slight, fair, nervous, shy, and spoke with a pronounced stutter. Their lives to that point could not have been more different. John Henry's father, Ebenezer Comstock, had disappeared after joining the California gold rush. Left to fend for herself, his mother, Susan Allen Comstock, became a live-in nurse to invalids. She was forced to abandon her young son, John Henry, and eventually left for California in search of her husband, who had, in fact, died of cholera en route to the West. John Henry endured a succession of often brutal foster homes and at the age of fifteen ran away and found a home for himself with a kindly ship's captain and his wife. "Ma" and "Pa" Turner supported the boy's dreams of an education, allowing him to earn school money in the summer as a cook on the Turners' ship on the Great Lakes. Earlier frequent beatings by an uncle, who was a "hellfire and brimstone" minister, left John Henry with an intense dislike for religion; so when he was warned to avoid the new "godless" university in nearby Ithaca, he promptly applied for admission.[5]

At Cornell John Henry earned his way through school as a lab assistant, construction worker, and chimesmaster. He had planned to study medicine, but his fascination with the natural world was encouraged by his professor of zoology, Burt Green Wilder. In 1873, while still an undergraduate, John Henry was asked to teach an entomology course for his fellow students. Upon graduation he was appointed an instructor in entomology and soon after was named chair of a separate Department of Entomology. He remained at Cornell for the remainder of his career, establishing the foremost center for entomology in North America.[6]

Despite their differences in life experiences, the serene Anna and high-strung Harry were drawn together, complementing each other's independent spirit and love of nature. John Henry built a small home on the Cornell campus overlooking Cayuga Lake where the newlyweds took up housekeeping. It was John Henry's first real home of his own. At Fall Creek Cottage, Anna assumed the traditional role of a university wife, running the household, supporting her husband's career, and providing a warm social environment for his students and colleagues. The Comstocks began a lifelong practice of inviting colleagues and students to take meals at their table and even live with them. Fall Creek Cottage, and later the Roberts Street house, became the zoology department's gathering place for dinners and lively evening discussions by the fireplace. President Andrew White and his wife were among the regular visitors to the lakeside cottage, which always offered an open door and warm hearth.[7]

From the outset the Comstocks shared their lives to a degree unusual for the times. John Henry helped Anna with household tasks, while Anna assisted John Henry with his teaching. When the hoped-for children did not appear, Anna began to devote more time to assisting her spouse, washing specimen bottles in the lab, drawing illustrations for his lectures, making stencils of his syllabi and classnotes, and preparing his correspondence.[8]

Anna's help became more crucial in 1879, when John Henry accepted the position of chief entomologist at the Bureau of Entomology in the U.S. Department of Agriculture. The young couple left their new home in rural Ithaca for the hustle and bustle of Washington, D.C. With a great vision for organizing the nation's work in entomology, John Henry plunged into the new job with typical enthusiasm. Anna had a servant in their city apartment and thus was free to devote even more time to assisting her husband. The commissioner of agriculture was so impressed with her work that in June 1879 he appointed her a clerk in the Bureau of Entomology. Reflecting their progressive approach to work, the Comstocks purchased the second typewriting machine in the USDA so that Anna could prepare John Henry's correspondence and notes. She also cared for the specimen collections and produced drawings of insect anatomy for his reports. She was learning about entomology, indeed, becoming something of an expert in her own right.[9]

The introverted John Henry lacked political savvy; thus, in the change of administrations after President Garfield was shot, he lost his appointment as chief entomologist. In 1881 the young couple returned to Fall Creek Cottage and university life. Taking her work more seriously now, Anna resumed her college studies, determined to receive the same science degree as her husband. To allow Anna time to study, the Comstocks hired a maid and took their meals at Sage College, a practice they followed at a number of busy times in their careers. John Henry took the physics course with Anna, and the two studied together. In 1885, upon completion of a thesis on *Corydalus cornutus* or dobsonfly, an insect of great interest to her husband, Anna was awarded the bachelor of science degree.[10]

To improve her skills as an illustrator, in 1882 Anna had sent away for a set of wood engraving tools and set about mastering that craft. She displayed talent, and the engravings she exhibited at expositions began to receive awards. In 1885 she studied in New York City with the noted wood engraver John P. Davis at Cooper Union. The Cooper Union pioneered the field of industrial art, training women in practical art skills and also developing the aesthetic sensibilities of its students. In 1859 it absorbed the Women's Art School of New York, and by 1870 had 231 women studying in its art school and 25 in its wood engraving department. Anna not only mastered the engraving skills needed to illustrate her husband's rather dry tomes, but in Davis's studio she further developed her talent as an artist. The period of separation while Anna was in New York was not unusual for the Comstocks. The couple corresponded regularly while apart and seemed comfortable with the distance.[11]

The production of publications grew to assume a central role in the Comstocks' personal and professional partnership. John Henry produced several important reports for the USDA and in the early 1880s set out to write the definitive textbook on entomology. He wished to apply Darwin's theory of evolution by natural selection to the study and classification of insects. His goals were to classify insects in a way that reflected their evolutionary history and to develop a new method for taxonomy that would facilitate these evolutionary studies. John Henry also planned to incorporate his innovative ideas about teaching evolutionary entomology through field and laboratory observations and comparisons. This task would occupy the Comstocks for well over a decade. By 1888 John Henry had published only part I of *An Introduction to Entomology*, in which the title page credits noted "with many original illustrations drawn and engraved by Anna Botsford Comstock." Anna's early wood engravings were skillful and accurate representations of insect anatomy. Her illustrations had become so important to the Comstocks' work that in 1893 John Henry and Simon Gage, a close friend and colleague, established Comstock Publishing Associates. As his own publisher, John Henry could control the stocking of his texts, the issuance of regular revisions based on new scientific research, and, most importantly, the quality and accuracy of their illustrations. Anna's work as lab assistant and illustrator continued to be an important asset to his career.[12] As Margaret Rossiter has demonstrated, male scientists whose wives assisted in their work were more likely to become leaders in their field than colleagues who were unmarried or whose wives played no role in their scientific endeavors. Anna's assistance gave her husband a significant advantage over his colleagues as he built the major center for the teaching of entomology in the United States.[13]

Anna was now a professional illustrator, a member of the Society of American Wood Engravers who had exhibited her work in Berlin, San Francisco, the Chicago World's Fair of 1893, and the Paris Exposition of 1900. She received awards at the New Orleans Exposition of 1885 and the Pan-American Exposition of 1901. As Anna matured artistically, her style evolved from textbook illustrations to aesthetic, yet equally accurate, interpretations of the natural world around her.[14]

As Anna grew more confident, her distinctive view of the natural world began to reveal itself in their joint projects as well. When the classic *Manual for the Study of Insects* appeared in 1895, it contained over eight hundred of Anna's original illustrations, and she wrote parts of the text as well. The title page read, "*A Manual for the Study of Insects* by John Henry Comstock, Professor of Entomology in Cornell University and in Leland Stanford Junior University, and Anna Botsford Comstock, Member of the Society of American Wood Engravers," showing how Anna had risen from the role of assistant to that of a full-fledged partner and coauthor. [15]

The influence of Anna's distinctive style of illustration and writing can be seen throughout this text. In addition to her straightforward insect drawings were a few plates, such as her *Moonlight Sonata*, that moved into an artistic realm of expression, demonstrating Anna's joy in the colors and patterns and interplay of plant and insect life.[16] The Comstocks' stylistic differences can be detected in the writing as well. John Henry wrote in the typical scientific prose of the day, as in the following description of bees:

> In the Apidae we find that the lower lip has been highly specialized for the procuring of nectar from deep flowers. Here the glossa is slender and greatly elongate, being longer than the mentum (Fig. 795); the basal segments of the labial palpi are also elongate.[17]

In contrast, Anna wrote in the same section:

> The clumsy rover, the bumblebee, is an old friend of us all. As children we caught her off thistle-blossoms and imprisoned her in emptied milkweed pods, and bade her sing for us. . . . And she has deserved all the attention and affection bestowed upon her, because she is usually good-natured and companionable. She is a happy-go-lucky insect, and takes life as it comes without any of the severe disciplining and exact methods of her cousin, the honey-bee.[18]

This lively, personal prose, accompanied by the dramatic and beautiful plates in the *Manual*, were a product of the interaction of the pair and not at all typical of textbooks of the day. Alpheus Spring Packard's *Guide to the Study of Insects*, their major competitor, was written in dry, technical prose. Its illustrations were often sketchy, copied from earlier works, and poorly reproduced. Both Comstocks were convinced of the importance of the new plates Anna produced for the *Manual*. They did not accept the common practice of using again illustrations produced a century or more earlier. Indeed, John Henry threatened to sue any colleagues who reproduced Anna's illustrations in their works! John Henry was a very visual thinker who believed that accurate illustrations were crucial for comparative anatomy and was determined that their books would be noted for their excellent illustrations as well as text.[19]

The *Manual* contained the results of John Henry's research program on the evolution of insects. Using innovative methods for analyzing how insects had changed and adapted to the world around them, John Henry proposed a new classification for all families of insects. This work established John Henry as a major figure in his field as both a researcher and a teacher. The *Manual* was rapidly adopted as a text in colleges across the country and became a standard reference.[20]

Although listed as junior author of the *Manual*, Anna does not seem to have played a central role in the theoretical work on the evolution of insects. There is no evidence that Anna conducted independent scientific work after writing her thesis for her bachelor's degree. In her autobiography, Anna credits the theoretical scientific work on the evolution and classification of insects solely to John Henry. She collected and prepared specimens, compiled observations on life histories, and prepared engravings that illustrated the anatomic points her husband wished to emphasize. However, she left no laboratory notebooks, nor did she ever publish any monographs or papers on insects. She was fascinated by the beauty, complexity, and harmony of the natural world, so she asked different questions about the natural world than did her husband. Although Anna continued to illustrate John Henry's publications after the *Manual*, she also began writing on her own, directing her writings to a more general audience to whom she could speak on more intimate and emotional terms, an audience that soon became very much her own.[21]

Had Anna wished to become a scientist, her close friend from their early Cornell days, Susanna Phelps, could have served as a role model and source of support. Susanna became a distinguished embryologist and comparative anatomist who conducted independent research. She married John Henry's close friend, Simon Gage, a microscopist who had lived with the Comstocks while a bachelor. Simon Gage and John Henry supported their wives' careers. Both men were charter members of Sigma Xi, the scientific honor society founded at Cornell in 1886, and at their urgings, Anna and Susanna were among the first five women elected members in 1888.[22]

But real career advancement at Cornell was not easy, even with a supportive spouse. Susanna Gage made such important contributions that she was one of the few women listed as a leader in her field in *American Men of Science*, but she was never granted a salaried faculty appointment by Cornell. There were no women professors at Cornell until the 1910s. Perhaps the realistic Anna decided to pursue a course less filled with obstacles. She was also genuinely more interested in the aesthetic in nature and seemed more comfortable in the more traditional roles for women. The role of illustrator was not uncommon for wives of nineteenth-century American scientists and, indeed, was viewed as more compatible with the feminine temperament than research.[23]

Anna's social skills also played an important role in her husband's career. Warm and outgoing, Anna was liked and respected by a succession of Cornell presidents, including Andrew D. White, Charles K. Adams, Jacob G. Schurman, and Livingston Farrand. The Comstocks were especially close to White, during whose tenure John Henry was able to build his Department of Entomology. Adams on the other hand was far less supportive of the entomology program. When Professor Comstock threatened to resign in response to Adams's planned budget cuts, Adams appealed to Anna to convince her husband to stay. Many observers saw Anna as a "power behind the throne" in securing institutional support for her husband. Because the shy John Henry was never inclined toward politicking, many colleagues attributed much of his success in building the most highly regarded department of entomology in the country in large measure to Anna's social and political skills in navigating their way through the shoals of university life.[24]

In 1906 Anna published her rather pithy observations of the social structure of university life as a novel, *Confessions to a Heathen Idol*, under the pseudonym Marian Lee. *Confessions* was in some ways a typical Victorian romance in which mismatched couples eventually sort themselves out and everyone lives happily ever after. But it also painted telling portraits of the personalities found on a campus such as Cornell. The protagonist is the young widow of a college professor who confides each evening to a teakwood "heathen idol" on her writing desk. Confronted with the possibility of remarriage, she hesitates, since she believes that a widow has the best of all possible worlds—the social status of a married woman and the independence of a single woman. Refusing to rush into remarriage to the wrong man, she gives up her independence only when her true love sues for her hand.[25]

The women Anna crafted in this novel—Marian Lee, her widowed mother-in-law, MaBelle, and unmarried friend Hilda—are mature, strong, and independent individuals who run their lives well. They have interests outside the home and are well aware of the many sacrifices they will be forced to make to ensure the success of their impending marriages. It is interesting to speculate how much of the novel reflected Anna's view of her own marriage, especially in view of the premature death of the young professor, Paul Lee. Given the closeness of Anna and John Henry, it is doubtful she wished him an early death as well. But with the heavy demands of his career, she may at times have felt like a widow herself. John Henry supported Anna's career and independence; thus her frustrations at the confining roles for women seem directed more at the university "system" than her husband. But the stresses she described in the novel probably did reflect the many small and large ways in which she bent and shaped her own life to conform to her husband's professional needs. Anna's use of a pseudonym may indicate that she knew she had stepped outside the bounds of propriety. John Henry reviewed the novel with the pithy remark, "For people who want this sort of thing, it is just what they want," but he did not discourage her from publishing it.[26]

In her marriage, Anna served as a calming influence on her high-strung husband. Although both were highly motivated workers, John Henry had a tendency to drive himself beyond his limits. The scientific problems he set out to solve were complex, and it took him over a decade even to begin to find the solutions. He set a grueling schedule for himself, rising at 4 A.M. so he would have several hours of quiet writing time before classes. He would retire at 8 P.M. unfailingly, often leaving Anna with a parlor full of callers. John Henry's drive finally took its toll in 1891, when he suffered a nervous collapse. He took a much needed rest, visiting a friend in Cuba, but Anna did not accompany him. She wrote, "It was quite imperative that I stay home and work, although I longed to go with him." Anna noted in her autobiography that she lacked nursing skills, especially patience, which may account for her remaining in Ithaca to complete illustrations at this critical juncture. John Henry recovered, and their combination of pursuing separate activities while holding common goals continued to serve the Comstocks well.[27]

Anna seems to have taken her husband's temperament in stride. While usually quiet, he could display a temper, such as the time he tired of the many cranberry pies his bride had baked and threw the remainder in the garbage. In her autobiography, Anna related such stories dispassionately and seems to have regarded it as her responsibility to maintain peace at home. Despite occasional stresses in their closely intertwined lives, the couple retained a genuine affection for one another.[28]

In the 1890s the Comstocks expanded their sphere of influence beyond the Cornell campus. Anna and John Henry were both invited to teach at the new Leland Stanford Junior University, headed by John Henry's former Cornell classmate, David Starr Jordan. Jordan tried to entice them to California permanently but settled for an arrangement whereby they taught in California during the winter and in Ithaca in the summer and fall. John Henry established Stanford's Department of Entomology, and Anna taught home extension courses.[29]

In 1893 Anna's career moved into a related but independent direction, nature-study, which combined her interests in nature, the aesthetic, writing, illustrating, and teaching. The nature-study movement arose out of the social crisis that followed an agricultural depression in New York from 1891 to 1893. Many rural New Yorkers left their failed farms for the city. Finding jobs scarce, many then wound up on public assistance. To provide relief, a group of distinguished citizens banded together to form the New York Committee for the Promotion of Agriculture. The committee's goal was to keep rural youth on the farm, in part by teaching them to appreciate the natural world around them.[30]

John Henry had become involved in agricultural issues through his work on insect pests. Anna often accompanied him when he lectured at farmers' institutes and had even given her own lectures to farm wives on rural education and the values of farm life. Because of her experience with rural education and natural history, in 1893 the Committee for the Promotion of Agriculture asked Anna and Professor James Rice to produce a curriculum for nature-study to be used in the Westchester County schools.[31]

Their work was well received, and in 1884 the New York State Legislature appropriated eight thousand dollars to the New York State College of Agriculture at Cornell for extension work, specifying that part of the money was to be used for a nature-study program. Asked by the committee to devote her considerable energies to developing such a program at

Cornell, Anna embarked on her own independent career as a teacher, public lecturer, and writer on nature-study. In 1896–1897 Professor Isaac P. Roberts of the School of Agriculture directed the program with the assistance of Liberty Hyde Bailey, a professor of horticulture, and Anna, who was appointed an instructor in nature-study. Bailey took over direction of the Nature-Study Program in 1897. Anna and Bailey set out to develop teaching tools for the new program, traveling by horse and buggy and by foot across the state, talking with rural teachers about current science education practices.[32]

They were especially impressed with the approach to science education at the Oswego Normal School, developed by a former Cornellian and friend, H. H. Straight. The Oswego School had imported from Europe the Pestalozzian method of education, which stressed learning as a result of sense impressions of nature. According to Johann Pestalozzi, children must be taught to observe and then express the results of their observations. In the Oswego science courses, students were encouraged to handle and study real specimens in a systematic way that would facilitate their forming generalizations about what they observed. Nature, not a dry textbook, was the teacher.[33]

In some ways, this approach was similar to the teaching methods Anna's husband had developed for his science students. John Henry stressed field observations of insects in their natural environment and had constructed an "insectary" building where insects could be reared and observed with their host plants. But as Bailey wrote in *The Nature Study Idea*, nature-study was not synonymous with natural history. Nature could be studied with either of two objectives: "to discover new truth for the purpose of increasing the sum of human knowledge" or "to put the pupil in a sympathetic attitude toward nature for the purpose of increasing his joy of living."[34] The goal of nature-study was the latter, and it allowed Anna to give full play to the almost mystical love of nature she had developed in her youth.[35]

Eventually, the Cornell Nature-Study Program reached a nationwide audience. Anna lectured at colleges across the country and at the chautauquas and farmers' institutes then in vogue. Teachers from across the state came to Cornell each summer to learn these new educational techniques. By 1900, of the 423 students enrolled in Cornell summer sessions, 111 of them were in the Nature-Study Program. Mostly women, the teachers who came to the summer courses were taken on field trips, introduced to laboratory work, and given hands-on experience. The Nature-Study Program and the Department of Home Economics it spawned were the first departments at Cornell to hire women faculty, although they were initially confined to lectureships. John Henry was appointed professor of nature-study in insect life, as well as entomology; indeed, until 1918, the Nature-Study Program was based in the Department of Entomology because of Anna's close ties to her husband's department.[36]

Expressing their lyric love of nature, Bailey and Anna wrote poetry, as well as prose, about nature. With Professor Stanley Coulter of Purdue University, Anna developed a course on nature in literature. She acquired another outlet for her interest in literature when she was appointed a poetry editor for *Country Life in America*.[37]

Anna's work in nature-study even influenced her husband's publications. In contrast to the more technical *Manual for the Study of Insects*, his *Insect Life: An Introduction to Nature Study* of 1897 was aimed at a general nature-study audience. John Henry wrote the text in a much more informal style than his scientific publications. In these illustrations, Anna finally merged her scientific expertise and artistic sensibilities. The insects are seen alongside their host plants, and, in addition to being technically accurate, the images have a grace that goes beyond pure scientific illustration, at times literally framing the descriptive language like an illuminated manuscript. Clearly aimed at a popular audience, the book was published by Appleton's and soon proved a favorite.[38]

In 1903 Anna produced her first book on her own, *The Ways of the Six-Footed*, a compilation of her essays on nature-study that introduced children to the insect world through the everyday habits of such familiar creatures as ants, butterflies, and cicadas. Illustrations by W. C. Baker and O. L. Foster supplemented engravings the busy Anna had prepared for earlier volumes. In 1906 she wrote a small volume, *How to Keep Bees*, at the request of Doubleday, Page and Company. The Comstocks were now in constant demand among national publishing houses as authors with a popular audience. The following year, 1907, Anna and John Henry coauthored *How to Know the Butterflies*, aimed at nature-study teachers. Although Anna was listed as second author, their collaboration seems to have reached a truly equal status in this volume. Such equality of contribution reached its richest quality in their popular publications.[39]

The demands on their time were now so great that they secured a small, secret room on the upper floor of White Hall to which they retreated to work uninterrupted. Anna wrote of these days, "Our writing was the thread on which our days were strung, despite a thousand interfering activities." As the demands of writing and teaching forced Anna to give up plans for new engravings, the Comstocks began to use photography and the services of other illustrators.[40]

Confident from these writing successes, Anna conceived a larger project, a *Handbook of Nature Study*, which would combine the best of her writings. Designed for an audience outside the classroom, it would encompass all aspects of nature-study, from biology to geology to meteorology. Anna's nine-hundred page manuscript was rejected by the State Department of Education and a succession of commercial presses. In desperation, Anna turned to the Comstock Publishing Company, and in 1911 John Henry reluctantly agreed to publish it as a favor to Anna. He was sure, however, that there was no audience for it and it would lose at least five thousand dollars. Much to his surprise and her delight, the *Handbook of Nature Study* turned out to be such a smashing hit that it remains in print today. It has been through numerous editions, translated into eight languages, and distributed on four continents. Ironically, the income from the *Handbook* kept the tiny publishing house afloat and financed many less popular productions![41]

The *Handbook of Nature Study* was in many ways the crowning achievement of Anna's career. In it she introduced her readers to the natural world just outside their doors, whether it be a moth or a weed or a sparrow or a rock. Natural beauty and fascination were right at hand to anyone who took the time to look and ponder. Anna asserted that there were two and only two occupations for a Saturday afternoon or forenoon for a teacher. One was to be out-of-doors and the other to lie in bed, and the first was best! Anna led the teacher through the lesson, beginning with a "leading thought" or goal for the lesson, followed by a section on method and a section on observations by pupils. The essays were well illustrated but primarily with photographs and ended with questions for the student of nature. Anna also included the nature poetry she so loved. In the *Handbook*, Anna was able to find her distinctive voice as a teacher, poet, artist, lover of nature, and woman.[42]

But these successes were accompanied by some bitter disappointments. In the summer of 1898, at the urgings of Liberty Hyde Bailey and with the support of her husband, Anna had the singular honor of being appointed assistant professor of nature-study, the first woman professor at Cornell. The Board of Trustees did not approve of President Schurman's action, however, and her title was returned to instructor of nature-study, although she retained the higher salary. Faculty and trustee opposition to women professors at Cornell remained high until the 1910s. In 1911, again at Bailey's suggestion, Margaret van Rensselaer and Flora Rose were appointed the first permanent female professors, with the understanding that such professorships would be confined to the Department of Home Economics. Anna remained an instructor, at least in title, until 1913, when she was again appointed assistant professor of nature-study. At the age of sixty-six, she was finally advanced to full professor, just two years before her retirement in 1922.[43]

Although very disappointed, Anna did not protest her demotion in 1898. She did not wish to jeopardize her husband's career and did not see herself as a feminist. Although she was a staunch egalitarian, she wrote that she did not support universal suffrage for women and was satisfied if women could vote in local elections. In her family, the men had engaged in political debates and the women confined themselves to more day-to-day concerns; thus Anna focused her energies on issues of curriculum and early education. Anna seems to have accepted such roles, despite her own demonstrated abilities, drive, and success.[44]

But the trustees' action had long-term consequences for the Comstocks' lives that she could not have foreseen. The blow was made doubly bitter when Anna applied for a professor's pension from the Carnegie Foundation for the Advancement of Teaching. The Carnegie Board ruled that Anna lacked the five years as a full professor that would have made her eligible. After the Comstocks' retirement, their publishing house had been poorly managed, leaving them with debts rather than income. In these financially straitened circumstances, full credit for the professionalism of Anna's decades of work would have gone a long way toward relieving their hardship.[45]

If real recognition for Anna's contributions was slow in coming from the Cornell trustees, she had established a national reputation based on her work in the nature-study movement. A noted lecturer, she also served as associate director of the American Nature Study Association and president in 1913. She was a contributing editor of the *Nature Study Review* from 1905 to 1917 and editor in chief from 1917 to 1923, and "Nature Study Review" editor of *Nature Magazine* from 1924 to 1929. She wrote two more popular volumes, *The Pet Book* in 1914 and *Trees at Leisure* in 1916.[46]

By 1923 Anna was so well known that she was voted one of the twelve most outstanding women in America in a League of Women Voters poll. A distinguished professional in her own right, in some ways she achieved national recognition beyond that of her husband. John Henry was proud of Anna's accomplishments, but the professional demands on Anna did place occasional stresses on the marriage. Despite John Henry's worries that Anna's summer teaching load was too taxing on her health, she maintained her strenuous schedule until days before her death. However, when he complained in 1904 about her frequent absences from home, she did cut back her chautauqua lecture schedule. Their intertwined careers at Cornell seemed to strengthen, rather than diminish, their partnership.[47]

After John Henry retired from teaching at Cornell in 1914 and Anna followed suit in 1922, both pursued their active professional lives as professors emeriti. John Henry remained active until he had a severe stroke in 1926. Anna wrote that "this calamity, for us, ended life. All that came after was merely existence." She noted, however, that the telepathy that had always existed between them reached "an amazing perfection" after John Henry lost the power of speech. Despite the burden of caring for her paralyzed husband and her own advancing cancer, Anna continued to teach nature-study courses; indeed she taught her last class in August 1930, just a few days before she died. John Henry died within the following year.[48]

Anna and John Henry's professional evolution is a fascinating and unusual story of scientific collaboration within the context of nineteenth-century life. Initially, Anna constructed her personal life as a housewife and helpmate within the traditional role of the faculty wife, assisting John Henry on the sidelines and creating a home that fostered his position in the Cornell academic community. When the lack of children left Anna with time to devote to her husband's career, she obtained her own degree and honed her skills as an engraver to meet his demands for the best possible visuals in his texts. Anna's story is an example of the "backdoor" way many nineteenth-century women entered science, and the tasks she initially undertook were typical of "women's work" in science of that day.

The Comstocks, however, are unusual in the way they grew individually and as partners. As they both began to take Anna's career quite seriously, she stepped out of the role of invisible assistant into a quite public and independent career as a leader in the nature-study movement. To achieve this, the Comstocks were willing to adopt unorthodox domestic arrangements to free Anna's time for professional work. Although a closely knit couple, they were, for the most part, comfortable with frequent periods of separation while one of them traveled for professional reasons. And very importantly, Anna was given full credit for her contributions on the title pages of their publications. He regarded her, and she regarded herself, as an equal member of their professional team.

Never a scientist, Anna's extraordinary talents took her in a related direction that allowed her to speak in a different voice, one more compatible with her own artistic and feminine spirit. She chose avenues traditionally open to women—art, literature, and children's education—where she felt more at home. She found avenues of expression for her abilities that complemented her husband's career. But she was a woman of great talent and drive, and went far beyond the role of assistant and faculty wife. She eventually achieved national prominence in her own right and left a rich legacy in the generations of children who have learned to love the natural world around them.

SYLVIA WIEGAND

Grace Chisholm Young and William Henry Young

A Partnership of Itinerant British Mathematicians

Grace Chisholm Young and William Henry ("Will") Young, internationally known for their contributions to the development of modern mathematical analysis, shared a stimulating life full of research, travel, and communication with other eminent mathematicians of their time. From 1895 to the 1940s, Grace and Will produced over 220 mathematical articles, several books, and six children. They traveled widely, always seeking a stable academic situation for Will that never materialized. In order to enhance Will's reputation and his chances of obtaining a professorship, they attributed most of their joint work to him. How significant was Grace's contribution? That is an unsolved mystery, but their six thousand letters and the opinions of family and scholars indicate that her part was essential and that Will and Grace were both important pioneering mathematicians. Theirs is an intriguing tale of the intellectual collaboration of two creative mathematical thinkers.[1]

Grace Chisholm was born in London on March 15, 1868, during the reign of Queen Victoria. Grace's parents were instrumental in her early intellectual development: her mother, a gifted teacher and musician, educated Grace and her siblings at home for a number of years, and her father, a retired government official, took her on memorable excursions such as to the British Mint.[2] Grace's early learning was hindered, however, by her frequent headaches and nightmares, which led the family doctor to recommend that Grace study only subjects that she specifically requested. The course of her life would be set by her choices: mental arithmetic and music.

Grace passed the senior examination for entrance into Cambridge University in 1885 at the age of seventeen, but higher education for women was unusual at the time, and her family was reluctant to send her. When Cambridge's Girton College offered her the Goldschmid Scholarship in

1889, however, her father was willing to match its value. Girton, England's first institution for educating women at the university level, had opened in 1869. Grace would enjoy her college years, making many lifelong friends—including her future husband.

Born in London on October 20, 1863, the son of a grocer, William Henry Young went to the City of London School. Will was considered the best mathematics student, as evidenced by his being chosen "mathematical captain" of the school in his final year. He then studied at Peterhouse College, Cambridge, under a major scholarship. According to the academic system at Cambridge, students attended lectures given by professors, but most of the students' time was spent in individual study with a tutor, preparing for the difficult tripos examinations taken upon completion of their studies. Honors and degrees were bestowed on students in accordance with their performance on the tripos. Will found the mathematics courses at Cambridge uninspiring, dull, and repetitive, and he hated the examinations. He had done well, but because success at the examinations was more determined by writing speed and powers of instant recall than by deeper knowledge and creativity, not as well as everyone had expected. Consequently he lost interest in pursuing research in mathematics. Upon the completion of his studies in 1885, he worked as a schoolmaster at several distinguished private secondary schools. In 1888 he was appointed lecturer in mathematics at Girton and fellow at Peterhouse, primarily to be a coach for the students. His personal experiences undoubtedly made him a better teacher; the Girton students found him stern but very effective in preparing them for the tripos examinations.[3]

From Will's reputation, Grace assumed that he emphasized memorization and test performance rather than learning for its own sake. Despite his rumored good looks, she was relieved to be assigned a different tutor.[4] Grace's opinion changed when she first saw him; contrary to the harsh accounts, he seemed "a mere boy . . . needing help."[5] Later, receiving temporary tutoring from Will, she was further surprised and impressed by his creative and enthusiastic teaching style.

In 1892 Grace completed her studies at Girton and scored the equivalent of a "first class" on the Cambridge tripos part 1 examination. (As women did not actually receive degrees from Cambridge at that time, the women's tripos scores were listed separately from the men's, as "equivalences.") As Grace's and another woman's first-class standings were announced, the entire student body of Girton College cheered enthusiastically. In response to a challenge, Grace then took the Oxford University examination unofficially and scored higher than all the Oxford students.[6]

In England women could not earn doctoral degrees, but an experimental program at the University of Göttingen in Germany gave Grace and two American women the opportunity to study there. Supported by a fellowship, Grace studied with the eminent mathematician Felix Klein at Göttingen. In 1895 Grace sought special permission to receive the degree from the government. Pleading her case to the Ministry of Culture was traumatic: "All my words forsook me; . . . what he [one of the officials] said I don't much know, and what I said I know less, but it all took about a minute, and he told me my petition was regarded with favour." Another official explained that "for them it was a very big matter because it was one wave in a very great movement, and one which they all and he in particular have much at heart." There was a mixup about the carriage to take her to her oral examination; the driver assumed that it must be a man who was about to receive a doctorate, and finding no men waiting for a carriage, he drove away. In Grace's words: "It is not so easy to do a thing till now supposed to be an exclusively male performance! . . . I had to go on my own legs as fast as I could, and of course I lost my way, . . . I got to the Aula very hot and five minutes late." At the examination itself, she realized she had prepared for different questions than were asked. Nevertheless, she was informed by Professor Kielhorn of the philosophical faculty that she had made the examination magna cum laude. Her reaction: "I was almost stupefied!" Later she and her friend May Winston "used the occasion to execute a war dance of triumph."[7]

Thus at the age of twenty-seven Grace became the first woman to officially receive a Ph.D. in any field in Germany.[8] Although Sofia Kovalevskaia had been awarded a doctorate in mathematics in absentia from Göttingen in 1874 after submitting a thesis, the rules for doctoral degrees had become stricter, and Grace was required to take courses and pass a difficult examination showing broader knowledge as well as prepare a thesis in order to receive her degree.[9]

Grace began a correspondence with Will by sending him a copy of her dissertation. Gradually their greetings to each other became friendlier and less formal. In one of his letters Will suggested they collaborate on an introductory book on astronomy (which never materialized), and added that he thought she had been looking "extremely well." Grace responded favorably regarding both the book collaboration and an invitation to dinner, "which would be very pleasant . . . if you were to invite one of the dons at the same time. . . . I hope you were not too late for your train . . . we had so much to talk about." When Will proposed marriage to her in 1896, she tried to explain that, due to her responsibility to care for her family, she had no intention of marrying him or anyone else. Most likely she was actually concerned that marriage would mean giving up mathematics. As for Will, somehow he did not hear her—he thought that she had accepted his proposal. Evidently she changed her mind—and she never told him that she had refused him. On May 7, she wrote, "Dearest love, I do so long for you, & hope, yet dare not hope, you will be able to come tomorrow."[10]

After their marriage on June 11, 1896, they settled into a routine with Grace working on research and Will supporting them both comfortably with his tutoring at Cambridge. Grace initially provided the impetus and the formal training for the couple to begin to do research together. But their professional relationship soon changed: "It is wonderful & delightful to me to feel how much I have helped & spurred him on," Grace wrote in early 1897, "but now he is often beyond my control & I am not able to work as I did with him & for him. . . . We have been working hard & enjoying it very much. We do collaborate in a delightful way."[11] Grace and Will agreed, however, that Cambridge was dead mathematically: "There was no mathematician—or more properly no mathematical thinker—in the place."[12] They decided to give up Will's coaching and become itinerant mathematicians: "At the end of our first year together he proposed, and I eagerly agreed, to throw up lucre, go abroad and devote ourselves to research," Grace explained. "Of course all our relations were horrified, but we succeeded in living without help, and indeed got the reputation of being well off."[13] This decision, crucial to their later achievements, required them to defy family, common sense, and the English mathematical community. They knew such a life would never make them rich, but the financial hardships lasted longer than they could have anticipated.

Dependent on savings, they set off for Göttingen in the fall of 1897, after their first child, Frank, was born. There, with the encouragement of Klein, Will wrote and submitted his first article, on the geometry of one-dimensional vectors in n-dimensional space. Grace had finished her first paper for publication, concerning astronomy, in February 1897.[14] The following spring they went to Turin, Italy, the center for n-dimensional geometry. This was a turning point in their professional relationship. Grace wrote to her family and friends enthusiastically to tell them that she was no longer dominating the couple's joint work, as Will was now developing his own ideas, beyond hers and beyond those of other mathematicians.[15] During the eighteen months in Turin, Will wrote three more papers, including one in Italian, and Grace, although occupied with the baby, wrote one paper on her own, also in Italian. Now they could both be considered productive mathematicians.

As the family grew to include six children, Grace's research productivity decreased. She managed the family well, with some help from relatives, and kept herself active in mathematical ideas by extensive correspondence and work with Will. The Youngs were usually able to hire helpers, some more satisfactory than others, for coping with the housework and their children. In addition, after the family moved to Göttingen, Will's sister Ethel came to stay with them and help with the household. When Ethel became ill in 1903, Will's other unmarried sister, Mary Ann ("Auntie May"), moved in and took over household duties; she became a second mother to the children.

Grace, the six children, and their aunt lived at Göttingen until 1908, while Will traveled to and from their home earning a living, coaching part of the year at Cambridge and other places, lecturing, and consulting with other mathematicians. Sometimes Grace left the children with their aunt and joined him. Many of the couple's verbal discussions when they were together and much of their extensive correspondence when they were apart concerned their research and papers. While Will was away from the family home for long periods, Grace managed the household, wrote to him daily (about mathematics and household details), and saw that his articles were published. Both of them worked intensely on mathematics, although Grace's family concerns left less time for research and travel. The children tried not to disturb her, however, and sometimes played quietly in the same room with her while she worked.

From 1898 to 1913, the period of their greatest collaboration, the major goal of the couple was a reputation and mathematical career for Will. Grace threw her energy into assisting with Will's research so that he could concentrate on research ideas and "flood the journals" with papers in his name. She helped to develop and clarify his ideas, filled in proofs, wrote results for publication, corrected mistakes, and checked everything carefully. This last was extremely important; errors tend to propagate, since a false theorem will invalidate subsequent papers dependent on it. Mathematicians can lose their reputations by publishing incorrect theorems. The nature of Grace's assistance is exhibited by Will's footnote to a paper published under his name in 1914: "Various circumstances have prevented me from composing the present paper myself. The substance of it only was given to my wife, who has kindly put it into form. The careful elaboration of the argument is due to her."[16] Grace assumed a role subservient to Will in most of their mathematical work and in their daily life, taking care of routine tasks in order to leave time and opportunities for Will to think creatively about mathematics. Will had no access to secretarial help and no funds to pay for any on his own, so Grace and later some of the children took on these duties. (The habit of using family members as assistants continued even after Will was established and had access to other help, as the family had developed a routine of cooperating as a team.)

More of their work should have been jointly credited; Will said as much in a letter to Grace (quoted in the "Letters" section of this chapter). But, as both felt that the family's fortunes would be better if Will could establish a reputation independently of Grace, they made an agreement to publish predominantly under Will's name—an agreement that fit with the social and economic mores of the period in which they lived.

Will was awarded a Doctor of Science degree from Cambridge in 1903, at age forty. In 1907 he was proclaimed a fellow of the Royal Society. In Grace's letter congratulating him, she referred to their earlier decision to devote themselves to research: "Dear old boy, I hope you feel very happy about the F.R.S. We all do. It is the fruition of your promise in the Regent's Park 11 years ago."[17]

In 1908 the family moved to Switzerland, where the same pattern of life continued. One reason for the move to Switzerland was the strained political relationship between Germany and England as World War I approached and another was that a good education for the children was cheaper there. Will was publishing at a fantastic rate now, still away from the family for long periods, and still unable to find a suitable professorship or any satisfactory employment at a university. Even the tutoring jobs dwindled, because English universities changed their procedures. Of course an academic position was out of the question for Grace because of her gender, despite her research credentials. By necessity, the family lived very frugally, supported by Will's occasional lectureships and several examinerships, which were all part-time. The University of Liverpool was willing to call his position "Associate Professor" in 1911 and "Professor" in 1913, but still paid him only £100 per year.

The main reason that Will could not find a satisfactory academic position was that the university system in England and the rest of Europe included very few distinguished professorships of the kind he sought, and the competition was keen. Moreover Will had followed an unconventional path: he had done no major research until midlife and lacked a formal degree until 1903. His numerous articles were not all of uniformly high quality, and he was pioneering new areas, areas that some established mathematicians looked upon suspiciously. His continental associations disenchanted the English, whereas his English background alienated continental universities. Still, some farsighted individuals wrote glowing letters about Will's work and attempted unsuccessfully to acquire him at their universities.

Will's abrupt, blunt manner may have further hindered his prospects for employment.[18] He was sometimes impatient, even with Grace, although she would interpret his actions in the best light and would never allow anything negative to be said of him.[19] In particular, Grace's family disliked Will from the start. One of her letters to Will written before their marriage referred to his social difficulties: "But you know, dearest, you are a curiously complex character & you ought not to be surprised if people often misunderstand you, especially when a good deal of their knowledge of you is drawn from letters. I expect I have got my work cut out for me to make Nellie [Grace's sister Helen] see you as you are."[20] Will's letters indicate that he was well-intentioned but demanding, outspoken, and critical as well as overly sensitive and especially paranoid about his difficulties in finding a good position. He believed that others would not give him enough credit for his achievements. For example, in several letters exchanged between Grace and Will, he despaired of ever being chosen a fellow of the Royal Society, because he thought certain individuals would block it, and he almost sent an angry letter of protest just before receiving word that the honor had come.[21]

While they were in Switzerland, their fortunes and situation worsened. Auntie May left to care for Aunt Ethel, who had become more seriously ill, their English money was devalued, and the family could not afford their large house and servants. Fortunately the children were resourceful and helpful with running the household: "As advised by the Swiss government, we bred rabbits and we grew potatoes: they were kept on our balconies in specially designed boxes, and we also supplemented our diet by puddings, made from starch, and used nettles and dandelions in place of spinach," according to their son Laurence. "All this my mother supervised, and each of us, except of course my father and my eldest brother, had our appointed domestic tasks to fit in outside our school hours . . . [the household chores] . . . had to be fitted in, most of the year, between school or university hours, and in the vacations between my father's requirements when he was at home, so that the bulk of the work was still done by my mother and generally with her characteristic look of intelligent sympathy, as if she was always ready to help someone else."[22]

Grace earned a small income from her vine culture, fruits, vegetables, and jam. These and other homemaking activities appealed to Grace, as did many subjects besides mathematics: medicine, languages, poetry, music, photography, early education, and her children. While the youngest children attended school, she began and completed the coursework in Switzerland for a medical degree—one of her childhood dreams. But she was never able to complete the internship, and her medical practice was therefore limited to the family. Grace taught her children six languages. She also communicated her love of music; each of the children played an instrument in informal family concerts. Another of Grace's hobbies was developing photographs with her son Frank.

Finally in 1913 the University of Calcutta offered Will a position, which carried an annual salary of £1,000 plus expenses and required only a few months of residence each year. When he arrived in Calcutta, Will found the Indian educational system to be depressing and hostile. In an effort to improve the situation, he planned a thorough study of universities all over the world, particularly of aspects relating to mathematics, so that the Indian university officials might learn from their foreign counterparts. Although he traveled through Europe, Japan, and the United States gathering information on higher education, he never finished his report.[23]

While Will was based in India and temporarily diverted from research, Grace, in Switzerland, was free to develop her own career, since Will was established and the children were older. She worked on new problems and published papers under her own name, in mathematics as well as other fields. One of the first women to do significant mathematical research, Grace was awarded the Gamble Prize (given by Girton College to distinguished alumnae) for her series of papers, appearing from 1914 to 1916 and containing her contributions to the Denjoy-Saks-Young theorem on the foundations of the differential calculus.

Will terminated his appointments with Calcutta and Liverpool in 1916, and set to writing up research ideas; during 1916 and 1917 he published over twenty papers. The London Mathematical Society awarded him the De Morgan Medal, a prize offered every three years for distinguished contributions to mathematics, and the University of Lausanne conferred an honorary degree on him. The Youngs' major tragedy came in 1917—the death at age twenty of son Frank while serving Britain as a pilot in World War I. In 1919, when Will accepted a professorship at the University of Wales at Aberystwyth, Grace remained in Switzerland with the younger children. Their daughter Cecily accompanied Will and assisted him as he began to build up the mathematics department at Aberystwyth. "He made mathematics exciting," recounted one of Will's students of this period. "Often one could not follow his leaping mind during the lectures, but afterwards, when one had read about the matter, the line of reasoning became clear. He spoke to us more as a learned society than as a collection of students. . . . It was one of the more memorable experiences of my life to sit at the feet of one who knew so much mathematics."[24]

In 1922 Will was elected president of the London Mathematical Society for a two-year term, and soon after resigned from his position at Aberystwyth, effectively retiring. In 1928 he was awarded the Sylvester Medal by the Royal Society for his contributions to mathematics. A year later he was elected the second president of the International Mathematics Union, begun in 1920. In the 1930s Grace and Will, still residing in Switzerland, undertook the writing of their autobiographies and other books. Will was elected to an honorary fellowship at Peterhouse College in 1939. In 1940 they became separated by World War II; when Grace accompanied her grandchildren back to England, their plane was the last from Paris to London until after the war, because of the fall of France. Will's lonely death in Geneva, Switzerland, on July 7, 1942, was a sad ending for their glorious partnership. The fellows of Girton College recommended that Grace be awarded an honorary degree, but she died in England on March 29, 1944, before the degree could be conferred upon her. (Recently the degree was awarded posthumously.) Four of the Youngs' children survived them, and remained active and intellectually acute even through old age.[25]

Major Achievements

Most of the Youngs' published articles and books list Will as sole author.[26] Among the articles attributed to Will are those concerning the (modern-day) Lebesgue theory of integration, one of the most influential discoveries in modern mathematics. As G. H. Hardy explained in Will's obituary notice: "Young, working independently, arrived at a definition of the integral different in form from, but essentially equivalent to Lebesgue's. If Lebesgue had never lived, but the mathematical world had been presented with Young's definition, it would have found Lebesgue's theorems before long." When Will realized that Lebesgue's work had preceded his own, he recognized Lebesgue "magnanimously, and set himself whole-heartedly to work at the further development of the theory," thereafter referring to the "Lebesgue integral."[27] Lebesgue included a tribute to Will's work in his book on integration.[28] Hardy also mentioned Will's significant contributions in Fourier series and in the elementary differential calculus of functions of several variables as "two [other] fields in which Young seems to me to show his powers at their highest."[29] What are now known as the "Young-Hausdorff inequalities" are among the results of Will's work on series.

In addition to the work in Will's name, twelve articles and two books are represented as jointly authored, and another eighteen articles are credited to Grace. In 1905 the Youngs published an elementary geometry book in both names. The couple's correspondence about the book and the fact that its German translation was published solely under Grace's name seem to suggest that she alone wrote the book.[30] In those days the study of geometry was normally limited to theorems for the plane, but the Youngs' book included paper-folding patterns for three-dimensional figures as well as pictures of the completed figures. Seven-year-old son Frank helped Grace with some of the figures.

In 1906 Will and Grace jointly published a second book, *The Theory of Sets of Points*, which was the first of its kind. Although set theory is now considered the foundation for every branch of mathematics and is studied by all mathematicians, the situation was different at that time: "In subjects as wide apart as Projective Geometry, Theory of Functions of a Complex Variable, the Expansions of Astronomy, Calculus of Variations, Differential Equations," Grace and Will commented in the original preface to the book, "mistakes have in fact been made by mathematicians of standing, which even a slender grasp of the Theory of Sets would have enabled them to avoid."[31] The mathematical community was not yet receptive to this theory, but at least the distinguished father of modern set theory, Georg Cantor, was enthusiastic about the Youngs' book.[32]

Two joint papers of Grace and Will and some signed by Will alone were important to the development of the theory of cluster sets and prime ends. A textbook written by E. F. Collingwood and A. J. Lohwater mentions these papers.[33] The couple's joint work on "crystalline symmetry" has also influenced modern analysis.

Grace's most significant independent mathematical work was her study of the existence of derivates of real functions—her part of the Denjoy-Saks-Young theorem. According to a once popular textbook on functional analysis, the theorem "is due to Denjoy and to Mrs. Young, who established it, independently of one another, for the case of continuous functions; then Mrs. Young extended it to measurable functions; finally Saks showed that the theorem holds for arbitrary functions."[34] The theorem gives a complete description of what possibilities can occur for the *derivates* of a function—the derivates are four quantities that are considered when deciding whether or not a function is differentiable. This is extremely useful to mathematicians who work on differential equations, because one need only eliminate two possibilities for the derivates in order to show that a function is differentiable.

Two important reference books mention other independent work: Klein's *Elementary Mathematics from an Advanced Standpoint* uses some of Grace's thesis work as an example, and E. W. Hobson's *Theory of Functions of a Real Variable* refers to several of Grace's articles, one of which is used for an estimate of an infinite sum.[35] The subjects of several of Grace's papers show her diversity of interests: higher geometry, algebraic groups, astronomy, Plato, and Pythagoras.[36]

In addition to mathematical works, Grace wrote two books for children, *Bimbo* and *Bimbo and the Frogs*, which were lessons on elementary biology directed toward children. These books involved a family much like the Youngs and included images of cell structure as seen under a microscope. Grace also worked for five years on an unfinished sixteenth-century historical novel, "The Crown of England," now being completed by Laurence Young. Grace wrote wonderfully vivid and humorous descriptions as a start toward an autobiography (never completed). Will's autobiographical notes were more terse and not as developed.

Letters and Opinions Regarding the Collaboration

Written almost every day when they were apart, the Youngs' letters to each other give evidence of Grace's assistance to Will.[37] They illustrate the couple's way of working and flow of ideas, as well as their devotion to each other. In several letters, Will mentioned his concern that Grace's medical studies were taking too much time away from mathematics, and he urged her to spend at least four hours a day on mathematics and at least half an hour a day on reading the literature. He cautioned in one letter, "[I] am very anxious your [medical] lectures should not prevent the progress of math. work which seems to me now of quite the first importance. The book just coming out [*Theory of Sets*] has taken far too long & I want some *higher* geometry books written before this time next year. eg. one on correspondence."[38] Will was always concerned that he and Grace were not accomplishing enough, frequently asking her to read the current mathematical literature and report to him in order to save time. Still, most of the books that he envisioned were never completed.

Will had high standards for Grace, and was often critical of her efforts to arrange the couple's ideas into a publishable form. In a letter to Grace concerning their book on set theory, Will set forth some of his objections; next he briefly sketched a skeleton for the book's preface that shows how few words were necessary for her to understand what he intended:

The 2nd Cor. on p. 60 ie. the last 3 lines, do not follow from the prop. alone. They require what is proved afterwards viz. that we can secure such set of intervals etc. viz. Theorem 32(b) old numbering on p. 62 . . . I can't think how you have managed to put that Cor. in its wrong place. This whole section pp. 49–64 strikes me as difficult. . . . By the way make a note for preface say that (1^0) We are entering in this subject with the holy of holies of mathematical thought. (2^0) Just as mathematics whets the intellect for other subjects eg. for Law, so sets of points does this for mathematics in general. No math. subject has ideas which do not seem easy after grappling with the sets of pts. (3^0) Absence of symbolism does not interfere with these facts.[39]

Will was overflowing with ideas for papers and books, too many to realistically pursue, and expected similar enthusiasm from Grace:

I am beginning away here to want you to do nothing but mathematics & to work regularly at that. . . . I am very fascinated with the mathematics that lies before us now that we know how to write books. We've got to write an introduction to Theory of Functions for one thing!! Let us write, I am sure that is the thing now, & don't let me do any teaching unless I have a decent post or *prospect* offered . . . we have both learned much. Let us use our knowledge. Keep me properly fed up! And let us have an early breakfast & good tea & I shall work like a Trojan. I am sure of it. The London university work just supplies the stimulus required. . . . As regards the Algebra as I call it, I mean the Galois, I think we must clear that up for we can't do diff. eqns. & various other things without. . . . If I had kept on at Cambridge much longer I should have relapsed into a useless hack . . . give me four hours/day.[40]

Grace had a hard time keeping the pace that Will demanded, and apologized, "I am *so* sorry you think I have been slack, I have tried to do the nearest I could under the circumstances to what we had arranged." However, she really enjoyed the work. Concerning the planned higher geometry textbook she wrote, "I got on a good bit with the Geom. yesterday, it is *most* interesting doing it."[41]

Will gave Grace advice for one of her important papers on derivates, and urged that she publish it in her own name:

Delighted to have your long letter. Write a small paper on your function and *prove* it is continuous & prove that it has a differential continuous [? hard to read] everywhere, *prove* that it has an upper right hand derivate distinct from lower right hand derivate at origin & that these lie between the upper & lower limits of the function defined by the differentials [?] in the open interval. Then send it to the Messenger in *your own name*. Can you not in this way get a monstrous function with no right hand derivate at an everywhere dense countable set or does that break down? Say: "So few examples have even yet been given of the possibilities that arise that the following one may be of interest." Then say what it is.[42]

Regarding his view of their collaboration and their attributing of credit, Will wrote:

I hope you enjoy this working for me. On the whole I think it is, at present at any rate, quite as it should be, seeing that we are responsible only to ourselves as to division of laurels. The work is not of a character to cause conflicting claims. I am very happy that you are getting on with the ideas. I feel partly as if I were teaching you, and setting you problems which I could not quite do myself but could enable you to. Then again I think of myself as like Klein, furnishing the steam required—the initiative, the guidance. But I feel confident too that we are rising together to new heights. You do need a good deal of criticism when you are at your best, and in your best working vein.

The fact is that our papers ought to be published under our joint names, but if this were done neither of us get the benefit of it. No. Mine the laurels now and the knowledge. Yours the knowledge only. Everything under my name now, and later when the loaves and fishes are no more procurable in that way, everything or much under your name.

There is my programme. At present you can't undertake a public career. You have your children. I can and do. Every post which brings an answer from you to my last request or suggestion gives me a pleasurable excitement. Life here is more interesting with such stimulants. I am kept working and thinking, too, myself. Everything seems to say we are on the right track just now. But we must flood the societies with papers. They need not all of them be up to the continental standard, but they must show knowledge which others have not got and they must be numerous."[43]

This letter tells much about the nature of their working relationship and the underlying assumptions that Will made. He was condescending to her, writing as though the children were her children alone; he made the decisions and he motivated her. On the other hand, this picture is not exactly accurate. She really made the choice that they would work the way they did, and he was devoted to her. She provided the "steam" to keep him going.

Their daughter Cecily, who—as their eldest daughter and as another collaborator with them—was the leading authority on the Youngs, wrote:

Another famous partnership, that of George Eliot and Lewes, can be taken as in many respects the counterpart of that of my parents. There it was the man who took the brunt of life off the woman's shoulders and spent his creative energies in fostering her genius. This my mother clearly appreciated.

When all is said, it remains that my father had ideas and a wide grasp of subjects, but was by nature undecided; his mind worked only when stimulated by the reactions of a sympathetic audience. My mother had decision and initiative and the stamina to carry an undertaking to its conclusion. Her skill in understanding and in responding, and her pleasure in exercising this skill led her naturally into the position she filled so uniquely. If she had not had that skill, my father's genius would probably have been abortive, and would not have eclipsed hers and the name she had already made for herself.[44]

Their son Laurence saw their working relationship somewhat differently:

The papers and books by W.H.Y. were *all* written up by G.C.Y, except at the end when Cecily sometimes helped with the writing. Mostly that was what *she* did, in addition to verbal discussions. W. H. had to earn a living—there were no research grants.

When her contribution was *substantial*, not merely routine, W. H. *managed* to put her name as joint author. She didn't really care; she would have preferred him to have the credit.[45]

This assertion that Will included Grace as a joint author when her contribution was "substantial" does not seem to fit with other evidence, including Will's above-cited acknowledgment in a paper that Grace had provided "the careful elaboration of the argument." The inconsistency of these two portrayals of the nature of Grace and Will's collaboration suggests that perhaps their son was more inclined to identify with his father, and likewise their daughter with her mother.

Still another viewpoint of Grace and Will's collaboration is provided by I. Grattan-Guinness, the author of several articles on the Youngs:

[GCY] became WHY's secretary and assistant, perfectly capable of making original contributions of her own but basically needed to see that the flood of ideas that was poured out to her could actually be refined into rigorous theorems and results.

When Will was at home he completely monopolised Grace's life and duties. He knew that he was making excessive demands on her, but he could not help himself and realized that one of the advantages of his travels was that it would give Grace periods of quiet and undisturbed work. (Yet when he went away again she usually suffered for three days with a headache!) Nevertheless, at all times it was really Grace who took the decisions of the household. . . . She would . . . decide the final form of the books and papers before they were sent off for publication.[46]

Conclusions

Who contributed the most? Were they collaborators, was he her mentor, was she his mentor, or did she do his work for him? Was she merely his scribe? A good case could be made for each of these descriptions. In fact it appears that their collaboration took on all these forms, at various times, although we cannot know precisely the division of labor and ideas between the couple. On the whole, since each helped the other's career tremendously and their joint results were far better than the combination of what each could have achieved separately, it seems reasonable to consider them as equal partners.

Certainly Will was not the sole originator of the Youngs' research; Grace was more than scribe and manager. Many mathematicians would argue that writing an introduction, carefully stating the results, and putting in the details, even if doing nothing else, is doing most of the paper! She did help in that way, but that description of their collaboration is far too simplistic. Grace definitely was a creative and sustaining force in their work. Grace was an original thinker herself, she provided a participatory ear and sounding board for Will, and she had the stamina and concern to keep everything running smoothly. In fact Will would probably have accomplished very little without her, and he realized it. Will had a profusion of ideas and great intelligence, but on his own he would have had neither the time nor the temperament to carry them through.

Will brought to the partnership a flush of ideas, command of the subjects, creative energy, and dedication to mathematics. Grace shared his knowledge and love of mathematics, and also contributed a creative mind and skill in seeing broad ideas through to their conclusion. Will's ideas came more rapidly, often in a raw, undeveloped form, and he spoke more about them, but sometimes after speaking, he would forget what he had said. Grace spoke less, but she captured his words, thought deeply about his ideas as well as her own, and presented them clearly.[47] Most mathematicians need a supportive spouse as well as stimulating discussions about mathematics—each filled these needs for the other. Both found an excitement and stimulation in the collaboration itself. This is often the case with a collaboration; results come as if by magic. When two mathematicians are working together on a problem, the work becomes more exciting and absorbing, and ideas often come more than twice as fast as usual. Each feels a part of a team; each is particularly reluctant to stop working because of concern about letting the other down.

Although Will was a demanding taskmaster for Grace, the reader should consider that he had a tremendous beneficial effect on her research production and knowledge. Will encouraged her to work on mathematics—not only to write up his papers or their joint work, but also to write independent papers that were to be in her name alone. Grace's dedication to Will helped her to achieve, because she was motivated to please him. It is likely that without Will's influence, Grace would not have continued to pursue mathematics. (It appears that the other women who studied with her at Girton and Göttingen did not continue to do research afterward.[48]) Avenues for women to pursue research careers were scarce, and Grace might have been distracted by her other interests, just as Will had been when they first knew each other. As an example of the climate of their times, consider the case of Emmy Noether, who received a doctorate from Göttingen in 1907 and is considered to be the best woman mathematician who ever lived. In the 1920s the mathematician David Hilbert unsuccessfuly pleaded with the faculty there to offer Noether a modest position as a minor lecturer: "After all," he entreated, "this is a University, not a bathhouse!"[49]

Mathematical research and personal ambition were more important to Will than to Grace. As a married woman she seemed more comfortable in emphasizing the family unit over her own identity, and concentrated on assuring recognition and respect for Will. As she explained to her sister Helen in 1905:

Will has insisted on my name appearing with his, which up till now, at my wish, we have not done. I am rather sorry, I liked being incog. to the outside world, & felt I had a perfect right to do so, husband & wife being one. I confess it seems to me a trifle "ordinaire" to put my name with his on the title page, I don't want to be mistaken for the modern ambitious female, ambitious for herself & her own glorification. Our work has just been our work, as our children are our children but I can understand & I respect his feelings & it is like him to have it & so I give way.[50]

Although due to a variety of circumstances and choices Grace received less recognition than Will during their lifetimes, she still fared well for a woman mathematician of her time. Both Youngs seemed satisfied with their personal and professional lives and with the choices they had made. As Will wrote to Grace in 1909 (referring to a possible position at Glasgow): "Of course there are some advantages in £1400 a year and a house. I wonder however whether people who have three times our income have really as good a time & whether they remain as young as we do!"[51]

PART III

A Spectrum of Mutually Supportive Couples

✦ 8 ✦

MARIANNE GOSZTONYI AINLEY

Marriage and Scientific Work in Twentieth-Century Canada

The Berkeleys in Marine Biology
and the Hoggs in Astronomy

The gendered history of Canadian science is a relatively new area of interdisciplinary inquiry, and many questions need to be investigated to understand how Canadian women's scientific careers differed from those in other countries. We know that science and scientific institutions in Canada developed later than those in the United States, that nineteenth-century Canadian science was highly utilitarian, and that, in the twentieth century, there was little private philanthropy to fund scientific research. There is also evidence that the convergence of two late-nineteenth-century trends in Canada—access to higher education for white, middle-class women of British (though not of French) background and the professionalization of science—opened up scientific careers for women. This was different from the situation in the United States, where, according to Margaret Rossiter, the professionalization of science led to the exclusion of women.[1] The small size of the Canadian scientific community, the increasing prestige of science, and the lack of graduate schools also meant that during the first half of the twentieth century foreign-trained scientists were solicited by universities and the federal government.[2]

We do not yet know how and to what extent women's presence in the paid workforce changed gender relations in Canadian science. From the biographical information available on hundreds of single women and more than a dozen married couples in Canada in the twentieth century, it is clear, however, that among the scientific couples women and men have had different career paths, advantages, and disadvantages.[3] We are beginning to see the gender implications of different scientific fields at different historical periods, and are in the process of identifying factors such as class and ethnic background, in addition to gender and marital status, that have influenced women's careers. It is evident that while single women could have more or less satisfying careers, those women who wished to have both a career and a family rarely succeeded.[4] Career opportunities depended on different institutional practices, the number of male experts in a field, and larger economic considerations. During the Depression, employed, married women were pressed to yield their jobs to unemployed men, and if they wished to continue with their own scientific careers, they had to resort to secrecy and other strategies.[5] Most married women scientists worked with their husbands, sometimes pursuing their own research but mostly helping their husbands as unpaid researchers. Collaboration with her husband could provide intellectual challenge for a woman and lead to private satisfaction, though hardly ever to professional recognition. In most cases such a collaboration resulted in the wife's invisibility within the Canadian and international scientific communities. Only rarely did married couples develop complete partnerships that allowed equal or better recognition for the female member of the team.

In this paper I will explore the collaborative partnerships and successful marriages of two famous Canadian scientific couples from different generations and different disciplines: marine biologists Edith and Cyril Berkeley and astronomers Helen and Frank Hogg. Edith and Helen were among the nineteen Canadian women natural scientists featured at a traveling exhibition organized by the National Museum of Natural Sciences in Ottawa in 1975,[6] and so they became known outside their respective scientific communities. Details of their options and experiences, including events that led to their midlife career changes, need to be documented to understand how their careers developed and were influenced by their partners.

Lifelong Collaboration: Edith and Cyril Berkeley in Marine Biology

The scientific relationship of Edith and Cyril Berkeley went through several stages from their initial meeting as students at the University of London in 1898 until Edith's death in 1963. The convergence of the Berkeleys' careers resulted from several factors: the relatively early stage of marine biology in Canada, the financial security of the couple, and the willingness of the husband to give up his own research to work with his wife.

Born on September 1, 1875, in Tulbagh, South Africa, Edith Dunington was the daughter of Martha Treglohan and Alfred Dunington, an English civil engineer, both of whom encouraged their daughters to follow their own interests. Dunington built bridges in many parts of the world, and Edith and her sister, Laura, lived in several countries before they reached school age and moved to England. In 1889 Edith entered Wimbledon High School, where the curriculum included mathematics and chemistry; there her interest in science crystallized into a determination to pursue university studies. In September 1895 she entered the London Medical

School, a part of University College, London, with a three-year Surrey Council scholarship (60 pounds sterling per year).[7] Although Edith started the medical course, for reasons that are unclear she began to read zoology at University College, London, under W. W. R. Weldon. Through him, she became interested in the anatomy of polychaetes, a complicated class of marine worms (phylum Annelida). She took courses in botany, chemistry, physics, and pure and applied mathematics and, in 1897, passed her intermediate exams with honors in zoology. Her good marks and "ability shown for chemical research" meant that Edith's scholarship was extended for a fourth year. Because of an unspecified illness, she had to postpone her final exams until 1900.[8] By the fall of that year she was teaching science at a girls' school in Kensington.

Cyril [Bergtheil] Berkeley, the son of Alice Collins and Louis Bergtheil, was born in London, England, on December 2, 1878. Cyril's parents divorced while he was a small boy, but he was influenced by his stepfather, who, like many other Victorians, was interested in science. Cyril studied science at St. Paul's School in London (1890–1895) and chemistry at the Industrieschule in Nuremberg, Germany (1895–1897). In 1897 he was "admitted to Professor Ramsay's chemical research laboratory at University College London," where he investigated the relationship between time and temperature in iodide-iodate reactions.[9] In the laboratory he met Edith Dunington, who was doing research with M. W. Travers, Ramsay's assistant. As Sir William Ramsay was known to encourage women science students and scientists, it is likely that he persuaded Edith to pursue research under Travers's supervision—though what she actually worked on cannot be ascertained.[10]

When Edith and Cyril decided to marry, he chose to obtain practical expertise rather than to pursue a degree in pure science and, in 1899, enrolled at the new Agricultural College at Wye, where he studied agricultural chemistry and bacteriology. In 1901 he obtained a position with the impressive title of "Imperial Bacteriologist" to work on the cultivation and manufacture of indigo in India. Like many other British citizens, Cyril and Edith welcomed the opportunity to travel and work in various parts of the empire, and, after their marriage on February 26, 1902, they sailed for India. Traveling first class, they scandalized the passengers by "fishing with an improvised net" for jellyfish while going through the Suez Canal.[11] During the next dozen years, in the Bihar district of Bengal Province, Cyril was working on indigo cultivation and manufacture. Edith, like many other financially dependent English wives in the "colonies," had to give up her own interests.[12] She was far from idle, however. In addition to running a large household, she served as a medical factotum to the surrounding district. While Edith and Cyril were occupied by different but complementary tasks, they developed a shared interest in nature on their sporadic excursions to the Himalayas. Both became fascinated by the large variety of plants, particularly rhododendrons. Influenced by Edith's knowledge about marine worms, Cyril also studied the vertical distribution of earthworms.[13] Eventually the hard work, harsh climate, and separation from their daughter, Alfreda (born 1903), who had been sent to live with relatives in England, took their toll. Edith was given a few years to live unless she moved to a drier climate. Fortunately for them, the Berkeleys had accumulated enough capital to take a long holiday while considering their future and, in 1912, left India. They were attracted by orchard cultivation in western Canada and, en route to England, visited the Okanagan Valley of British Columbia. There, they bought a small property including a hog farm, and in 1914 Edith, Cyril, and Alfreda embarked on the next stage of their lives. They soon realized, however, that they were not "cut out for the farmer's life."[14] With Edith's health restored, they gave up ranching in 1916, changed the family name officially to Berkeley (because of anti-German sentiment in Canada), and moved to Vancouver. There both Cyril and Edith found positions at the relatively new University of British Columbia.

Around World War I, science was still a new and underpaid profession in Canada, and few Canadian men chose "to dedicate their lives to research science when other professions held greater reward."[15] Those trained in Britain were actively solicited by Canadian universities. Cyril was hired as assistant in the Department of Bacteriology, at one hundred dollars per month. Women scientists, like Edith, also replaced men during the war years at many institutions. F. F. Wesbrook, the university's president, actually recommended that Edith, with her training at University College, London, should apply for the position of laboratory assistant in zoology. She was hired to help A. H. Hutchinson in "starting courses in Zoology" (in the Department of Botany and Zoology) at thirty dollars per month. It is not clear whether she taught or was in charge of the student laboratory.[16] At a time when women students outnumbered the men, Edith was the first woman to be employed at the University of British Columbia in a scientific position.[17]

The Berkeleys spent the summer of 1917 on Vancouver Island collecting simple marine organisms for laboratory work in their respective departments. They discovered excellent possibilities for biological research at the Pacific Biological Station (near Nanaimo), one of two stations founded by the Canadian government in 1908 (the other one was at St. Andrews on the Atlantic coast). The congenial atmosphere of the Biological Station, the beauty of the coast, and the rich research opportunities offered by the complex world of the sea reawakened Edith's original interest in polychaetes. There were numerous species of these marine worms in the Pacific, but little was known about them or their taxonomic relationships to each other.

Cyril, who did not like supervising students, was at the time searching for other jobs and went to work for the Hercules Power Company in San Diego, where he developed a process for kelp fermentation for the large-scale production of acetone. Edith remained at the university. When war work in acetone production came to an end, Cyril rejoined his family in Canada. Alfreda continued her studies in Vancouver, but in 1919 Edith and Cyril moved to the Biological Station to do research as volunteer investigators. This was made possible by the peculiar state of marine biology in Canada as well as their own financial security. Although marine biology was a relatively new and rapidly expanding field of great economic and theoretical importance, the Biological Board of Canada (established in 1912) had no full-time employees until the mid-1920s. The stations provided, however, research facilities and room and board for "volunteer investigators"—that is, academic scientists and their students, who studied different aspects of marine life.[18]

It is ironic that, at a time when married women scientists found it difficult to obtain paid positions, Edith gave up hers, as a laboratory assistant, to become a volunteer investigator. Cyril, who did not want to return to teaching, supported her decision. With no immediate prospects for employment, he was delighted to move to Vancouver Island. Though details of their financial situation are not available, it is clear that the Berkeleys no longer had to work for a living.[19] This enabled Edith, at age forty-three, to embark on her major research, the laboratory investigation of polychaetes.

Cyril had hoped to work on the "chemical activities" of several species of marine bacteria, but there was practically no equipment for such work at the station. Influenced by Canadian biochemist J. B. Collip (of insulin fame), who spent summers at the station, Cyril began research on biochemical problems in marine life. Between 1920 and 1933, he published seven papers on oxidizing enzymes and another dozen on marine biochemical subjects. During this period, Edith became an authority on polychaete taxonomy.[20] Between 1923 and 1930, she published six, and from 1930 to 1961 five papers on her own, establishing one new genus, fifteen new species, and eight new varieties of polychaetes. From 1930 to 1964 Edith and Cyril jointly published thirty-four papers on polychaetes, establishing four new genera, forty-three new species, one subspecies, seven varieties, and one new name as the result of their productive collaboration after 1930.[21]

Edith's involvement in polychaete taxonomy occurred at a time when there were few polychaete experts in North America. She was inundated with samples and requests for identification from many researchers. Collecting, identifying, analyzing, and describing thousands of marine specimens is arduous work. Though fieldwork on the Pacific shores was enjoyable, peering through microscopes for hours on end is hard on the eyes, and writing detailed descriptions takes time away from actual laboratory work. Cyril and Edith had been discussing polychaetes from the late 1890s and collecting them since 1917. By 1930 she was overwhelmed by "new material sent to her . . . from both coasts of Canada and the U.S.A.

with requests for identification."[22] Cyril gave up most of his own research to assist Edith with hers, collecting with her in the field and taking on some of the microscopic work.

Edith preferred research to writing and left no personal account of her life and work. From Cyril's privately printed *Autobiography*, which recounts both their lives and gives full credit to Edith's accomplishments, it is clear this change in direction was due to his desire to help his wife. He assisted her with writing up the results and so "left her free to do most of the . . . dissection involved, and all the drawings."[23] While the two divided up the work to suit their interests, Edith was the senior author on all their joint publications, which appeared in journals such as *Nature*, *Proceedings of the Zoological Society*, *The Canadian Field-Naturalist*, and *Contributions to Canadian Biology*. During the same period (1930–1964) Cyril, a prolific writer, continued to publish his own papers.

Although scientific work took up much of the Berkeleys' time and energy, they had a variety of other interests. Foremost among these was a shared love for their rhododendron garden, with nearly a hundred species and varieties of these magnificent plants reminding the Berkeleys of the Himalayas. In bad weather and in the evenings the two spent time reading on geography and botany (Edith) and philosophy (Cyril). They were friendly and "never too busy to talk to young scientists and their wives . . . it was easy for the Berkeleys to bridge the gaps between several generations."[24] Working during a period when marine biology was becoming professionalized in Canada, the Berkeleys, without advanced degrees, contributed greatly to this scientific field and were at the center of a wide network of North American marine biologists. Their knowledge and genuine enthusiasm influenced their friends, colleagues, and summer students.

During their long lives, and since their deaths in 1963 and 1973 respectively, Edith and Cyril Berkeley received several types of recognition. Four species of polychaetes were named after the Berkeleys collectively, and in 1971 an issue of the *Journal of the Fisheries Research Board of Canada* was dedicated to them. Because of his contributions to agricultural chemistry, in 1919 Cyril became a fellow of the new Canadian Institute of Chemistry. In 1937 a species of earthworm was named after him. In 1968 he received an honorary LL.D. from the University of Victoria, British Columbia, which, he maintained, belonged equally to his late wife. Edith's contributions were also honored individually. In 1958 a rhododendron hybrid, "Edith Berkeley," was registered by the Royal Horticultural Society.[25] In 1969 the Department of Zoology at the University of British Columbia inaugurated the annual Edith Berkeley Memorial Lecture.

"Tied . . . to Astronomy for Life"—Helen and Frank Hogg

The careers of Helen and Frank Hogg, who lived and worked a generation later than the Berkeleys, represent a different pattern of collaborative marriage. Helen Sawyer Hogg (1905–1993) was one of Canada's best-known scientists, an astronomer of world renown, whose career developed in two distinct phases. During phase one, from 1931 to 1951, Helen and Frank Hogg worked in the same institutions, "Mrs. Hogg" in globular clusters and "Dr. Hogg" in spectroscopy. He had the prestigious posts; she, in spite of the recognition of her peers, was employed at a lower level.[26] In phase two, after she became a widow in 1951, Dr. Helen Sawyer Hogg developed a prestigious career of her own.

Astronomy, as a discipline, had long been part of Canadian government investigations, but it was peripheral in Canadian universities. Graduate work had to be pursued outside the country, and most astronomical work in Canada was conducted in the federal government's observatories. Astronomers, like other government scientists, had to conform to rigid regulations, and women scientists, like all women in the civil service, could not remain on the payroll after their marriage.[27] Although there were numerous local astronomical organizations, by contrast with the United States and Britain, avocational or amateur astronomy was underdeveloped.[28]

Born on August 1, 1905, in Lowell, Massachusetts, Helen Sawyer was the daughter of Carrie Sprague, a former teacher, and Edward Everett Sawyer, a banker. At a young age she developed a keen interest in nature, was introduced to star watching, and saw Halley's comet. While a student in Lowell High School, Helen heard enthusiastic reports of Mount Holyoke College from a cousin, and in 1922 she decided to study chemistry there. Brought up and influenced by strong, educated women (her mother and aunt), in college Helen met several forceful women scientists.[29] In her junior year she took Anne Young's astronomy course, which included a trip to Connecticut to view a total eclipse of the sun on January 24, 1925. More than half a century later Helen recalled, "the glory of the spectacle seems to have tied me to astronomy for life."[30] In her last year at Mount Holyoke, she gave up chemistry to concentrate on astronomy.

On January 8, 1926, Helen noted in her diary, "Miss Young asked me to luncheon with Miss [Annie Jump] Cannon." The three discussed scholarships and graduate work at Harvard, and subsequently Young and Cannon wrote to Harlow Shapley, director of the Harvard Observatory, who, in turn, offered Helen the Edward C. Pickering Fellowship of six hundred dollars. "I am so excited it's hard to put my mind to work," Helen noted on February 24, 1926. She accepted the offer and, in June 1926, graduated from Mount Holyoke magna cum laude. Having developed a "fondness" for globular star clusters, groups of stars that contain some of the oldest stars in our galaxy, she was delighted with the prospect of working with the world's authority on this subject.[31]

The Harvard Observatory was used for research by professors and students from both Harvard and Radcliffe. Helen studied photographic plates in the observatory's collection (taken at Bloemfontein, South Africa, and other astronomical stations connected with Harvard) to identify variable stars, which give out varying amounts of light. The December 1927 meeting of the American Astronomical Society at Yale University had a major impact on her career. Helen knew about the various astronomical societies from Young, and was encouraged to go to the 1927 meeting by Shapley, who even drove her there. A talk by Jan Schilt of Columbia concerning the lack of meaningful material on "the period-luminosity curve for Cepheids [variable stars] in globular clusters"[32] gave new directions to this young scientist, who was "still searching for a research mission in life."[33]

Helen Sawyer's work on photographic plates (taken by others) was in the tradition of "women's work in astronomy"—painstaking, detailed, repetitive work traditionally done by women,[34] some of whom she actually met at the observatory. With Shapley's prompting, Helen branched out into an area that led to a career of her own and at the same time involved research that refuted her own adviser's (Shapley's) theories. Helen's more active involvement with direct photography using the seventy-two- and seventy-four-inch telescopes at Victoria, British Columbia, and Richmond Hill, Ontario, respectively, came after her marriage to Frank Hogg and their subsequent move to Canada.

Frank Scott Hogg (1904–1951), the son of Ida Barber and James Scott Hogg, M.D., was born on July 26, 1904, in Preston, Ontario. He studied honors mathematics and physics at the University of Toronto and developed his interest in astronomy in Clarence A. Chant's course. With a gold medal from the Royal Astronomical Society of Canada, and with the encouragement of his parents and professors, twenty-two-year-old Frank Hogg went to Harvard University in 1926. He was Cecilia Payne's first graduate student and worked with her on stellar spectroscopy. Payne was a recent graduate, and there is no evidence that she particularly mentored Frank or in any way predisposed him to support the scientific aspirations of his wife and other women. But by the time Frank Hogg met Helen Sawyer, there were several astronomical couples at Harvard, and this may have influenced both Frank and Helen.[35]

In 1929, after becoming the first person to obtain a Ph.D. in astronomy from Harvard, Frank Hogg went to Europe on a Harvard Travelling Fellowship. Helen and Frank were married on September 6, 1930, and spent the first year of their marriage in New England. Her marital status did not prevent Helen from obtaining a teaching post (astronomy) at Mount Holyoke while finishing her doctoral dissertation at Radcliffe. Frank did research at nearby Amherst College and, before long, had two choices: to teach at Harvard or do research at the Dominion Astrophysical Observatory (DAO) in western Canada. Helen refrained from influencing him, but the opportunity to return to Canada and work with the biggest astronomical telescope was irresistible.[36] In August 1931 the Hoggs drove a Model A Ford across the North American continent to Victoria, British Columbia, where the DAO was located. Frank worked with the observatory's director, John Stanley Plaskett (1856–1941), on the radial velocity and rotation of the Galaxy. Plaskett was lucky to obtain funds to hire Frank Hogg at twelve hundred dollars per year. He could not employ Helen, because of antinepotism regulations,[37] but was delighted to have her work at the DAO, provided her with a desk in Frank's office, and offered her access to the seventy-two-inch telescope. She did not know at the time that she was the first woman to use the telescope.[38] Only seven years previously Plaskett did not hire Cecilia Payne, a single woman, because, as an old-fashioned man, he worried that "there would be difficulty about the observing end of it with a woman in this isolated place." As a married woman without a salary, Helen was in a better position to work there, because Frank was her "built-in chaperon."[39] She had already published (from 1927 to 1932) thirteen papers at Harvard, seven with Shapley and six on her own, and remained convinced into her old age that it was because of her "own standing" at the Harvard Observatory that Plaskett "offered" her both the "use of the telescope for . . . research on variables in globular clusters and office room with Frank."[40] Plaskett was in a good position to assess Helen's work, because his own son, H. H. Plaskett, was a professor at the Harvard Observatory. It is more likely, however, that by offering Helen the unpaid research opportunity, he ensured that Frank would accept the position at the DAO.

Helen was the first person to use the telescope for direct photography, instead of spectroscopy. This was quite a challenge. At Harvard, Helen had done no direct observation, though at Mount Holyoke she had used an eight-inch Alvan Clark refractor telescope. She now found that photography with "an instrument the size of 72 inch was a totally new experience."[41] It took Frank Hogg and the observatory's technical assistant several weeks to convert the telescope for direct photography. Finally, on September 22, 1931, Helen Sawyer Hogg exposed photographic plates and recorded globular clusters for the first time. Frank, who had experience with the sixty-inch reflecting telescope at the Mount Wilson Observatory in 1930, and the observatory's night assistant, Frank Hutchinson, were there to help. Nearly fifty years after the event Helen recalled her routine: "I stayed all night at the top of the dome and after each exposure, let the plate holder down in an old leather handbag to observers on the floor below, who swiftly in the darkroom replaced the exposed plate with a fresh one, and I hauled the bag up again."[42] This phase of her research lasted until October, when the revolution of the earth precluded further direct photography. During the winter Helen helped Frank with his work on spectroscopy; this was "behind-the-scenes help . . . like typing manuscripts, not the kind you acknowledge."[43] The rest of the time she studied short-period Cepheids by using previously collected plates from the Southern Hemisphere. Helen returned to night observation only at the end of July 1932, carrying a basket containing her five-week-old daughter, Sally. "As I was nursing, this meant that she had to come to the dome with us for the night."[44] The baby's presence at the observatory did not last long. Dr. Plaskett received a grant of two hundred dollars per year from the Gould Fund of the U.S. National Academy of Sciences, which he turned over to Helen. With this unexpected income she hired a capable live-in maid who freed her from housework and childcare.[45]

In 1934 C. A. Chant offered Frank a position as lecturer in astronomy at the University of Toronto. In addition to teaching, he was to conduct research using the new seventy-four-inch telescope at the newly opened David Dunlap Observatory at Richmond Hill, fifteen miles north of Toronto. The proximity of the Sawyer and Hogg families, the opportunity to work on the radial velocities of stars using the best telescope in Canada, and the pay, two thousand dollars per year (although an average salary for the times), made the new position attractive. Frank became an assistant professor in 1936, an associate professor in 1942, and a professor of astronomy and director of the observatory in 1946. He was president of the Royal Astronomical Society of Canada (1941–1942), published numerous scientific and popular articles, and invented the two-star sextant.

Thus, in 1935 Frank was well launched on an illustrious career, whereas Helen remained unemployed. In their early days in Toronto, they shared an office at the University of Toronto. Helen later recalled that the arrangements "were made with Dr. C. A. Chant who had great respect for me and my ability and it was just taken for granted that I would share Frank's office." While waiting to move to suburban Richmond Hill, she organized the astronomical library of the university. Later, she continued to work without pay at the observatory, but on her own research program on variables and star clusters.[46] This was the beginning of a highly productive period for Helen. In 1936 R. K. Young, the observatory's director, hired her as a research assistant for three hundred dollars per year. The Hoggs' sons, David and James, were born in January 1936 and September 1937 respectively, but household help was available and affordable, and Helen enjoyed three years of intensive research. The pay was poor even after she became a research associate in 1939; in fact until the early 1940s she never earned more than five hundred dollars a year.[47] She worked on her favorite topic, however, and published a number of prestigious papers on variables in globular clusters in addition to her first catalog of globular star clusters. In 1941 Helen was appointed a lecturer of astronomy at the University of Toronto for about eleven hundred dollars per year. Later she recalled that she was paid less than other university lecturers because she was a woman and the wife of a higher-ranking faculty member.[48] It is unlikely that she would have been offered this post if it were not for World War II! Many faculty members were absent because of the war, and Frank, who had a heart condition as a result of childhood rheumatic fever, taught most of the undergraduate and graduate courses in astronomy. At night he worked at the observatory, commuting from the campus in downtown Toronto to suburban Richmond Hill in the subzero winter nights. Helen taught two courses, conducted research, spent time with her children, and discussed science with her husband and their colleagues.

They were a compatible couple, as Helen recalled: "tied together from Harvard days. We liked to talk about astronomy and astronomers."[49] They had a common interest in science, but as they worked in different areas of astronomy, Helen and Frank became internationally recognized experts in variable stars and spectroscopy respectively. Their different research areas precluded competition but ensured continued intellectual discussions. These, in turn, strengthened their personal relationship. From the time of Helen's first foray into stellar photography in Victoria in 1931, Frank was a constant presence while she worked on the telescope. The nature of Helen's work required a two-person team: "As long as he was alive and able," he helped her. While she observed nightly from the top of the telescope, he remained on the platform below to receive the exposed photographic plates. Helen also helped Frank, first with measuring plates of some of his observations, later typing and editing his popular astronomy column in a Toronto newspaper.[50]

Although Helen remained a lecturer while her husband was alive, with his encouragement she spent most of her time on research. Thus, despite her junior position, Helen became a well-known astronomer—a seeming anomaly in the 1990s, but not unusual in an incompletely professionalized science. In 1940–1941 she was president of the American Association of Variable Star Observers; in 1946 she became a fellow of the Royal Society of Canada (the first woman in the physical science division), which embarrassed her, as her husband was not elected; and in 1950 she received the prestigious Annie Jump Cannon Prize of the American Astronomical Society. Her colleagues at the university treated her respectfully, particularly after she was awarded the Cannon Prize.[51]

The first phase of Helen's career came to a sudden end when Frank died on January 1, 1951. Ironically, her husband's premature death improved her career. Becoming a widow "made all the difference in the world." The university president promoted her to assistant professor of astronomy in 1951; she became associate professor in 1955, and full professor in 1957. She was well aware at the time that this was an unusual career path for a woman. In 1991 she remarked, "I don't know if I ever would have become full professor being married to a scientist."[52]

Recognition for her achievements took many forms. In 1968 the government of Canada named Helen Sawyer Hogg an officer of the Order of Canada. She became a companion of the Order of Canada in 1976, an honor she greatly valued. She was the recipient of numerous honorary doctorates and prizes, including the 1983 Klumpke-Roberts award. In 1984 a minor planet was named Sawyer Hogg after her. In June 1992 the University of Toronto dedicated "its 61 cm telescope on Las Campanas, Chile in honour of Professor Emeritus Helen Sawyer Hogg to celebrate her lifetime of accomplishment."[53] Later that year she received the "Commemorative Medal for the 125th Anniversary of Canadian Confederation," an award for those who "have made significant contributions to Canada."[54] An emeritus professor since 1976, Helen was active to the end. In spite of illness, she published articles on astronomy and the history of astronomy and worked on star clusters and her fourth catalog of variable stars. Her last contribution was to prepare a tape on "the role of women in science" two days before her death on January 28, 1993.[55]

Conclusion

The two couples discussed in this essay had successful partnerships for a number of reasons: they were compatible, had complementary scientific interests, supported each other, and the husbands encouraged their wives' research. There were many factors that contributed to different collaboration styles and career paths of the two couples. First among these was the historical context. The Berkeleys were products of the British colonial era; they first sought employment in India and moved later to Canada, then still part of the British Empire. Although both attended university, they had no need for advanced degrees and did not aspire to academic careers. A generation later, the Canadian Frank Hogg and American Helen Sawyer Hogg obtained doctorates in astronomy at Harvard, began research at a Canadian government institution, and later combined teaching and research.

During the first half of their adult life, the two women's "careers" were secondary to those of their husbands. Edith Berkeley, like other British wives in the empire, gave up her own scientific activities to support her husband's career in India, remained financially dependent on him, and was even deprived of mothering their daughter. Helen Hogg depended on Frank financially but did not have to give up either her research or her children. Both women obtained university posts because of the wartime shortage of male scientists. Antinepotism regulations, which usually prevented married couples from working at the same institutions, were waived, although Edith's and Helen's salaries were much lower than their husbands' and those of other men.

Cyril and Frank both had early and successful careers, while Edith and Helen developed careers of their own in their middle age, but for very different reasons. The Berkeleys were financially secure, which enabled Edith to become a volunteer investigator in marine biology. Helen's independent career began when she became a widow, but at a great personal cost. Perhaps because astronomy grew out of the craft tradition, where wives often succeeded their husbands, or because of the urgent need for teachers in a growing department (at a salary lower than Frank had commanded), Helen was given rapid promotion. It is the only known case of a wife replacing her husband in Canadian science. By contrast, Edith's scientific career benefited because Cyril gave up his own investigations in bacteriology and biochemistry to assist her research. Their congenial married life and the years of scientific collaboration brought them great satisfaction, honors, and recognition. The Hoggs' collaboration was different because Helen's work with a huge telescope was based on a two-person team. Frank willingly formed part of that team both from a genuine desire to promote Helen's research and because throughout their married life she never earned enough to hire an assistant.

Although most women scientists in Canada had inferior careers to those of equally qualified men, a few collaborative relationships among married couples, such as the Berkeleys and Hoggs, transcended the factors that, on paper at least, determine successful careers. While Edith Berkeley was a generation older than Helen Hogg and they were born and educated in different countries, it was in Canada that they had scientific careers. Creative strategies enabled them to work without pay, or for a pittance; their close relationships with their supportive husbands and clear assessment of their options helped them to weather the difficult years. Edith Berkeley and Helen Hogg were productive scientists because of their own ability. In spite of economic and institutional constraints, each in her distinctive way managed to have a career of her own.

SYLVIA W. MCGRATH

Unusually Close Companions
Frieda Cobb Blanchard and
Frank Nelson Blanchard

Frieda Cobb and Frank Blanchard arrived at the University of Michigan in 1916 to pursue graduate study in botany and zoology. Already knowledgeable scientists with a love of the natural world, Frieda and Frank met at Ann Arbor, found that they shared many interests, eventually married, and became, in Frieda's words, "unusually close companions." Frank, a herpetologist, became widely known for his field- and taxonomic work and his definitive life-history studies of several species of snakes and salamanders. Frieda's research on *Oenothera* (evening primroses) was important in the understanding of mutations that was developing in the early twentieth century. After her marriage to Frank, she extended her genetic studies to garter snakes. The two collaborated in studies of those animals. Their work provided the first demonstration of Mendelian inheritance in reptiles.

Frieda Cobb was thirty-two and already a professional biologist when she married Frank Blanchard. She was born in 1889 in the era that Margaret Rossiter has described as one of developing professionalization of science, an era in which women became professional scientists but usually held subordinate positions, accepting (perhaps unconsciously) a pattern of segregated employment and underrecognition.[1] Her introduction to science came while the pattern Rossiter discussed was being set. Then and later Blanchard pursued research opportunities that her skills and opportunities opened for her, doing so because of her interest in the work itself rather than from a desire to promote women scientists. She continually sought to understand more about the natural world and shared her delight in nature with those with whom she came in contact.

Blanchard's strength as a scientist and her attitudes toward marriage and family developed out of an unusual family background. Her father, Nathan Augustus Cobb, born in Massachusetts in 1859, was a pioneer plant pathologist and nematologist whose career included research on three continents. He was Frieda's first scientific mentor. Like many American scientists whose careers began in the late nineteenth century, Cobb studied both in the United States and abroad, earning a Doctor of Philosophy degree at the University of Jena in Germany in 1888 with a dissertation on parasitic and free-living nematodes.[2] In 1889 he moved to Australia, where he soon became the plant pathologist of the Department of Agriculture in New South Wales, the first full-time plant pathologist in the British overseas government.[3]

Frieda Cobb, born in Australia on October 2, 1889, grew up in a closely knit, scientifically oriented family. Her mother, Alice Vara Proctor, had graduated in the first class (1876) of the state normal school in Worcester, Massachusetts, and taught for several years before marrying Nathan Cobb in 1881. The Cobbs took their three small children to Europe and had three additional children, of whom Frieda was the first, in Australia. Alice Cobb shared her husband's love of science and helped him in his work. He always had a home laboratory where he did some of his research and crafted tools that he needed. When possible the Cobbs employed domestic help, which gave Alice Cobb time to devote to her husband's research. As she told her sister, "I write, draw, mount specimens for microscopic work, set type and print (rubber type) cut labels, clean slides, cover slips, etc. make trays for holding slides, etc. etc." Nathan and Alice Cobb shaped the attitudes of their children toward science. The young Cobbs shared their mother's devotion to their father's work as well as his enthusiasm for science.[4]

One of Cobb's many experiments, in which his children helped him, involved a new method to analyze the internal structure of wheat cells, in order to estimate the protein content of flour produced from various varieties of wheat. He compared the amounts of cell wall and cell cavity in the aleurone layer, a special external layer in cereal grains. First he prepared and photographed cell layers. He then projected magnified images from the negatives on large sheets of paper and carefully traced those images. The next step was to cut the cell wall lines of the traced images with a sharp steel tool, to separate the cell walls and the cell cavities. That last task Cobb assigned to his daughters.[5]

Frieda Cobb recalled that after school she and her three sisters sat around the dining table carefully cutting the papers. "Though we did not know just why, there was something tremendously important, almost sacred, about the aleurone layer. The breeze was shut out of the room; if one 'cell' accidentally was lost, or blown into the wrong plate, probably our father's scientific career would be wrecked—or, anyway, we would have failed him and he would be disappointed in us, an equally great calamity."[6] The Cobb children also helped make indexes for some of Nathan Cobb's publications and assisted with other routine tasks. More important, they internalized the belief that their father's scientific work was crucially important and the household must be organized to support that work. They loved and admired him, as he did them. Only much later did they understand their mother's central role in helping her husband and in creating a supportive home atmosphere.[7]

Frieda Cobb spent most of her childhood in Australia, but in 1898 Alice Cobb and the children returned to Massachusetts for a two-year stay while Nathan Cobb traveled and studied in Europe and the United States. Frieda attended public schools in both Australia and the United States and had some lessons from her older sister.[8] However, her informal education at home was at least as important as her formal schooling in shaping her interests. Not only did she learn the importance of the scientific work ethic, but she also gained a love of plants and animals.

In 1905 Nathan Cobb resigned his position with the Australian government so that the children could continue their education in the United States. He moved first to Hawaii, where he designed and organized a laboratory to study the diseases of sugar cane, and then to Washington, D.C., where he spent the rest of his long, productive career working for the U.S. Department of Agriculture. He described over a thousand species of nematodes, grouping them taxonomically. He shaped the new science of nematology, developing methods of study still used in the 1990s. Included in every edition of *American Men of Science* published before his death in 1932, Cobb received stars in the 1921 and 1927 editions. Nematologists consider him the "Father of American Nematology."[9]

Frieda was fifteen when the Cobbs left Australia. She worked in her father's laboratory in Hawaii for over a year, learning additional laboratory techniques and helping with photography, which became one of her major lifelong interests. When the Cobbs moved to Washington, she finished high school there. In 1909, when she was almost twenty, Frieda entered Radcliffe College. There she studied science, taking primarily botany and zoology. She joined the Science Club, serving as president in 1911–1912.[10] After three years at Radcliffe, Frieda interrupted her college studies to help her father with marine nematode work in his home laboratory and served as his assistant full-time from 1912 to 1915.[11] In 1915 she entered the University of Illinois, where her older sister Margaret was doing graduate work in psychology. She received her B.S. in general science from Illinois in 1916 and then spent a summer assisting her father in nematode research at Woods Hole, Massachusetts. She collected nematodes, prepared and studied microscope slides, and found her work "simply fascinating."[12] Thus, by her midtwenties Frieda Cobb had a sound education in general science, considerable skill in laboratory techniques, and experience working in or near some major research centers.

In the fall of 1916, when Frieda went to Ann Arbor, Michigan, she did so at the urging of Harley Harris Bartlett, who was doing pioneering work in plant genetics. Bartlett became her second scientific mentor, and Ann Arbor remained her home for the rest of her life. Though certainly well qualified to work with Bartlett, Frieda received Bartlett's offer primarily because of her family connections. Bartlett and Frieda's brother Victor had been classmates at Harvard College. By 1909 Bartlett was working as a chemical biologist for the Bureau of Plant Industry in Washington; he often visited the Cobb home, joining Victor and his sisters for family picnics and canoe trips. In 1915 the University of Michigan offered Bartlett an assistant professorship in the Department of Botany. Attracted by the potential of a new botanical garden being developed at the university, Bartlett accepted the position. However, he wanted one of Cobb's daughters, all four of whom had learned laboratory techniques through working with their father, to go with him as an assistant. Frieda agreed that she would do so after she had completed her undergraduate work at Illinois.[13] She became Bartlett's graduate student, earning her Ph.D. in 1920.

Shortly before Frieda moved to Ann Arbor, Frank Nelson Blanchard arrived there, also as a graduate student. Like Frieda, Frank came from a New England family based in Massachusetts.[14] Born December 19, 1888, he grew up in Somerville, near Boston, but enjoyed summer visits to an uncle's farm. He became an enthusiastic gardener and keen observer of nature and also began a lifelong interest in photography. In 1909 he entered Tufts College, serving as an assistant in botany during his senior year. Based on his work there, he published his first professional paper; that article, on algae, was his only botanical paper.[15] All his subsequent work was in zoology. After graduation from Tufts in 1913, Blanchard accepted a position as an instructor of zoology under Professor Clarence E. Gordon at Massachusetts Agricultural College in Amherst.

The three years at Amherst helped make Frank Blanchard a zoologist. There he developed teaching and laboratory skills and did extensive fieldwork. He taught classes, tutored students, and prepared study specimens; when Gordon took a leave of absence in the spring of 1915, he placed Blanchard in charge of all the work in zoology and geology. On the day that Gordon left, Blanchard taught both the geology class and the invertebrate zoology class. He noted in his diary, "Thus begins my teaching experience. I hope it may continue as well as it starts." He spent most of his time in class preparation, studying, exploring the countryside around Amherst, and playing the piano for relaxation. During part of the summer of 1915, the year before Frieda Cobb was there, he studied at Woods Hole. Gordon encouraged Blanchard to seek graduate fellowships, and after several letters of inquiry, he applied to the University of Illinois and the University of Michigan. Michigan offered him a three-hundred-dollar fellowship. Delighted, Blanchard wrote in his diary, "It makes me happier than I have been for any time since my appointment to my M.A.C. position. I can now take up graduate work at the place I have wanted to."[16]

In June 1916 Blanchard traveled to northern Michigan, where he spent the summer at the University of Michigan Biological Station on Douglas Lake. He thoroughly enjoyed the classes, field trips, swimming, boating, and picnics. At the end of the formal session he helped close the camp, then remained in the area to collect specimens, primarily tiger beetles. It was the first of many summers spent at the station, beginning as a student, later as a faculty member. On September 24, he returned to Ann Arbor, received the keys to an office in the Natural Science Building, and met with some of the zoology professors.[17]

Frieda Cobb also arrived in Ann Arbor that autumn and received an office in the same building as Frank. She too began classes at the university. The zoology and botany departments were both small; students encountered one another frequently. The zoologists usually attended meetings of the botany journal club, and the botanists did the same for the zoologists. Frank's first mention of Frieda in his diary was that "Miss Cobb" spoke at the botany journal club. While Frank, a zoologist, had long been interested in botany, Frieda, though studying botany, also liked zoology. She had a small pet snake, a *Storeria*, acquired when she first moved to Ann Arbor, and was sad when the tiny creature escaped from his dish and was crushed. She noted, "I always had a hankering to take up herpetology as a hobby." Not long after that she wrote to her father that a man in the zoology department working on king and milk snakes needed specimens from the Washington area. She asked him to send some, saying that it would help a fellow student and that she herself was interested in the subject. Later Frieda recalled that she and Frank began working together as graduate students. However, at first, immersed in their own research projects and with different circles of friends, they spent little time together.[18]

After another summer at the Biological Station in 1917, Frank returned to Ann Arbor, where he continued to explore much of the surrounding area, both alone and with others, collecting specimens of snakes, salamanders, frogs, turtles, and beetles. In May 1918 he went to Washington to study specimens at the National Museum; he returned to Michigan in August but resigned his fellowship in November to take a position at the National Museum in the Division of Reptiles. At first he missed Ann Arbor and found it hard to continue writing his dissertation, on the taxonomy of king snakes, but he did complete the work and received his Ph.D. in June 1919. In October, while he was considering other permanent positions, he received an offer of an instructorship at the University of Michigan. At the urging of his major professor, he accepted and returned to Ann Arbor.[19]

Meanwhile Frieda was doing doctoral research in plant genetics with Bartlett. When she began working for him, scientists were attempting to understand inheritance and variation in nature in the light of Mendelian genetics. Working with peas in the mid-1800s, Gregor Mendel had discovered mathematical laws governing inheritance of dominant and recessive characteristics. In the early 1900s many zoologists and botanists helped develop a new field, genetics. In Holland Hugo de Vries introduced the concept of mutation based on his work with the evening primrose, *Oenothera*. Mutations, sudden variations in characteristics that continued to appear in subsequent generations, provided scientists with a way to study and test Mendel's laws by studying the new characteristics over many generations.

However, botanists soon found that the genus *Oenothera* did not appear to follow conventional Mendelian laws. Its genetic behavior became a major puzzle that geneticists tried to solve. That work continued over several decades in both America and Europe. In Washington Bartlett had begun growing *Oenothera* for study purposes and corresponding with de Vries. De Vries visited the United States to study wild *Oenothera* populations, and Bartlett went with him on a trip to Alabama. Together they published an account of their findings.[20] Bartlett, who remained convinced of the correctness of de Vries's basic theories, hoped to develop a major center for *Oenothera* research at the University of Michigan.[21] Choosing Frieda Cobb as his assistant proved a major factor in his success in doing so.

The University of Michigan had created a botanical garden in the late 1800s, but as the university grew, the old site proved inadequate; the university purchased land for a new garden. The first greenhouses and laboratory at the new site were ready for use in February 1916. Bartlett moved his experimental work from Washington to Ann Arbor and began building the new garden as a research center, especially for *Oenothera* studies. When Frieda arrived later in 1916, she quickly became involved in that *Oenothera* work. One task was to plant the seeds, keeping accurate records of all the seeds planted. On October 9, for example, she planted 1,800 seeds from six different collections. Later in the month she helped Bartlett sort out all the seeds from that year's *Oenothera* garden into two or three hundred "nifty little seed envelopes." Then she counted out about 30,000 seeds for additional planting. By the next spring, busy with her graduate classes, she felt she was behind in everything, "except the *Oenothera* work at the gardens, which I never neglect. That *has* to be done, and what's more it has to be done, when it has to be done. It is some work to keep track of it all, and check up after the potting, and record all the plants saved and discarded."[22]

As the work grew, gardeners helped with the actual planting, cultivating, and harvesting, but Frieda became responsible for more and more of the record keeping and supervision at the Botanical Gardens. In 1917 Bartlett agreed to go to Sumatra to study rubber plants and arranged with the United States Rubber Company to provide funds to pay Frieda and another graduate student to carry on his work at the Gardens. He was gone throughout 1918. During 1917 the garden staff had raised 15,000 to 16,000 seedlings for research purposes. In 1918 the total rose to over 30,000. Of those 11,000 were for Frieda, 1,000 for Bartlett's other graduate student, and the remainder for a pharmacy professor who was raising medicinal plants for drugs for the war effort. When Bartlett returned in 1919, he became director of the Botanical Gardens. As one of the conditions of his appointment he insisted that Frieda be named assistant director at a salary of twelve hundred dollars per year, a position considered the equivalent of an instructorship. Frieda thought her appointment was "a remarkable thing for the University to do. . . . fits in with my research work and studying better than any I could have and I am pleased that it went through." In March 1920 Bartlett reported that the administrative scheme he had proposed was working successfully. "The Assistant Director, Miss Cobb, has taken entire charge of the Garden routine, and has relieved the Director to such an extent that no reduction of his teaching has been necessary." He also proposed raising Frieda's salary so that it would remain the equal to that of an instructor, noting "Miss Cobb has given very excellent service to the Garden."[23]

The pattern set by 1919 remained in effect until the mid-1950s: Bartlett continued as director until his retirement in 1955; Frieda remained assistant director until her retirement the following year. Frieda became Bartlett's research colleague in the *Oenothera* studies, and she took over the active administration of the Botanical Gardens, leaving Bartlett free to pursue other interests, including the chairmanship of the Department of Botany, which he assumed in 1923. Frieda kept the Gardens running smoothly, maintaining facilities for scientific research and an atmosphere conducive to such research. Her presence made it possible for Bartlett to branch out into other areas of botanical fieldwork and ethnobotany. He conducted botanical studies in Asia and Central America, published in many fields, and became an internationally recognized botanist who received stars in *American Men of Science*. Warm, outgoing, and generous, he helped and encouraged several generations of Michigan students who knew him as Uncle Harley. Those students served as his extended family, for he never married. He became especially close to and supportive of Frieda, Frank, and their children, but, because Bartlett retained the title of director of the Botanical Gardens, Frieda's position was always technically assistant director. She never received adequate monetary or academic recognition for the value of her administrative and research work.[24]

When Frieda first began working with Bartlett in 1916, she was also a graduate student. She found most of the coursework interesting, fought to overcome a fear of speaking to seminar groups and presenting research papers, and became engrossed in the *Oenothera* research. She understood that she had chosen a long-term project, for she needed to study the structure and chromosomes of plants resulting from various crosses of pure strains over several generations, and it required a year to produce each new generation. However, she found her work fascinating. Based on the work in progress, she and Bartlett prepared a paper, "On Mendelian Inheritance in Crosses between Mass-mutating and Non-mass-mutating

Strains of *Oenothera pratincola*," for the December 1917 meeting of the Botanical Society of America. Bartlett wanted her to present the paper, but she declined. She did, however, "work up the data for him," and when the paper was published in 1919 in the *Journal of the Washington Academy of Sciences*, her name appeared first. Later she presented papers and published regularly.[25] Bartlett encouraged her, as he did many other students, both male and female. Soon she was doing the major share of their *Oenothera* research, while Bartlett turned to his other interests.

During 1917 and 1918 her father made several attempts to persuade her to choose a nematode project for her doctoral research and to return to Washington to work with him. Though acknowledging that she remained very interested in nematodes, she respectfully and tactfully declined his offer.[26] Instead she continued to study *Oenothera*. She earned her Ph.D. in 1920, and her dissertation was published in *Genetics*. In it she demonstrated that Mendelian inheritance did occur in certain strains of *Oenothera*, an important addition to the discussion of plant genetics and mutation theory. She continued her *Oenothera* research, but by the late 1920s scientists were turning to other organisms to study genetic problems because the genetic structure of *Oenothera* is extremely complex and the growing of many plants over many generations is labor-intensive and thus expensive. Frieda published her last *Oenothera* paper in 1929 but continued to collect research data, believing, as *Oenothera* specialists still do, that there remained interesting genetic problems which those plants illustrated.[27]

After Frank returned to Ann Arbor late in 1919, he and Frieda began to take field trips together and became close friends. They shared a love of outdoor life as well as enthusiasm for research in natural history. Frank had often gone camping and canoeing with friends, as had Frieda. Frieda explained to her worried parents that she was safe in doing so: "Here girls do go out of doors and enjoy life so much more than where you are that such things are not conspicuous." After a trip collecting salamanders with a woman friend, she wrote, "I love to go out in the woods with zoology in mind." Frank and Frieda explored a nearby river, watching red-winged blackbirds and finding nests. Frieda wrote, "I went with Mr. Blanchard, of the zoology department. I like to go out with zoologists and learn where to look for different animals and what they are when found. And although I don't know this zoologist very well I have always liked him as far as I have known him (he came here the year I did). He was very kind and thoughtful at the time of my examination."[28]

When Frank went to the Biological Station for the 1920 summer season, Frieda cared for his study animals. She reported to him, "The creatures are all doing well except the garter snake that the king snake ate. I am not sure that you will ever get the king snake back again; he is a great favorite here. He has had five mice." Always helpful to others, she also took on the *Oenothera* work for a colleague who was ill and for another who had left for a position in Washington. She told Frank, "I seem to have a habit of acquiring abandoned work!" Later in the year, when she was in Washington helping a sister with a new baby, she reported to Frank that she had seen "many good earthworms and slugs and flies going to waste. Hope you are not having too much trouble finding food for the animals. I'll soon be there to take care of them again."[29]

Together Frank and Frieda searched the woods for rare four-toed salamanders, one of Frank's research interests, and worked to prepare their dissertations for publication. After a day of reading page proofs, Frieda told her family, "It has been a race to see which of us would have our thesis out first, but I think mine is sure to be now. His comes from the government printing office. It is on king snakes and about ten times as long as mine." They also bought a car together. "The Ford is great fun! . . . It is mine only half of the time, the other half of the time being Dr. Blanchard's. He very much wanted one for field work but did not think he should afford one when there is so much of the year that he could not use it. I had decided to get one anyway, but thought it would be just as well to try a half one first." Frieda had learned to drive during the war, when the Young Women's Christian Association (YWCA) had offered a special program to teach women to drive so that they could, if needed, take the place of men who were overseas. They used the car for field excursions and picnics; Frieda, who for her first five years in Ann Arbor had walked or used a bicycle for the four-to-six mile round trip to the Gardens, enjoyed being able to drive to work.[30]

On June 12, 1922, in a quiet ceremony in Ann Arbor, Frieda and Frank were married with Bartlett and two others as witnesses. Neither had wanted a large wedding, but Bartlett, who thought there should be some kind of a celebration, arranged a picnic following the ceremony. At first there was little change in the pattern of their lives. Both returned immediately to work, Frieda at the Botanical Gardens and Frank preparing to teach ornithology at the Biological Station. Frieda told her mother that they did not plan to "keep house" in the fall. Both had been used to rooming and taking some meals at boardinghouses. They expected to spend most of their time at work, so Frieda did not wish to do much housekeeping. However, they found a new apartment, near the Gardens, which Frieda felt would make the housekeeping easy. By the late summer of 1923 they had purchased a house and were shopping for used furniture. Though the new house added to Frieda's responsibilities, she assured her mother that Frank was very good about "doing his part." Furthermore, she prepared simple meals and hired help with cleaning and washing. They tried briefly to do all their own housework but found it too difficult. They preferred to "stop doing the things we care less about doing—let someone else do some of them, and let the others go altogether." Occasionally company came for dinner. "By both helping we can enjoy having company, when otherwise it would be too hard." They did most of their entertaining during the winter months, "for after that we will have to be in the woods again all the time we can."[31]

To the Blanchards, their most important activities were their studies of plants and animals, Frank's teaching, and Frieda's responsibilities at the Gardens. During Frank's first years at the university there seemed little hope of advancement, as there was another herpetologist on the faculty, but the Blanchards stayed in Ann Arbor because, Frieda told her family, it was a good place for Frank's research and she had a position there. Frank developed a reputation as a well-prepared teacher "whose interest in his students extended beyond the classroom and throughout their careers." He effectively taught laboratory, field study, and research techniques and launched several graduate students on solid professional careers. The publication of "A Key to the Snakes of the United States, Canada, and Lower California" helped establish his reputation as the leading authority on North American snakes.[32] In addition to supervising and coordinating *Oenothera* research at the Gardens, Frieda served on several doctoral committees and gave occasional lectures and seminars in botany. She also directed the growing of plants for class use and campus decorating; prepared budgets; accessioned plants and seeds; kept extensive plant records; and supervised the gardens, greenhouses, and employees, whom she also hired. She arranged exchanges, primarily of seeds, with other major botanical gardens.[33] Though she never achieved the national recognition that her father, Bartlett, and Frank had, she did sound scientific research and was listed in every edition of *American Men of Science* from 1933 through 1960.

Frank and Frieda helped each other with research, discussing ideas, keeping detailed records, and revising and proofreading papers. Frank could do little to help directly with the *Oenothera* research, which took place entirely at the Gardens, but Frieda, who was a talented scientific artist, made drawings of the patterns of the scales on the heads of different species of snakes for some of Frank's publications and helped prepare diagrams and illustrations for papers that he presented and slides for his classes. She arranged their home life to facilitate his research as well as her own, and whenever possible she accompanied him on his out-of-town collecting trips; however, Frieda usually could not leave Ann Arbor if Bartlett was taking one of his frequent long-duration trips, because then she was entirely responsible for the Gardens.[34]

The Blanchards also worked together studying salamanders and snakes. For over a decade they collected and studied Michigan salamanders, taking photographs and detailed notes. Frank published most of the salamander papers coming from those long-term studies, but Frieda published one, and they published one jointly. However, it was in studying garter snakes that they combined their specialties, with Frank concentrating on life history and Frieda on genetics. In 1922, shortly after their marriage, they happened upon a rare melanistic (black) male Eastern garter snake (*Tham-*

nophis sirtalis sirtalis). They began experiments using that snake, breeding it with normal striped garter snakes. Frieda's interest and her position made it possible to conduct the research at the Botanical Gardens, where she designed and had built enclosures for the breeding experiments. The snakes were in partially covered cement-walled pits with natural ground bottoms about four feet deep and five by six feet in area. Greenhouses surrounding the row of fourteen pits helped keep out predatory animals. The door to the research courtyard was locked, because moving or harming a snake could ruin years of research. Frank and Frieda both spent much time observing, photographing, and caring for the snakes. In 1924 Frieda reported to her mother that they had another brood of hybrid garter snakes. "The year-old broods are flourishing. The F_2 will be so exciting! Maybe they will be born two years from now." Eventually they acquired other melanistic garter snakes from Ohio and Ontario; the experiments continued for almost two decades. When finished, the research project provided the first demonstration of Mendelian inheritance in a reptile and important additions to understanding the life history and breeding habits of the Eastern garter snake.[35]

The Blanchards' devotion to research and love of nature shaped the pattern of their lives. The birth of their first child, Dorothy, in October 1925, altered but did not basically change that pattern. At first Frieda feared that she would not be allowed to keep her job at the university; she applied for a leave with pay but expected to be fired. "The attitude of the dean toward women holding positions in the University, especially those whose husbands do, is such that, with this added complication, there may not be any use in my trying!" But, as she told her mother when the baby was about two weeks old, she did not want to stay home all the time. "I could not stand it, I fear—not even with such a lovely little baby for company." Bartlett, supportive and needing Frieda at the Gardens, asked the dean for a two-month leave of absence for her. The dean approved; the assistant to the president passed the request on to the regents, saying, "The request for a leave of absence for Mrs. Blanchard is made because a child was born to her only this week. It remains to decide whether she should have leave of absence with or without salary. Perhaps this question should be decided by the full Board as the circumstances are so unusual. Mrs. Blanchard's services are very valuable to the University and she has been for many years most faithful and enthusiastic in her work." The regents, in an unusual move, approved the leave with pay; Bartlett brought the Blanchards the welcome news.[36] Though Frieda would be allowed to keep her position at the university, the Blanchards still had to plan for childcare. Frieda thought of her own mother, who had been isolated in Australia when Frieda was born, and realized that she herself was fortunate to be surrounded by friends in Ann Arbor. Still she found it hard "to see Frank go off without me to work, after having gone with him for three years, and being there with him longer than that. . . .

I hope that I may soon go with him again." They called on the family for help. Frank's mother and grandmother came to live with them for six months. After they left, the Blanchards hired additional household help for cleaning, childcare, and some meal preparation. Frieda arranged the schedule so they could have time with their daughter, for both were loving parents. Frieda bathed the baby before leaving for work in the morning and kept her up in the evenings. "I know," she wrote, "all the books would frown upon the plan of having her awake from five til nine in the evening, but we can not bear to miss her altogether." She enjoyed being with her daughter, reporting, "It is perfectly plain that she is the loveliest tiny girl that ever was."[37]

Though the Blanchards sometimes had problems getting and keeping good help, their childcare arrangements worked reasonably well, permitting Frieda to continue research and prepare her *Oenothera* and other papers for publication. The ways in which they cared for their children remained essentially the same following the birth of their second daughter, Grace, in 1927, and their son, Frank Nelson, in 1931. They spent part of each day with the children and included them on some nearby collecting trips. Once, after taking the girls to the campus, Frieda wrote that she always felt she should apologize to the university for having babies, "so I take pleasure in showing anyone on the campus what remarkably fine ones they are, and therefore justifiable." Frieda left detailed instructions about the care of the children for the hours when she was at work. She regretted not having more time to spend with the children, but she also believed that she probably had as much opportunity "to work and play with them and enjoy them as I would if I were at home all the time and doing the work there." Occasionally they left the children with hired help so that they could attend professional meetings or take trips together.[38] Growing up in the Cobb family had helped give Frieda the confidence and self-assurance she needed to combine a family and a professional career.

In 1927 the Blanchards both applied for sabbatical leave for 1927–1928; they planned a collecting and research trip to Australia and New Zealand. Such a trip would be possible only if they could both obtain leave with pay. Bartlett strongly endorsed the request, pointing out that Frank would be collecting specimens for the university museum and Frieda seeds for the Botanical Gardens and specimens for the herbarium, including specimens "of great exchange value in building up our collections." He also urged that the leaves be considered together, since "the two Blanchards would be of great help to each other in the field work which they contemplate." The chairman of zoology heartily approved, and the dean endorsed the request, using Bartlett's arguments and adding that "it means virtually a University of Michigan expedition to the other side of the world at practically no expense." The regents approved the request. Both Blanchards obtained leave for a year at half salary.[39]

The Blanchards planned the trip as a combination study, collecting, and vacation trip. They decided reluctantly to leave two-year-old Dorothy with Frieda's sister in Washington, D.C., but took with them their new baby, Grace, who was six weeks old when they sailed. In New Zealand, among other adventures, they located, on a not-easily-accessible island, the rare Stephen Island frog (*Liopelma hamiltoni*) and the tuatara (*Sphenodon punctatum*), an ancient type of reptile. Frieda later wrote, and illustrated with their photographs, an article about that part of the trip for *The National Geographic Magazine*. From New Zealand they moved on to Australia, including Tasmania, where they located the "lost frog of Tasmania" (*Crinia tasmaniensis*), previously known only by an 1864 specimen in the British Museum. Although they spent some time attending professional meetings and studying museum collections, they found greater satisfaction in their trips to remote areas. Carrying Grace in a special "sling" that Frieda had designed, they went by foot or horseback into the interior. There they collected research specimens and seeds from native plants, which they shipped back to Ann Arbor. They returned home, after picking up their older daughter, almost exactly a year after their departure. The trip, which gave them an opportunity to study nature together in distant parts of the world, was a high point of their lives.[40]

Seven years later, when the time for another sabbatical arrived, Bartlett was in the Philippines, and Frieda could not leave the Gardens. Thus, Frank took one of his students on a five-month collecting trip in the southwestern United States. So now, Frieda reported, "we must wait another seven years for a big trip." That opportunity never came. In 1937, while at the Biological Station, Frank developed an infection and high fever. After over a month in the university hospital, he died on September 21, 1937, of bacterial endocarditis. Frieda had been at the hospital daily; Bartlett had also visited regularly, often stopping at the Blanchard home to read to the children. The support of family and friends, the needs of her children, and continuing her work helped Frieda cope with her tragic loss.[41]

The research that Frieda and Frank had started together did not end with his death. She gently prodded other herpetologists to finish projects they had begun with him; she herself carried on the garter snake research. In May 1937, shortly before Frank's fatal illness, they had presented together at the annual meeting of the American Society of Ichthyologists and Herpetologists the first of several papers they had planned on the garter snake work. As a widow, Frieda finished that project. She gave papers at professional meetings and published four garter snake papers, listing Frank as coauthor for the first three. Though she put his name first as a tribute to him, she completed the unfinished research and writing. "Pulling on alone is dreary," she reported, "but I can do it." Her goal was not only to finish their snake research but also to bring up "his three precious little children . . . as he would wish it done."[42]

Frieda published the last garter snake paper in 1942. She also worked with one of Frank's students to finish a long-term project on Eastern ringneck snakes (*Diadophis punctatus edwardsii*). That project and writing a biography of her father were her major research activities in the late 1940s and 1950s. She raised her children, who followed careers in biology and geology; helped numerous university students, some of whom lived in her home; cared for her mother; continued as the active administrator at the Botanical Gardens; and remained a supporter of scientific research, conservation, and the study of nature. She died, at age eighty-seven, in Ann Arbor on August 29, 1977.

Frieda Cobb Blanchard had a long and productive career in the natural sciences, beginning with her training as an assistant to her father. Originally a graduate student under Harley Harris Bartlett, she soon became a scientific administrator for the University of Michigan Botanical Gardens and Bartlett's research colleague. It was, however, her husband, Frank Nelson Blanchard, who became her closest scientific partner. They shared a happy, productive partnership, uniting their research interests to collaborate as geneticist and herpetologist on a long-term, groundbreaking study of garter snakes. After Frank died, Frieda continued and published that research. Shortly after his death, she reflected on their life together: "for twenty-one years, beginning as graduate students, we had our offices in the same University building. Coming and going together, usually conferring at some time during the day, often working together, we consciously enjoyed the unusually close companionship thus allowed us." With that warm, loving companionship they had strengthened one another. In their creative partnership they shared ideas and skills as well as home responsibilities. They had been a team, studying together the natural world in which they lived.[43]

MAUREEN M. JULIAN

Kathleen and Thomas Lonsdale
Forty-Three Years of Spiritual and Scientific Life Together

Kathleen Yardley and Thomas Jackson Lonsdale met in 1922 as graduate students at University College, London, where they were among the new wave of young scholars invited to help restart the European scientific laboratories after the interruption of World War I.[1] Nineteen-year-old Kathleen was studying X-ray crystallography with Sir William H. Bragg, the inventor of the Bragg X-ray ionization spectrometer and the winner of the 1915 Nobel Prize in chemistry. Twenty-year-old Thomas was working on the elastic properties of metal wires with Alfred W. Porter.

Through a career stretching across the next half-century, Kathleen Lonsdale would establish herself as the first woman crystallographer to attain a worldwide reputation not only as a scientist but as a crusader for peace and prison reform as well. Her scientific achievements and honors were many. She showed the planarity of the benzene ring, and demonstrated the reality of *sigma* and *pi* electrons and their representation by molecular orbitals. She did groundbreaking work on the magnetic anisotropy of crystals, diffuse scattering, temperature factors in diamond, and solid-state reactions. At the age of almost sixty she began an exhaustive study on the crystallography of body stones. Her careful derivation of complex crystallographic formulas led to her editorship with others of the *International X-ray Tables*. She promoted the cause of women in science both by her example and in her writings and lectures. She was one of the first two women to become a fellow of the Royal Society and the first woman president of the British Association for the Advancement of Science.

More into practical science than his wife, Thomas Lonsdale would initially earn the major income for the couple and their growing family, first as a scientist in the silk industry and later as an engineer in the Department of Roads. As Kathleen's star rose, Thomas consciously assumed the support role by retiring early at the age of sixty to assist in Kathleen's enormous worldwide correspondence for peace and prison reform and to take on home duties.

Kathleen was the youngest of ten children born to the Londoners Jessie Cameron and Harry Frederick Yardley. Jessie had worked in an Italian restaurant before being apprenticed to a shirtmaker; Harry had become first a telegraph boy and then a postman. Twenty-three-year-old Jessie and thirty-one-year-old Harry were married in 1889 in the Baptist Church at Dalston Junction. Their first nine children, three girls and six boys, were born in England, where two sets of twins died in infancy. Unfortunately Jessie's strict Baptist beliefs conflicted with Harry's agnosticism and drinking habits. Family life was so unhappy that Harry Yardley left his job with the post office in England and joined the army to fight in the South African war. After completing his service, he was assigned to Newbridge, Ireland, as the postmaster for the British garrison stationed nearby. In an effort to revive his marriage, Harry invited Jessie to join him in Ireland, where their youngest child, Kathleen, was born on January 28, 1903, in the family quarters above the post office.

When Kathleen was five years old, her parents permanently separated. Feeling threatened by the unrest in Ireland and burdened by her marital problems, Jessie took the children to Seven Kings, a small town east of London that turned out to be on the zeppelin route during World War I. Kathleen was pressured by her mother and by a fear of being damned to make a "confession of faith." Her formal baptism into the Baptist church followed, but did not give her peace of mind. Kathleen later wrote: "I had jibbed first at *having* to believe anything, at being obliged to accept the truth of any statement upon authority only. Perhaps my skepticism was precocious; it was certainly involuntary. I would have preferred to be credulous; it would have been far less strain."[2] It was in Seven Kings also that Kathleen's pacifism was awakened as, amidst the terror of the zeppelin attacks, her mother cried for the young German men dying in the airships that had been "shot down in flames."[3]

Although not living with her father during her formative years, Kathleen believed that he influenced her in many ways. "I became a lifelong teetotaller largely because he was not. I have never even attempted smoking principally because he smoked so much. But he was a great reader and a natural mathematician. Our house was full of encyclopaedias to dip into and of books worth reading; and my scientific turn of mind I believe I owe to him."[4] Her father's agnosticism, moreover, contributed to Kathleen's difficulty in accepting the Baptist faith the way her mother did.

At sixteen Kathleen enrolled in Bedford College, a small women's college of the University of London. She started an honors program in mathematics, but after the first year switched to physics. Her headmistress advised against giving up mathematics because, as Kathleen remembered years later, "in physics there would be far more competition from men and . . . I would be a fool to think that, with my comparative lack of background of any practical skill and knowledge, I would be able to compete in a man's field."[5] Two years later, in 1922, she headed the University of London list of all students in physics with the highest score on the written examination in ten years. Sir William Bragg was among her oral examiners and, as a result, invited her into his laboratory at University College with a grant of £180 per year. By contrast she probably could have secured a position teaching mathematics and science at a middle school for about £240 a year. Kathleen continued to live at home and contributed a substantial part of her salary to her mother.

Thomas was named after his paternal grandfather, Thomas Lonsdale, whose hobby was dating New Testament manuscripts from the Greek. Thomas's father, James Jackson Lonsdale, held a bachelor's degree from the University of London and a doctorate in science from the University of Durham. He became senior science master at the Sloane School, Chelsea, and in 1910 published work on ionization produced by the splashing of mercury.[6] Thomas's maternal grandfather, John Porter, was a shoemaker, whose family consisted of eleven children, including the fourth daughter, Nora Porter, who married James Jackson Lonsdale and bore two sons, James as well as Thomas. James eventually worked in the automobile industry.

Thomas was a gifted linguist who spoke fluent French and took great pleasure in Latin and Greek. He had begun piano lessons at five years of age and read music as soon as he read story books. For the rest of his life, he remained fond of music, especially Bach. Originally Thomas had wanted to be a medical doctor, but found that he had taken a sequence of subjects more appropriate for the physicist than the physician. Like his namesake grandfather, Thomas was philosophically inclined and loved to argue about religion. When Thomas and Kathleen began to spend their Sundays together, they attended sermons in different London churches.

The Carey-Forster Laboratory where Thomas and Kathleen worked was a small cottage. The young scientists discussed religion and politics as well as science, and played international table-tennis tournaments in the basement. They called themselves Bill Bragg's group, and soon everybody jokingly went by the name of Bill. When Kathleen arrived, she was called "Jane," the group's idea of the feminine equivalent of Bill. Kathleen worked in the southerly downstairs room, where by the beginning of 1923 she had a large table with parts of her own ionization spectrometer on top. On one occasion she needed to do some soldering, and Thomas offered to do it. Kathleen refused but instead asked to be taught. Years later, when they were both being interviewed by Jean Metcalfe for the BBC program *Women's Hour*, Thomas was asked what attracted him to Kathleen. He replied that it was her skill as a mathematician.[7]

Kathleen's specialty was crystallography, in which the basic experimental technique is X-ray diffraction. A crystal is mounted in the path of an

X-ray beam, and the positions and intensities of the diffracted beams are measured. This experimental data consisting of hundreds of reflections is used to determine the molecular structure of the crystal. To do this, crystallographers perform many complicated mathematical operations.

For her master's dissertation, Kathleen analyzed crystals of succinic acid, succinic anhydride, and succinimide.[8] Another project she began as a graduate student was a derivation of a set of space-group tables that translated crystal symmetries into mathematical language.[9] This practical paper led beginning crystallographers through the mathematical minefields of X-ray analysis and was one of the few papers reprinted by the *Philosophical Transactions*. Throughout her entire career much of her professional energy was dedicated to the enormous mathematical and editorial task of producing these crystallographic tables. Versions and later editions of the *International Tables for X-ray Crystallography, I, II, III* are found on the desk of every practicing crystallographer today.[10] Dorothy Hodgkin, the winner of the Nobel Prize for chemistry in 1964, commented: "Some may feel Kathleen Lonsdale's work on the tables hampered her own achievements in research. It is difficult to judge. Her work has helped so many others to success and her own researches are, in spite of lost time, so considerable, and so permeated by her knowledge of symmetry."[11]

In 1923 Professor Bragg was offered a better position at the Royal Institution, and he invited Kathleen to come with him.[12] Meanwhile Thomas, too, left University College and was working as a physicist with the British Silk Research Association, which was located in the Textile Department of the University of Leeds. There he studied the elastic properties and tensile strength of silk filaments.[13] To perform these measurements, he invented an extensometer to measure extension and load at the elastic limit.[14] This research was presented at a joint meeting of the Faraday Society and the Textile Institute held in 1924.

When separated, Thomas and Kathleen wrote nearly every day, often about religious and philosophical problems. Finally Thomas wrote to ask Kathleen to marry him; they were engaged four years. Although Kathleen intensely loved her work and would even run the last few feet to the laboratory, she first intended to retire and equally throw herself into wifehood and motherhood.[15] They planned to have four children. Thomas knew how much she loved her work and how good she was at it, so he would not consider her giving up science. He was not marrying to get a free housekeeper; they would have their family and their science both, somehow.[16]

Kathleen and Thomas were married at the Baptist Church in High Road, Ilford, Essex, on August 27, 1927. At first they had temporary lodging in Leeds with a Congregational minister, the Reverend Plowright. His daughter was married to a Quaker named England, and they all became dear friends. England and his brother had been jailed during World War I as conscientious objectors, and the brother had died as a result of the hard prison life. The Englands impressed Thomas and Kathleen with their Quaker work, especially in depressed areas of South Wales, which were in severe economic crisis following World War I.[17] This was the first intimate contact that the Lonsdales had with the Quaker people and Quakerism, which eventually became a central life force for both Thomas and Kathleen.

In her last days at the Royal Institution before marriage, Kathleen had worked on a series of eight similar ethane compounds that all crystallized in the same space group with similar crystal repeat distances. In part I of the series, she carefully tried to deduce the carbon valences from the molecular symmetry. Some of the compounds were very difficult to work with because they evaporated very quickly. Part II of the series extended the study to other similar ethane compounds with different space groups. By the time her papers on this research were received by the Royal Society on December 15, 1927, she was already married, so she used the byline Kathleen Yardley, M.Sc. (Mrs. Lonsdale), as a wedding announcement to the scientific community.[18] In a tribute to Kathleen's scholarship, Bragg invited her to skip the Ph.D. and submit this research as her thesis for the more prestigious D.Sc. degree. She had completed all her requirements and was waiting to be called for her oral examinations. Since she and Thomas were planning a vacation trip, Thomas suggested she phone to see what was holding things up. Her degree had been awarded without an oral. Thomas teased her that the examiners were willing to give up their four-guinea fee to avoid facing her long mathematical formulas.

Kathleen was welcomed to the Department of Physics at the University of Leeds by Professor Richard Widdington, who arranged a part-time demonstratorship to supplement the Amy Lady Tate Fellowship that Bedford College awarded her from 1927 to 1929. Her next paper on the anisotropy of the carbon atom was a theoretical outgrowth of her ethane studies.[19] She did not need any X-ray equipment to do it.

Then the Royal Society gave her an apparatus grant of £150 that she used to buy an X-ray tube, a new ionization spectrometer, and an electroscope for £113 17s.[20] So equipped and looking for a new crystal to study in 1928, Kathleen received some large crystals of hexamethylbenzene from Christopher K. Ingold, a professor in the Chemistry Department at Leeds.[21] These crystals had molecules that contained the benzene ring, the foundation of much of organic and industrial chemistry. Bragg had predicted that the benzene ring would be puckered like diamond. Benzene is a liquid at room temperature and contains four molecules to the unit cell. Since the hexamethylbenzene crystals are triclinic, the structure of the crystal does not have any special symmetry, making the calculations particularly tedious. In the evenings Kathleen worked on her calculations for hexamethylbenzene while Thomas worked on his Ph.D. He was measuring the torsional strengths of metals. Using apparatus set up at home, he did experimental work in which he monitored the changes, produced by torsion, in the dimensions of wires made of silver, gold, aluminum, nickel, and lead.[22]

Sometimes they took a break and, for example, made gooseberry jam. Every week they shopped together at the Leeds Market. In the morning they would prepare vegetables for dinner before walking over to the university. Kathleen would then allow herself an extra half-hour in the late morning to finish the preparations for dinner, which they ate at noon. Soon she had a repertoire of meals that took thirty minutes to make, and she cooked regularly because they found eating at home cheaper than at the university.

She continued to collect data on the hexamethylbenzene crystals. She recorded the intensities of over one hundred X-ray reflections, where each reflection represented a different bending angle of the X-ray beam by the crystal. After correcting for the polarization of the X-rays, Kathleen noticed that the data had sixfold symmetry. Since the benzene molecule had six sides, she had discovered a powerful clue to the structure. Another clue came in the way this new data arranged itself in layers. Further calculations showed the benzene ring was flat and not puckered like the carbon atoms in diamond. She calculated the dimensions of the ring. Bragg immediately sensed her great achievement although it contradicted his own theories of the benzene ring. In October 1928 he wrote: "I think your new results are perfectly delightful; many compliments on it! I like to see the benzene ring 'emerging.' "[23] On November 24, 1928, Kathleen published a brief account of her results and then followed it with a more detailed paper.[24] Later in life she said this was the most satisfying work of her whole career.

The Lonsdales were now expecting a baby. Kathleen, who had begun investigating another structure, did the calculations in bed to save her strength. When their baby girl was born, they named her Jane after Kathleen's laboratory name. Kathleen found it impossible to care for Jane, do all the household chores, and still have time for scientific studies. She wrote to Bragg in London, and the managers of the Royal Institution gave her a grant of £50 for a year. The Lonsdales used this money to hire a Mrs. Snowball, who came in every day to wash and clean. Kathleen cared for Jane, did the cooking, and, when the baby fell asleep, continued her crystal studies.[25]

Thomas's work on the tensile properties of silk filaments[26] came to an unfortunate halt. His grants depended on the internal policies of India as one of the major buyers of English spun silk. On January 26, 1930, India celebrated its first Independence Day. Payment of the British salt tax was refused, and soon foreign (which meant mostly British) goods rotted on shelves throughout India. The outcome for the Lonsdales was that Thomas had to leave his job in the silk business. Kathleen, too, had to leave the experimental equipment that she had carefully purchased and assembled in Leeds. At that time Thomas's salary was £350 per year, and Kathleen was earning about £150 at the university. They were now expecting their second child. Thomas was fortunate in finding a new job at the Testing Station of the Experimental Roads Department of the Ministry of Transport outside London at no loss in salary. In October 1930 the Lonsdales moved back to London. Their second daughter, Nancy, was born in Windsor in 1931.

All along Kathleen continued to work at home on the calculations on the structure of another crystal with a benzene nucleus, hexachlorobenzene. These calculations, which she had started in Leeds, were long and tedious, for she had to do them by hand using logarithm tables for the Fourier series. For example, each point had eighty terms added together, and she calculated 450 points, or 36,000 terms. This was the first structure solved with the Fourier method.[27] On November 2, 1931, Ingold, who had given her the crystals, wrote: "Ever so many thanks for the reprint of your wonderful paper on hexachlorobenzene. . . . The calculations must have been dreadful but one paper like this brings more certainty into organic chemistry than generations of activity by us professionals."[28] New insights like Kathleen's work on the planarity of the benzene ring brought credibility and then prestige to crystallography.

After the birth of the Lonsdales' third and last child, Stephen, in 1934, Kathleen returned to Bragg at the Royal Institution and remained with him until his death in 1942. At first, after the disappointing news that no X-ray set was available, she used a large electromagnet to study magnetic anisotropy.[29] She demonstrated that the orbitals representing the *sigma* electrons were of atomic dimensions and that those of *pi* electrons were of molecular dimensions because the diamagnetic susceptibilies of aromatic compounds are greater perpendicular to the aromatic ring than in the same plane. This experimental work was the first detection of the *sigma* and *pi* electrons and validated their representation by molecular orbitals.[30] In a separate study, Ellie Knaggs, a fellow research worker, did the X-ray analysis of cyanuric trioxide, and Kathleen supplemented the work with magnetic measurements.

One of the scientists to whom Kathleen communicated her results on cyanuric trioxide was K. S. Krishnan, who wrecked his laboratory in Calcutta testing these somewhat erratic and explosive crystals.[31] When Krishnan later visited London, he had difficulty getting lodging, and the Lonsdales invited him to stay with them. Since Krishnan did not eat meat for religious reasons, the Lonsdales decided to observe a vegetarian diet while he was with them. They made yogurt from the culture that he brought from India, long before the product became commercially available in England. Kathleen, who had a nervous stomach, felt better on this meatless diet. Eventually she became a vegetarian both as a health measure and as an extension of pacifism to the animal world.

This pacifism fits with the teachings of the Society of Friends that the Lonsdales joined in 1935 as adults by convincement. Like many parents, they had reexamined their own beliefs in trying to decide what to teach their children, before officially embracing Quakerism. Soon the whole family was active in the local Friends Meeting at Uxbridge.

In 1938, as Nazi Germany and England were preparing for war, Thomas was reassigned from the construction of roads to the building of air raid shelters. A little later, the Lonsdales moved to West Drayton, a town about halfway between Harmondsworth and Uxbridge, which promised better schooling for the family's children. The move was in vain because, with the coming of World War II, the local school building was converted to wartime uses. Eventually Jane, Nan, and Stephen went to Quaker boarding schools for the duration of the war. Also at this time Kathleen's mother, Jessie Yardley, moved in with the family and stayed until she died thirteen years later in her eighties.

Kathleen's lifelong interest in thermal vibrations began when she observed some diffuse (or non-Bragg) spots on a Laue X-ray photograph of benzil, on the structure of which she and Knaggs published in 1939.[32] The diffuse reflections in diamond led to her work on natural and artificial diamonds beginning in the late 1930s.[33] Applying the techniques of divergent X-ray beams and noticing a particularly fortunate coincidence for diamond, she measured the distance between carbon atoms to four decimal places.[34] In 1966 Clifford Frondel of Harvard University suggested that a rare hexagonal form of meteoritic diamonds be named "lonsdaleite."[35] In her reply to Frondel, she wrote, "it makes me feel both proud and rather humble that it shall be called lonsdaleite. Certainly the name seems appropriate since the mineral only occurs in very small quantities (perhaps rare would be too flattering) and is generally rather mixed up!"[36]

Although Kathleen peacefully pursued her research during the early years of the war, she eventually became caught in a struggle between her pacifism and England's all-out war effort. By law she was required to register as a fire watcher. She was already voluntarily performing the functions of a fire watcher, but registration marked a definite and positive official participation in the war. She argued with herself:

> I know I am a fool, that I am risking my job and my career, and that one isolated example could do no good, that it was a futile gesture since even if I did register my three small children would exempt me. I had wrestled in prayer and I knew beyond all doubt that I *must* refuse to register, that those who believed that war was the wrong way to fight evil must stand out against it however much they stood alone, and that I and mine must take the consequences. The "and mine" made it more difficult, but I question whether children ever really suffer loss in the long run *through having parents who are willing to stand by principles.*[37]

Eventually Kathleen's letter refusing to register for civil defense duties came to the attention of the authorities. The local magistrate summoned her and imposed a small fine of two pounds. Refusing to pay the fine, Kathleen was sentenced to one month at Holloway jail. On January 22,

1943, just six days before her fortieth birthday, she left the Uxbridge Police Court in a police van. She was so scared that imprisonment was almost an anticlimax. Permission was granted for her to have some scientific instruments and papers so she could work in her cell in the evenings.[38]

During the summer of 1943 Kathleen received an invitation to go to her first international scientific meeting—at the Institute of Advanced Studies Summer School in Dublin. The Dublin Institute had been founded by Eamon de Valera, the mathematics scholar who became the prime minster of Ireland.[39] Theoretical lectures were given by Max Born and Paul P. Ewald, a German crystallographer who was then at the Queen's University, Belfast, Northern Ireland. These talks were an introduction to the experimental work described by Kathleen. The chairman was Erwin Schrödinger, the president of the Institute. After the meeting Kathleen visited the post office at Newbridge, where she was born.

When the blitz began, Thomas stopped building air raid shelters and became part of the air raid investigation unit. He was first with the Home Office and later with the Ministry of Home Security. The unit traveled continuously as needed to Plymouth, Coventry, Sheffield, Leeds, Birmingham, Doncaster, and Newcastle in a huge truck towing a trailer. As soon as they arrived in a town, they reported to the police and worked closely with them. Often they worked through the night and seldom could visit their own homes. When a cut on Thomas's left thumb turned septic and sent him to the hospital, the doctors were more concerned about his near total condition of exhaustion than the thumb. He was kept in the hospital for two weeks and then sent to a convalescent home.

Kathleen was now an established scientist, but unlike many of her scientific contemporaries from the Royal Institution, she had not been invited to fellowship of the Royal Society.[40] After World War I, parliament had passed legislation that made it legally possible for a woman to be elected. William Astbury, who had been a graduate student with and collaborator of Kathleen at the Royal Institution with Bragg, had put Kathleen's name forward soon after his own election in 1940. Finally in 1944 Sir Henry Dale consulted Kathleen about forwarding her nomination. After he assured her that there would be no dissension among the fellows, her name was put forward. Marjory Stephenson, a chemical microbiologist of the Medical Research Council, was nominated to represent the biological sciences. On March 22, 1945, they were elected along with J. Monteath Robertson, another former member of Bragg's group at the Royal Institution. On May 17 Kathleen signed the Charter Book. Dale enjoyed teasing Kathleen by commenting that her marriage made her alphabetically the first woman elected into the Royal Society! She was the first married woman elected and of course the first woman with children. Kathleen was pleased with her election and always wrote F.R.S. after her name. Years later in a speech titled "Women and Science" at the Royal Institution, she mused: "Looking back over a quarter of a century, I sometimes wonder do you know, whether my going to prison as a conscientious objector in 1943 and my election to the Fellowship of the Royal Society in 1945 were all that unconnected."[41]

After Bragg's death in 1942, Kathleen stayed on for three years at the Royal Institution. Then in December 1946, after living on year-to-year grants until she was forty-three years old, she finally got her first permanent position and founded her own crystallography research group at University College, London, where she and Thomas had met almost a quarter of a century before. Although she would have preferred a separate department of crystallography, her group became part of the Department of Chemistry.[42] A special chair was created for her at Ingold's initiative.

First she took a six-month leave of absence to gain experience on Ralph W. G. Wyckoff's electron microscope and Geiger counter spectrometer at the National Institute of Health in Washington, D.C. When she returned to University College, London, Kathleen entered a new phase in her academic life. The carefree days of all research and no administration were gone. She worked along with J. D. Bernal in preparing slides and teaching an M.Sc. intercollegiate course, which was organized a few blocks away at Birkbeck College. The lectures were written up in her book *Crystals and X-rays*.[43]

In 1949 Judith Grenville-Wells (later Milledge) came from South Africa for doctoral work on diamonds with Kathleen. At first, in exchange for room and board at the Lonsdales, Milledge did some secretarial work on the *International X-ray Tables* and checked the mathematical equations. Milledge, befriended by Thomas as well as Kathleen, spent some summer holidays motorcycling through northern England and Scotland with the Lonsdales; Thomas drove, and Kathleen and Milledge alternated riding in the sidecar and on the seat behind Thomas. Upon obtaining her doctorate, Milledge joined the faculty of University College with the rank of reader in crystallography. The joint studies of Kathleen and Milledge included natural and artificial diamonds,[44] diamond inclusions,[45] minerals at high temperatures and pressures, and the mechanisms of solid-state reactions. The most important example of the latter is the detection of the X-ray diffraction patterns of the intermediate products in the conversion of the photooxide of anthracene into a mixed crystal of anthrone and anthraquinone.[46] Eventually Milledge became Kathleen's literary executrix and continued the diamond and solid-state reaction work at University College.

In 1962 D. A. Anderson, a urologist and chief medical officer of the Salvation Army, showed Kathleen his extensive collection of body stones. She began a crystallographic study of the collection and received many body stones from all over the world, her favorite being Napoleon III's bladder stone.[47] D. June Sutor, a New Zealander, worked with Kathleen on this project and eventually took it over and extended it.

Kathleen loved traveling, and everywhere she went she relentlessly crusaded for peace and prison reform in addition to furthering her scientific interests. For example, in 1951 she visited Russia on a mission of peace organized by the Society of Friends. She asked to be taken inside an ordinary nonpolitical prison as well as to the Russian Institute of Crystallography in Moscow. To the prison official's surprise, Kathleen asked penetrating questions about the 885 men and 15 women who were incarcerated there. Kathleen later commented: "Our woman interpreter was laughing as we came away and when I asked her why, she said that the prison governor had asked 'How is it that such a nice lady knows so much about prisons?' "[48]

In January 1956 Kathleen simultaneously became a grandmother and a Dame of the British Empire; that same year she emerged as an antinuclear weapons activist. By 1956 the atmospheric nuclear testing by Russia, the United States, and Great Britain was so extensive that the standard carbon-14 dating system needed revision. In response Kathleen interrupted her work, and in a six-week marathon she wrote her antiwar tract, *Is Peace Possible?* In the foreword she stated that the book was "written in a personal way because I feel a sense of corporate guilt and responsibility that scientific knowledge should have been so misused."[49] For the rest of her life she devoted much time and energy to writing antiwar pamphlets and giving lectures at schools, prisons, and churches. She was also president of the British section of the Women's International League for Peace and Freedom. Thomas looked after the correspondence.

After World War II Thomas had returned to the Road Research Laboratory. He worked on the testing of motorcycle crash helmets, methods for making roads and airstrips more resistant to skidding, apparatus for recording the flow and speed of traffic, and the design of road signs that were easier to read.[50] Thomas retired early, in 1960 at the age of sixty, because he felt he would be more useful assisting in his wife's enormous worldwide correspondence in the causes of peace and prison reform. He took on the domestic chores and enjoyed quite a reputation for bread making. In 1965, anticipating Kathleen's retirement in 1968, they moved to Bexhill-on-Sea, a tiny, quiet coastal resort town two and a half hours by train south of London. Their three children and ten grandchildren visited often. Kathleen loved giving the grandchildren stamps from her correspondence.

At the end of the day after her long train commute from London, Kathleen was extremely tired. "She would return home at 8:30 P.M. and Thomas would put her straight to bed, bringing her up a cooked supper and tea and cheese, a brown roll and yogurt. She slept badly at night, waking at 4:30 A.M. and rising at 5:30."[51] In December 1970 she became ill with leukemia. After a brief visit to Bexhill to celebrate Thomas's seventieth birthday, she died in the hospital on April 1, 1971. Thomas continued to live by himself at Bexhill. The house looked bare with the bookshelves empty of Kathleen's vast library. Almost eight years later, on March 10, 1979, he died suddenly while visiting his daughter Jane in Bearsted, Kent. He was seventy-seven years old.

Marriage was central to Thomas and Kathleen's lives, and theirs was a marriage built deeply in religion and in a faith in each other. At first Thomas's career dominated, when she followed him to Leeds and then back to London after his first job collapsed. Quakerism and the peace movement were the foundations upon which their spiritual lives were built. Thomas was the strong support at home while Kathleen was the interface with the world. While she was in prison, he suffered alone and at home. They both took up prison reform; she actively served on several prison boards, and he did the supporting correspondence. She traveled all over the world for the cause of peace; he entertained her international visitors and again did the behind-the-scenes paperwork. He was proud of her scientific achievements and professional successes. In addition to being a fellow of University College and a fellow and governor of Bedford College, she became the first woman president of the British Association for the Advancement of Science in 1968 and president of the International Union of Crystallography in 1966. She was awarded honorary doctoral degrees from the Universities of Wales, Leicester, Manchester, Lancaster, Leeds, Dundee, Oxford, and Bath. In contrast, Thomas took early retirement and concentrated on giving Kathleen whatever editorial and emotional assistance he could.

The Lonsdales combined a love of and excellence for science with a deep religious conviction and a humanitarian conscience. In a speech "Women in Science—Why So Few?" Kathleen wrote:

> For a woman, especially a married woman with children, to become a first class scientist she must first of all choose, or have chosen, the right husband. He must recognize her problems and be willing to share them. If he is really domesticated, so much the better. Then she must be a good organizer and be pretty ruthless in keeping to her schedule, no matter if the heavens fall. She must be able to do with very little sleep, because her working week will be at least twice as long as the average trades unionist's. She must go against all her early training and not care if she is regarded as a little peculiar. She must be willing to accept additional responsibility, even if she feels that she has more than enough. But above all, she must learn to concentrate in any available moment and not require ideal conditions in which to do so.[52]

Kathleen was, indeed, "a first class scientist," and Thomas, "the right husband."

PART IV

Couples Devolving from Creative Potential to Dissonance

JOY HARVEY

Clanging Eagles

The Marriage and Collaboration between Two Nineteenth-Century Physicians, Mary Putnam Jacobi and Abraham Jacobi

The American woman doctor Mary Putnam Jacobi, writing in 1882 for the *North American Review*, posed the question "Shall women practice medicine?" For this pioneering woman physician, who had been in practice for ten years and had studied and worked in the field of medicine for a total of twenty years, the answer had to be affirmative. "The taste for medicine when profound and genuine," she wrote, "is certainly peculiar enough to establish a decided variation on the most conventional type of women."[1]

She went on to demonstrate that not only women but married women could practice medicine. Many American women practicing medicine had married, but it was not known to what extent either their households or their practices may have suffered by the combination. She insisted "women . . . most likely to succeed in medicine have often also marked capacities for success in marriage . . . because their ability to perform such work as that involved in the practice of medicine and demanding high organic vigor, tends to be increased after marriage and the possession of children."[2] Quoting a fragment from Tennyson's poem "The Princess," she continued:

> The crane, I said may chatter with the crane
> The dove mate with the dove, but I
> An eagle clang an eagle in my sphere![3]

Putnam Jacobi held an ideal image of marriage in which the partners functioned both as a couple and as dedicated professionals bringing their strengths together, much like eagles. She saw them extending this strength to a complex community around them and from that to the society at large. She had seen such a marriage in France between Elie Réclus and his remarkable wife Noémie, who included in their extended household a number of his brothers, including the remarkable, gentle anarchist, the geographer Elisée Réclus.[4] In 1891 Putnam Jacobi extended her study of women in medicine to describe the collaboration between married physicians, using as her model the Gleasons, husband and wife "practising medicine in harmonious partnership" at the Elmira Sanatorium in New York. She went on:

> There is something idyllic in this episode. Here in western New York was realized, simply and naturally, the ideal life of a married pair as was once described by [Jules] Michelet [the French historian], where the common interests and activities should embrace not only the home circle but also professional life. . . . By Mrs. Gleason's happy career, the complex experiment in life which was being made by the first group of women physicians was enriched by a special and, on some accounts, peculiarly interesting type.[5]

She may well have believed that she could properly number hers among these special marriages. Mary Putnam Jacobi, married to the New York physician Abraham Jacobi, had cheerfully overcome many obstacles on her way to becoming an outstanding physician. Her attempt to establish an effective and new style of marriage created other problems for her that she would find more difficult to resolve.

In a description that paralleled her own life history, she had written in the early eighties of the career of a typical woman physician: "She is ready for practice at twenty-seven, marries at the same time or a year later. Her children are born during the first years of marriage thus also during the first years of practice, and before this has become exorbitant in its demands. The medical work grows gradually in about the same proportion as imperative family cares grow lighter."[6] One reason, she added, that women could continue to be professionals after marriage was due to the "rearrangement of domestic work" that the growth of modern industries had brought about.[7] Putnam Jacobi believed herself to be a pioneer in marriage as she had been in medicine. This paper will assess the success and failure of this attempt at marital and professional collaboration.

Mary Corinna Putnam Jacobi was born in 1842 in England, where her American parents were living. Upon his return to America, her father, George Palmer Putnam, built up a large and important publishing house. A bright, inquiring girl, Mary Putnam was touched by the sight of poverty and illness she encountered among the Civil War soldiers and former slaves when, at twenty, she went to nurse her soldier brother in New Orleans. She decided to pursue a medical career. She had already studied at the New York College of Pharmacy in 1863, which awarded her the first degree presented by that school to a woman. The next year, she went on to obtain a medical degree at the Woman's Medical College of Pennsylvania. Desiring hospital experience, she moved to the New England Hospital for Women but found the conditions unsatisfactory.[8] In 1866 she went to Paris, center of hospital medicine, where she first attended a sequence of hospital clinics on which she reported in articles signed PCM in the *New York Medical Record*. After some struggles, she was admitted to the Ecole de Médecine at the end of 1867, opening a door through which three other women entered in early 1868.[9] Delayed by the chaos of the Franco-Prussian War and the Paris Commune, which radicalized her, she received her degree in 1871, earning a medal for her thesis on fatty degeneration of the liver.[10] Then, urged by Dr. Elizabeth Blackwell, she returned to New York to teach materia medica at the New York Woman's Medical College of the New York Infirmary, established by Elizabeth and Emily Blackwell.[11]

Her future husband, Abraham Jacobi, was born of middle-class Jewish parents in Westphalia in 1830. He would outlive his much younger wife by many years. He studied medicine in three German universities, Greifswald, Göttingen, and Bonn, a virtue he attributed to the decentralized German system. He was fortunate to have had excellent and famous teachers, including Rudolf Virchow, who remained his model of the liberal politician and physician.

Jacobi had been involved in the 1848 revolution in Bonn, which remained the great event of his life. During this time he met another young revolutionary, Carl Schurz, who would make a great name for himself in politics in America and who later would become his closest friend. Shortly after Jacobi received his medical degree in 1851, the Prussian authorities threw him into prison for almost two years for his revolutionary involvement. For Jacobi this revolutionary experience was one of the high points in his life, as Putnam Jacobi's experiences of the Réclus family and the Paris Commune were for her. In 1853, after his release, he went first to England, following one of his notable professors. He tried Boston and then New York, where he found a warm welcome within the strong German-American community and soon established himself as an important clinician and a tireless participant in the growing number of medical societies.

By 1857 he was lecturing on children's diseases at the College of Physicians and Surgeons in New York. He established a number of pediatric clinics in New York and became professor of clinical pediatrics in 1870, a position he held for nineteen years. He was a major figure behind the development of Mount Sinai Hospital and the New York Academy of Medicine. His professional activities included organizations on the New York and national levels. These included the presidency of the New York Academy of Medicine and the presidency of the Pediatrics Section within the American Medical Association (AMA). A leading figure in the establishment of the American Society for Pediatrics, he also was a member of the New York Board of Health, and worked for the improvement of New York politics through civil service reform. In his old age, he became the president of the AMA, the "Nestor" of American Medicine.[12]

When the couple met in 1872, Mary Putnam Jacobi had just completed seven years in the Paris medical clinics. She was admitted to the New York Pathological Society in 1872 during Abraham Jacobi's presidency. A record exists of the lively professional exchange between the two doctors at meetings of the society that year. She later credited him for her ready admission to a number of other New York medical societies, including the Neurological Society, the State Medical Society and the New York Academy of Medicine, of which again Jacobi was president.[13]

She and Abraham Jacobi married in 1873. She had been engaged to two other professionals but had backed away from marriage. In the first case she canceled an engagement to her professor at the College of Pharmacy because she doubted the strength of her feelings. In the second case she reconsidered marriage to a young French doctor following the events of the Paris Commune, partly because she would not have been able to teach or practice medicine in France.[14] Jacobi had been married twice before to conventional women who died in childbirth.[15] During the first decade of their marriage their two children were born, and both husband and wife published a great many medical articles while teaching and practicing medicine, and participating in political reform movements. For Putnam Jacobi it was a "golden decade."

Their first surviving child (there had been an earlier stillbirth) was a son, Ernst, born in 1875. His birth was doubly welcome since Jacobi's earlier marriages had resulted in no surviving children. Putnam Jacobi wrote to her mother about her husband's delight in his child, his crooning nonsense to his child over the cradle, and his finding imaginary flaws in the infant to goad his wife into passionate denials. She wrote, "However this present delight is certainly an exquisite part of my own happiness, just as the hope of seeing him grow up with a fundamental likeness to his father in body and soul modified in a few details by my own share in his inheritance and education is certainly a most important part of the happiness of the future."[16]

When the baby was one year old, Putnam Jacobi had traveled to Switzerland with her husband and her younger sister to see her old friends the Réclus in Zurich and show off her child while Jacobi visited relatives in Germany. She reported to her mother that she and the Réclus had talked all day long about the future of relationships between men and women. Elie Réclus suggested that passionate and sentimental marriages would fade away and marriages would be formed on friendship or social interest with a restricted number of children.[17]

The medical collaboration of the Jacobis flourished in those early years. During the period between 1875 and 1880, they seem to have assisted each other in publications. The first of these was published just before the birth of their son. Putnam Jacobi edited a little book on infant diet taken from Jacobi's lectures, which emphasized the importance of breast-feeding newborn children for the first months. She chose a popular style with the intention of educating young mothers about the topic. The second collaboration was her contribution to the medical literature in the well-researched book published by Abraham Jacobi on diphtheria in 1880. Only a brief mention of some of her chemical experiments on albumen recognizes her collaboration, but the careful review of the French as well as the German, English, and American studies marks her interests and knowledge.[18]

Her male medical colleagues were impressed by her presentations at meetings, which were "always the equal and even superior to those of the men who attended." As one member of the Pathological Society later wrote, her knowledge of both pathology and the literature was so wide and her criticism so "keen, fearless and just that in our discussions we felt it prudent to shun the field of speculation and to walk strictly in the path of demonstrated facts."[19] She was making her name beyond the New York medical societies. She had won the Boylston Prize in 1876 offered by Harvard Medical College for an essay defending the capacity of women to function effectively during the menstrual period. This was a task she had shouldered at the urging of the feminist community.[20] Between 1879 and 1882 she published twenty-two papers on a wide range of topics emphasizing diseases of women and children and research in neuropathology.

She, like all women physicians of the time, was barred from serving as a consulting physician in the major hospitals in New York, although she taught and worked at the New York Infirmary for Women. She started an outpatient pediatric clinic at Mount Sinai, where both Jacobi and his close friend Dr. Ernst Krakowizer were important staff members. A year later the dispensary was incorporated into the regular hospital services, and a woman was on its staff for the following ten years, but this did not change the exclusion of women physicians from inpatient care.[21]

In 1882, when the New York Post-Graduate School of Medicine was organized, Mary Putnam Jacobi briefly joined the faculty of that body, lecturing to both male and female physicians on children's diseases. She received no pay, and her name and the name of one other initial woman lecturer have dropped out of the history of that body. Her opening session was remembered as a remarkable event with diagrams and case histories during which she urged further study of the pathology of the fetus and the physiology of the living child.[22]

A rivalry may have begun to develop between husband and wife as both were working in the new specialty of pediatrics, and as Putnam Jacobi actually exceeded what she had termed the "organic vigor" of the new professional women. She was simultaneously lecturing at the Woman's Medical College, practicing medicine at the New York Infirmary, and seeing private patients, not to mention continuing a high publication rate. With two young children at home, the increased load upon the young woman must have pushed her to the breaking point, even with domestic help, while her husband was similarly overextended as he often worked fifteen hours at a stretch. Although the Jacobis shared social and political ideals, many fundamental differences lay beneath the surface. They were both freethinkers, but he was from a Jewish background, she from a strongly religious Protestant one. Philosophically, he had adopted a Haeckelian monism; she leaned toward Comtean positivism.[23] Both were socialists, but she had been radicalized by the Paris Commune and trusted democratic chaos working from below, while he believed in a state socialism in which improvement was imposed from a centralized state above.[24]

A growing strain, originally stemming from their overcommitment to medicine and politics, may have been aggravated by the introduction of a third party, the politician Carl Schurz, Abraham Jacobi's oldest and closest friend. He had come to New York to work in local politics after an outstanding career as an officer during the Civil War, as a senator from Missouri, and as secretary of the interior under President Rutherford B. Hayes.[25] While Jacobi was a taciturn man, Schurz, like Putnam Jacobi, was an ebullient personality, needing sympathetic friends with whom to share ideas. When Jacobi came to America in the 1850s, Schurz, who had preceded him, gave him excellent advice about where to set up his practice. After Schurz retired to private life in the early 1880s and moved to New York City, he and Jacobi became an inseparable team, working for political reform, sharing drafts of all their writings, constantly attending the same meetings, and eventually bringing their families to spend every summer as neighbors at Lake George, New York.[26] The growing collaboration between Schurz and Jacobi negatively affected the more experimental collaboration between husband and wife.

The only surviving letter from Jacobi to his wife, written in the spring of 1883, detailed a serious quarrel over an invitation to Schurz's birthday celebration. When Jacobi came home exhausted in the evening and found his wife had left for Schurz's house, he refused to go himself. He quarreled with her the next night at supper, saying that he, as Schurz's oldest friend, and not she had been invited. Someone, he said, had to stay home with the children (although they had a nurse). Putnam Jacobi, perhaps feeling she had been put in an impossible situation, went off in a rage with a woman colleague who had come to call. She wrote to him from an address in the city, and he returned the letter saying that she could "speak but not write." After describing his own account of the quarrel, he added:

> If you had less temper, and more patience and a less exaggerated conviction of what is owed to you, you might have recognized the fact that I have done a great deal for you which might have enabled you to find a quiet and rational modus vivendi. I trust you may after a while, for your own sake and that of the children—my own claims are none. . . . You will find that I shall not show you by sign or word that I bear you a grudge for this hasty step of yours. I shall not mention it at all and the children will be glad to see you though child-like they do not miss you.[27]

After this quarrel, Putnam Jacobi returned home. Jacobi presumably said no more about it, as he promised in the letter. But the unhappiness generated by this quarrel may have been aggravated by the fact that their two children came down with diphtheria in the summer of 1883, probably soon after this letter was written. The four-year-old daughter, Marjorie, recovered, but the beloved eldest son, Ernst, not quite eight, died. Childhood death and disease were not unusual in the late nineteenth century, but in this case a number of factors combined to make it traumatic for both parents and to make them grieve for the rest of their lives. Putnam Jacobi's mother had borne nine children, and all except one had survived into productive adulthood. On the other hand, Abraham Jacobi's first two wives had lost a series of infants.

A major factor that must have aggravated the loss was that Abraham Jacobi was a known authority on the diagnosis and treatment of diphtheria, the disease from which his son died. This disease was one in which both diagnosis and treatment were about to change completely following the new therapeutic revolution in bacteriology. Jacobi had opposed this revolution both in the preface to his book on diphtheria and in a talk celebrating the great German pathologist Rudolf Virchow. He had condemned as worthless the work of the bacteriologist Edwin Klebs, the rebellious disciple of Virchow.[28] Klebs had (it turned out) correctly identified the diphtheria bacillus (the Klebs-Loeffler strain), which would soon lead to the development of a truly effective treatment, through the use of antitoxin serum, tried on a large scale by Emile Roux in the Paris hospitals ten years later. Jacobi could never publicly admit his error in opposing this research, and, I must add, initially his was not an unreasonable position, as Evelyn Hammonds has pointed out.[29]

Jacobi and his wife continued to grieve throughout their married life for the loss of this son.[30] Jacobi never publicly worried that he himself may have introduced the contagious disease inadvertently into his home, although as a regular practitioner of tracheotomy he constantly faced the threat of bacterial contagion. As a major authority on the disease, he had treated many cases of children suffering from diphtheria in 1883. He later reported on these cases in his clinical lectures. Perhaps because some doubt remained in his mind, he searched for a scapegoat in the figure of the old German nurse who cared for the children. In one of his most widely quoted papers on diphtheria in adults in 1884, he described a conscientious nurse who had in her care the two children of a physician who had continuous sore throats, which she hid from the physician. Jacobi believed she had passed the disease on to the children. Children's nurses and many other adults were carriers of the disease and passed it on to children, he argued, but no physician had been known to spread the disease. In spite of Jacobi's claim, he had regularly performed tracheotomies on children with membranous diphtheria, during which infectious matter could easily have been sprayed onto his clothing. The German nurse who had tended the children was dismissed, and Jacobi's paper became an important citation on the topic of disease carriers. On his behalf it might be added that although he never admitted he had misread the bacterial revolution, later, when the diphtheria antitoxin was developed and proved in clinical trials, at least one source identified him as one of those behind the decision of the New York Board of Health to introduce this system of immunization into the free clinics and hospitals for the poor.[31]

The death of her son was a great blow to Putnam Jacobi, perhaps even more than for her husband. She blamed not the nurse but herself for being unable to save her son and went through agonies of self-contempt. In a letter written to an acquaintance she described in the most heart-wrenching manner her belief that "the whole world should hoot" at her as a physician who could not save her own child. She seriously considered experiments in bacteriology as a consequence, a subject, she wrote to a sympathetic friend, "on which I have been reading all summer and expect to devote myself to. It has a horrible fascination for me. I cannot get away from it. But I never should feel that any knowledge now gained or even any lives saved would be in any way a compensation for Ernst."[32]

Putnam Jacobi never pursued significant bacterial research herself, but she urged one of her bright postdoctoral students, Anna Wessel Williams, from the Woman's Medical College of the New York Infirmary for Women, to begin this study. She helped Williams obtain a position working with William Park, the New York physician and public health researcher who made bacteriology an important aspect of diagnostic and preventative medicine in America.[33] In 1894 Williams discovered what is still termed the Williams-Park strain of diphtheria bacteria, from which much more effective diphtheria antitoxin could be derived.

Perhaps the death of that child forever called into question the view that each physician/marriage partner had of the other. She may have regretted the loss of the double personal and professional communication with Jacobi in which she had earlier delighted, as the friendship between Schurz and Jacobi expanded as they shared their political interests, their German language, and the stories of what Putnam Jacobi termed their "Trojan war," their revolutionary youth. Almost twenty years after Ernst's death, Putnam Jacobi confided in her future son-in-law that through confidence in her husband's professional opinion, "knowing his great love for our little son," she had failed to insist on taking precautions about his health that "I [know (crossed out)] now think. might have sufficiently increased his power of resistance when the trial came, to have enabled him to escape as Marjorie did."[34]

Jacobi has been called the father of American pediatrics, first founding the pediatric wing of the Obstetrical Society and then being crucial in establishing the *American Obstetrical Journal*, which initially carried articles on pediatrics. With Osler he founded the American Pediatric Society in 1892. He had been a president of the Pediatric Society of the AMA and eventually a president of the national organization. Jacobi failed to use his influence to bring his wife into these national organizations as he had twenty years earlier brought her into the New York medical societies. Putnam Jacobi had complained quite bitterly about the exclusion of women from the International Medical Congress held in London in 1888. Jacobi did not lead a discussion or a debate on this issue, nor did he, when president of national medical organizations, make an endorsement of a future intent to admit competent women. Certainly Mary Putnam Jacobi could have claimed some right to a place in the American Pediatric Society, founded by her husband. She had published over thirty papers on some aspect of pediatrics alone, about one-third of her total output. Most of these were very clearly written diagnostic analyses, often with pathological follow-up. Many of these were presented before the pediatric section of the New York Academy of Medicine and appeared in major medical journals. Yet she is not even mentioned in a recent book on American pediatrics that gives considerable space to her husband.[35]

Mary Putnam Jacobi thought that women must compete with male physicians, and she took pains to say so, a position in contrast to that of Elizabeth Blackwell, as Morantz Sanchez has so vividly demonstrated.[36] In 1883, in a commencement address to the Woman's Medical College, she had deplored the "colonial" role of women in American medicine, which had kept them from accomplishing at the appropriate level. She insisted that if women physicians did not compete directly with their male counterparts and act actively and independently, they would be shunted into a second-class position.[37]

Putnam Jacobi fought for women's social and political rights with the same energy she brought to medicine. She studied the health problems resulting from the long hours of shop girls and prepared reports for the New York Consumers' League. She spoke before the New York Legislature in 1894 asking for the vote for women. Knowing in advance that the legislation would fail, she suggested and developed the early programs for a League for Political Education. In her book *Common Sense Applied to Woman's Suffrage* (written as an expansion of a speech she had given before the New York State Legislature on behalf of women's suffrage), she compared the arguments about women's capacities to those arguments she had heard throughout her life directed at women physicians.[38] She also discussed the historical relationships between men and women. "The Patria Potestas is gone; a man has lost, first the right to kill his own son; then the right to order the marriage of his daughter; then the right to absorb the property of his wife. Nevertheless he survives and the family shorn of its portentous rights bids fair in America to remain the happiest of all conceivable natural institutions." The unwillingness to give women the vote was possibly based, according to Putnam Jacobi, on the fact that "one class remains over whom all men can exercise sovereignty—namely the women. Hence a shuddering dread runs through society at the proposal to also abolish this last refuge of facile domination."[39]

Her husband's friend Carl Schurz dismissed the arguments in her book, as she told him she expected him to do. She insisted that they approached the issue "from starting points which are essentially unlike."[40] For Putnam Jacobi, women's rights seemed natural and inevitable. Her own brothers had become warm advocates of women's suffrage, as she reminded Schurz, without any special pleading on her part. "But where men say either that the women of their families are quite incompetent to have opinions or as you probably think from your own experience that it would not occur to them to take any interest, then to claim [the same] of other women seems eccentric and absurd."[41] She had earlier twitted Schurz for having formally met with and praised his old enemy Bismarck and suggested that the old revolutionary of 1848 no longer existed.[42]

In 1899, just before the marriage of her daughter Marjorie, Putnam Jacobi urged her future son-in-law to try to persuade the reluctant Marjorie to live in an extended household with them, sharing their large house. Attributing this desire on her part to her "habituation to collectivism," she added that it would be "a subtle and far reaching personal triumph for me to see Dr. Jacobi enjoying Marjorie's social relations where he has often objected to sharing the responsibility of mine as being too much trouble for a hard working man. I feel as if what he had refused to one part of myself, he had after all accorded to another part."[43] She was distressed that her daughter did not adopt her own view of changing marriage roles; she deplored the "childishness" and conservatism she detected in her daughter's view of marriage, referring jokingly to the marriage ring as that "time honored badge of servitude."[44] "Marjorie has inherited from her father more of his racial conventionalism than of his individual and youthful tendency to revolutionary innovation—[a] tendency which attracted me and which my own innate tendencies affiliated— more so I might say than he ever understood."[45]

From 1901, a developing brain tumor was to decrease Putnam Jacobi's incredible energy and incapacitate her, resulting in her death in 1906. Her comment on Carl Schurz's death, which preceded hers by only a few months, was articulated with painful difficulty to her family: "The heaviest blow has fallen" referred not to her own loss but the blow to Jacobi.[46]

Even after Putnam Jacobi's death, Jacobi rarely publicly recognized his wife's work in medicine. Although his wife had written authoritatively on the history of women in medicine, his own venture into this subject was an article on women physicians for a German journal that appeared in English translation only after the death of his wife and did not cite her much earlier and far more analytical paper. While she had named and credited him for his support, he indirectly referred to her but not by name, in the course of mentioning the successes of women physicians. German physicians, he observed, should not fear competition from the women.

"If they achieve but little—habeant sibi—so much the better for their male competitors."[47] In the thirteen years before his death in 1919, he did not include recollections of his wife's truly unusual accomplishments, even in his lengthy statements about his medical career made to his biographer. One might have expected something more from a man so determined to establish his own place in medical history and to memorialize his friends.[48]

Perhaps I have picked up this problem from the wrong end. The question may not be why there was not more collaboration between husband and wife but whether there was a realization by that clear-thinking woman, Putnam Jacobi, that any collaboration between the two physicians would be credited as the work of Jacobi. Her one opportunity for independent recognition as a physician may have depended upon separate practice and publication.

By the turn of the century, it must have been clear to Putnam Jacobi that she could not rely on the goodwill of even the best male physicians to promote the interests of their women colleagues. By 1900, she issued a call for women physicians to form strong coalitions to overcome prejudices against the admission of women to hospitals and to make them more visible in research medicine as well, adding that "as there are no resources at hand to overcome [prejudice] by force, it must as on other analogous occasions be bought off. . . . [T]he women physicians of New York should claim to stand on their own feet, to work for their own interests in the same resolute way in which for instance groups of foreign physicians have developed themselves into powerful organizations."[49] In order to be seen as equal partners with male physicians, women had to improve their lot as a group. "We are first physicians and then women physicians," she insisted. "We are doubly members of a class and therefore set apart to support each other."[50]

Putnam Jacobi may also have mistaken the nature of the romantic idealism of Schurz and Jacobi's generation. Men and women inspired by the 1848 revolution had advocated new sexual relationships outside marriage, often complex triangular relationships with close friends of a married couple. The collaborations of the Romantics were inspirational, rarely practical, and based on highly charged and often destructive triangular sexual relationships. This kind of male-female relationship was in marked contrast to the ideals of the Réclus and their friends, who saw marriage as a unified, cooperative, and creative enterprise.[51] This later nineteenth-century ideal had as a definite social purpose, which included but extended beyond childbearing, incorporating the skills of both marriage partners. Caught between the two competing ideals, Mary Putnam Jacobi's dream of a marriage of two happily collaborating physicians may have been in truth a clanging of eagles.

✦ 12 ✦

LINDA TUCKER AND
CHRISTIANE GROEBEN

"My Life Is a Thing of the Past"
The Whitmans in Zoology
and in Marriage

Charles Otis Whitman (1842–1910), one of the most influential figures in the development of American biology, belonged to that nineteenth-century vanguard who organized science and research in the newly formulated American universities.[1] Emily A. Nunn (1843–1927) presented early challenges to all-male bastions of science, gaining a degree of access but often facing frustration and rejection.[2] After becoming Mrs. Whitman, Emily ceased to pursue a career other than that of wife and mother. In marriage, her identity as a scientist was lost, and her support of her husband's work was consistent with the role of wife. The Whitmans present not the ideal portrait of interactive synergy but a case in which gender, marital, and parental roles overrode the wife's personal and professional development, to the detriment of the creative potential of the couple.

Born on July 2, 1843, in West Suffolk, England, Emily was a descendant of smallholding farmers and artisans.[3] Her maternal grandfather, Robert Sculthorpe Kendall, was an artist. His wife, Mary Towler Kendall, had forced her way into classes at the college where her father was a professor. "To the scandal of the faculty," she finished her courses, thereby becoming the first recorded "college woman."[4] This extraordinary woman, who lived with Emily's family, set the precedent for higher education for daughters as well as sons. Mary Towler's daughter, Miriam Towler Kendall, married her second cousin Charles Robert Nunn in 1839. The Kendalls were Baptists, and Charles Robert Nunn adopted the faith on his marriage to Miriam. With their seven young children, the couple emigrated to America in 1852 for religious reasons. The third, Emily, was nine years old. The family lived in Ohio, first renting a farm near Medina and later buying one at Peru, twelve miles south of Norwalk. By 1860, four more children had been born.

Little is recorded of Emily's life on the farm, to which she remained strongly attached throughout her life, or about her education. The first available information on Emily dates from 1857, when her name was entered in the tiny journal kept by her father. "Received of Emily June, 1857 50.00," he wrote of his thirteen-year-old daughter.[5] The Nunns were a close-knit, businesslike family in which sharing was accompanied by strict accounting and repayment with interest, and where girls as well as boys earned income. In 1867 Emily, her sister Miriam, and her brother Lucien Lucius (L. L.) moved to Cleveland. The sisters taught public schools while L. L. attended a private academy. Emily ran the small household, and L. L. made the furniture.

In 1869 Emily joined her older sister Ellen in Dresden to study German. The training stood Emily in good stead; in 1871 she taught languages in Chicago, making as much as fifteen hundred dollars a year. From January 1874 to June 1875 Emily studied at Newnham College, Cambridge University, founded for women in 1871.[6] She attended the lectures of physiologist Michael Foster, who had recently introduced such progressive reforms as laboratory instruction in elementary biology and physiology.[7] There are no formal records of courses, examinations, or other activities. Emily's interest in zoology was further evidenced by her studies at the University of Zurich under Heinrich Frey, a professor of anatomy and zoology.[8] Official records there do not mention Emily, who presumably attended lectures as a special student ("Auditorin")[9] and did not maintain a residence in her own name.[10]

Before the fall of 1877 she also resided in Cambridge, Massachusetts,[11] and associated with Boston-area scientific circles. Samuel Scudder and Alpheus Hyatt of the Boston Society of Natural History (BSNH), as well as Elizabeth and Alexander Agassiz, figured in her later scientific life. A voluntary association of amateurs, the BSNH inaugurated its program of teachers' classes in the natural sciences, taught by the faculty of Harvard and the Massachusetts Institute of Technology, in 1870.[12] In 1879 Emily's special contributions to a particularly successful class were noted.[13] Women became eligible for BSNH membership in 1876, and in December 1881 Emily was elected an associate member.[14]

Emily came to the attention of Henry F. Durant, president of Wellesley College, near Boston, who hired her to teach zoology.[15] Envisioning a "female Harvard," Durant selected a highly trained female faculty. With Durant's approval Emily traveled to Baltimore in the fall of 1877 for laboratory instruction in the new Johns Hopkins University. Serving primarily male postgraduates, Hopkins offered unparalleled resources for laboratory work in the sciences. As a woman, Emily was welcomed in the Teachers' Class in Physiology, a Saturday class conducted by biology professor Henry Newell Martin, whom she may have met at Cambridge. Unsatisfied, Emily requested regular student status in laboratories and classrooms. "These advantages are so exactly what I need, & since they are not found elsewhere in our country, if they are denied to me, I suppose there will be nothing else left than to cross the ocean. . . ."[16]

Emily assumed such privileges as "after hours" laboratory instruction with Martin, attending Ira Remsen's chemistry lectures, and a room with desk space in the library.[17] President Daniel Coit Gilman disapproved of her aggressiveness and her profiting from confusion about policies, but was sympathetic to her needs. However, the trustees, loath to mix the sexes in biological laboratories, restricted Emily to the teachers' class.[18] She dropped out during Christmas break.[19] Emily did in fact "cross the ocean," returning to Cambridge to work with Foster and Francis Balfour.[20] She attended at least through the summer of 1878.[21] During this sojourn, a fellow American student reported home that Emily was "doing some very good work,"[22] resulting in her first scientific paper. Her histological study of the effects of poisons on frog epidermis appeared in volume 1 of Foster's *Journal of Physiology*.[23]

Emily left Cambridge in the fall of 1878 to begin her tenure at Wellesley College. The number and variety of science courses offered there were impressive. Emphasizing independent experimentation, Wellesley maintained well-stocked laboratories and the latest apparatus. In 1878–1879 Emily offered a one-year elective course, Zoology. Subsequently she organized a two-year course, Biology, with a choice between Advanced Physiology and Comparative Anatomy in the second year.[24]

In the summer of 1879 Emily joined a select group of eleven in the second session of the Chesapeake Zoological Laboratory (CZL), a summer research program of The Johns Hopkins University.[25] Emily, the only woman, studied the development of the ctenophore *Mnemiopsis*, getting "interesting results regarding the changes which accompany the fertilization of the egg," according to CZL director William Keith Brooks.[26] The outdoor expedition was uncomfortable in the extreme. Barges served as transport, laboratories, and dormitories. Mosquitoes made Crisfield, Maryland, unbearable, and the party relocated to Fort Wool, Virginia. One participant, Kakichi Mitsukuri, recalled that Emily had "disgusted everybody" on the expedition.[27] However, the reports of assistant director Samuel F. Clarke, though laden with ethnic stereotypes, indicated no awkwardness concerning the presence of a woman. In Clarke's accounts, Emily did not distinguish herself as a feminine weakling, nor was she the obvious substitute when the cook quit.[28] The following October Emily presented her research at a meeting of Hopkins' Scientific Association.[29]

Emily taught two more years at Wellesley. Durant, who insisted on teachers "of the highest Christian character . . . above suspicion or reproach,"[30] terminated Emily's contract in the summer of 1881—because of her agnosticism, according to her successor, Mary Alice Willcox (1856–1953). "She came into collision with Mr. Durand [sic] in various ways. I remember his saying to my father [a Congregational minister] that in bible class she discussed the story of the prodigal son, explaining him as probably insane. I understood that that was the last straw." A man who collected specimens for Wellesley gave Willcox evidence of Emily's eccentricity. "[Emily went] collecting to pitcher plant swamp clad only in a tight fitting garment that reached from neck to ankles. She fell into a pool and insisted on drying herself by running along the road in front of the carriage which had brought her. 'She looked just like a monkey.' "[31]

By the end of 1881 Emily had returned to England, where she worked diligently, according to zoologist Thomas Henry Huxley (1825–1895), as an investigator in Huxley's South Kensington laboratory.[32] In June 1882 Huxley presented her results on the development of the enamel of vertebrate teeth to the Royal Society.[33] Foster wrote to his friend Anton Dohrn, founder of the famous Stazione Zoologica in Naples, that he and Huxley strongly recommended giving Emily access to its facilities for marine zoological research. She would study the "nerves of *Actinia* going over [Richard] Hertwig's work."[34]

She arrived at Naples on November 22, 1882, the first woman and the second American among the 223 guest scientists to that date. The Stazione hosted experienced workers who conducted their own research with assistance from Dohrn's staff. Emily would have to prove herself equal to the new surroundings, and as Dohrn wrote to Huxley, "she realizes how difficult it is to work on something coherent. I have repeatedly said that she should not publish anything in the next 2 years in order to publish something sensible. She wants to stay now for 2 years. If she succeeds, all the better."[35]

Instead of staying for two years, Emily suddenly left the Stazione in May 1883, after only six months. Her reason for leaving is not clear, but may be related to an embarrassing and far-reaching incident. Without permission, Emily ingenuously took from the Stazione bad copies of group photographs and sent them to her sick mother.[36] Dohrn was furious. Edmund Beecher Wilson, who knew Emily from the CZL, was in Naples at the time and spread the story in the United States.[37]

On her return to the United States Emily spent the summer of 1883 as a guest at Alexander Agassiz's Newport Marine Zoological Laboratory, studying early stages in the development of Brachyura. Founded in 1877, this private laboratory extended its research privileges very selectively. There she met her future husband. Also about forty years of age at the time, he had a background and experiences that were similar to Emily's in many respects.[38]

✦ ✦ ✦

Charles Otis Whitman grew up in Maine, the son of a carriage maker. To finance his education, he began teaching at an early age. He entered Bowdoin College in 1865 at age twenty-three, earning a bachelor's degree in 1868 and a master's in 1872. Like Emily, he came to advanced studies as a teacher interested in continuing education. An informal course, Louis

Agassiz's Anderson School of Natural History at Penikese Island (1873–1874), strongly influenced him. Charles was thirty-three years old and secure in his position at Boston English High School when he decided to make a serious commitment to zoology. After three years of preparation in Leipzig under Rudolf Leuckart, Whitman earned the doctorate with the dissertation "The Embryology of Clepsine" in 1878.[39] Emily lacked these advantages, and the only evidence of any degree is the notation "M.A." behind her name in a Johns Hopkins publication.[40] Her scattered efforts did not provide the focused discipline of the Ph.D., which was virtually unattainable for American women of her generation.

Despite this critical difference, similar achievements and associations of the next five years carried the promise that both Charles and Emily would teach science at some level and with any luck continue to conduct research. In 1878 Charles left Leipzig for Boston to teach one last year at English High School, while Emily returned from England to begin teaching at Wellesley. For the academic year 1879–1880, Charles was offered but refused a fellowship to The Johns Hopkins University. At this point Emily was returning from her summer on the Chesapeake Bay to teach two more years at Wellesley. Charles spent the next two years as Professor of Zoology at the Imperial University in Tokyo. He resigned in protest against administration policies in August 1881, the same year Emily was dismissed from Wellesley.

Soon after these unpleasant events, the two converged on Europe from opposite ends of the globe. Charles took his place as the first American investigator at the Stazione Zoologica in Naples in November 1881, by which time Emily must have been in London. Leaving Naples in May 1882, six months before Emily arrived, Charles spent several months in Leipzig before sailing to the United States in September, with no position and no firm plans.[41] Although there is no evidence that Charles and Emily were previously acquainted, they probably knew of each other through mutual friends and colleagues within the international elite of science.

Before the year was out, Charles was hired as Associate in Zoology at Harvard College's Museum of Comparative Zoology (MCZ). Emily left Naples in May 1883, and both spent the summer in Newport. Charles was doing research on early stages of pelagic fish eggs, in collaboration with Alexander Agassiz, director of the MCZ. In June 1883 Emily published an article about the Naples station, the first contribution to the new journal *Science* by a woman. Charles followed with his own the next month.[42] An MCZ associate awaiting Charles's return from the summer at Newport recalled that "when he and Miss Nunn got off the train I was introduced to Miss Nunn and told of their engagement."[43] They were married on September 4, 1883, at Newton, Massachusetts, where Charles was living.[44] "I presume the enclosed cards may be a surprise to you," Charles wrote to Dohrn's assistant Paul Mayer regarding his marriage announcement. "[I]t was almost a surprise to me."[45]

At the time of their marriage, Charles and Emily had shared the experiences of teaching, informal teachers' classes, independent research, traveling for better opportunities, and instability in employment. Both had shown tenacity and dedication. There were as yet no indications that Emily would be lost in Charles's shadow. But although he was hardly secure in his new profession of zoologist, Charles was better prepared than Emily to take advantage of the few opportunities available in American institutions.

For a short time after marriage, Emily maintained some involvement in science. The summer following her marriage, she continued her work on Brachyura in Newport with her new husband.[46] She was elected to corporate membership in the Boston Society of Natural History in January 1884.[47] In 1886 she published a second article on the Naples station.[48] Emily never published another scientific paper, however, in spite of her work at Chesapeake Bay, Naples, and Newport. Charles, on the other hand, had no teaching or museum duties to distract him from his chief responsibility, research. He wrote and published papers based on research begun in Leipzig, Naples, and Tokyo. He edited the microscopical section of *The American Naturalist* and wrote a number of technical articles. In 1885 he wrote his only book, a manual of microscopical techniques.[49]

In March 1884, Bryn Mawr College dean Martha Carey Thomas began the search for her first faculty. Emily was among the first four candidates considered: "Biol. Miss Nunn—ask at Wellesly [*sic*]."[50] The same year, Thomas discussed Charles with William T. Sedgewick of the Massachusetts Institute of Technology. "Whitman excellent—one of the best zool[ogists] in country—Splendid man only wife objection. She is notorious has pestered Huxley nearly to death. Behaved badly at Naples—All Americ[ans] ashamed of her[.] Certainly not desirable woman for womans [*sic*] college."[51] Two years later Thomas confirmed her favorable judgment of Charles and her reservations about his wife, whom she had heard to be in an asylum. Emily later referred to her treatment in a mental institution near Boston in a letter to her brother.[52] Neither Whitman ever held a position at Bryn Mawr College.

In mid-1886 Emily followed Charles to a second unpromising position, as research assistant to Edward Phelps Allis, a wealthy amateur naturalist, and manager of his private laboratory in Milwaukee, Wisconsin. There in 1887 Charles announced the new *Journal of Morphology*, backed financially by his new employer.[53] As editor, Charles gained exposure and influence that prepared him for future leadership. On May 15, 1887, Emily gave birth to a son, Francis Nunn Whitman, at the age of forty-four.

The following year Charles took a position that was truly national in scope, director of the newly incorporated Marine Biological Laboratory (MBL) in Woods Hole, Massachusetts. He accepted no pay for this work, which required residency in summers only. Established jointly by the BSNH and the Women's Education Association of Boston, the MBL welcomed women from its inception.[54] Charles was poised to make his mark after ten years as an underemployed zoologist. In 1889 he was appointed professor of animal morphology at a new graduate-level research institution, Clark University in Worcester, Massachusetts. Within three years he assumed even greater responsibilities as head professor of biology at the University of Chicago. Founded by John D. Rockefeller in 1892, Chicago was a comprehensive university with prospects of liberal funding. The dual commitments to Chicago and Woods Hole marked the remainder of Charles's professional life. The next intense decade of organizing, teaching, administration, editing, and professional activities inevitably affected the Whitman family.

From 1888, the pace of life must have been maddening. Charles was one of the organizers of the American Morphological Society (later American Society of Naturalists) in 1890, and president from 1891 to 1894. The summer before the move to Chicago, Charles declined an invitation from a friend who had taught high school with him in Massachusetts. "I have the organization, equipment and planning of a laboratory, and the preparation of courses for next year on my hands, in addition to my usual duties in connection with Clark and Woods Holl [now Woods Hole] Laboratory. . . . After June 20, I expect to be at Woods Holl. I shall then return here to pack up and move to Chicago. Imagine all that stares me in the face."[55] The following month his former employer Alexander Agassiz angrily broke off relations with Charles for neglecting to publish their joint research results, then several years old.[56] Charles also left incomplete most of the researches begun at Allis's Lake Laboratory in Milwaukee.[57] Not to be deterred from new ventures, Charles launched another publication, the *Zoological Bulletin* (later the *Biological Bulletin*) in 1898.

While professional work swamped Charles, Emily was consumed by the need to care for her young children and aid her siblings. A second son, Carroll Nunn Whitman, was born in Worcester on March 24, 1890. Emily was forty-seven. The winter of 1890 brought "a series of illnesses" to her and three-year-old Francis. Emily also cared for her "invalid brother here from London" and made arrangements for another sick brother in a Boston hospital while she was in the final stages of pregnancy.[58]

Emily had followed her husband from Cambridge to Milwaukee, Worcester, and Chicago within six years. On arrival at Chicago she was nearly fifty and had two sons, aged five and two. In the winter of 1894, she suffered six weeks of illness, "stricken with withering weakness."[59] Anton Dohrn wrote during his visit to the MBL in the summer of 1897 that "one can see that Mrs. W. is not well and talking to her is a pain, talks as much as ever, complaining about him."[60] The following spring, Emily wrote to Dohrn, "Our home has been a hospital this winter," mentioning that this time Charles too was ill. Between 1899 and 1901, Emily's energy went into caring for her older son Francis, whom she took to Europe for respiratory problems.[61]

Around 1900, roughly midway in Charles's tenures at the MBL and

Chicago, the pace of his life slowed dramatically. At age fifty-seven, he unofficially semiretired from the university. From his home near the campus, Charles conducted research, held conferences with graduate students, and extended informal hospitality to guests, but lectured only once a week, to graduate students. He delegated administrative duties to Frank R. Lillie, who was appointed associate professor of zoology and assistant director of the MBL in 1900. This behavior sparked resentment in the Department of Zoology.[62] Charles also distanced himself from MBL affairs after 1902, postponing actual retirement until 1908. Reasons for his withdrawal are not clear. Some documents dating from this period suggest that domestic problems contributed to Charles's reclusion, while the MBL crises undoubtedly took their toll.[63] Whitman's absorption in his pigeon work also indicates the middle-aged scientist's desire to leave behind a magnum opus treating fundamental problems in biology.

By 1900, Charles had been performing experiments with pigeons for more than five years and had established a colony of several hundred at his own expense. Pigeons surrounded and filled his house. In 1904 he spoke of his "columbarium" as the first installment of a biological farm for the experimental study of evolution.[64] The expense of the pigeon work was staggering; observations, experiments, and daily care were demanding. Whitman took the birds to Woods Hole every summer, but after 1903 began to remain in Chicago in part because of the risk and expense of shipping them.[65] A former student recalled that Charles spent most of his seven-thousand-dollar salary on the birds.[66] Emily shared her husband's fanaticism to the extent that she joined him in cashing in life insurance policies and mortgaging their home to raise money for Charles's work.[67]

Besides spending money on his own research, for years Charles had performed such personal charities as paying for reprints, pigeons, and cages needed by students, and reaching into his "own poor pocket to settle the $600 unsettled salaries" at the MBL. Charles's wealthy Milwaukee employer, Edward Phelps Allis, believed that Charles "had to contribute something himself" in order to produce his final volume of the *Journal of Morphology* in 1903.[68]

Such liberality was made possible in part by a wealthy brother-in-law. Just about the time of his sister's marriage, L. L. Nunn (1853–1925) was becoming a wealthy man in Western mining and electrical power.[69] A lifelong bachelor, L. L. estimated that he spent nearly ten thousand dollars a year on his family from 1883 to 1909.[70] L. L. came to the aid of Charles's scientific ventures time after time. He contributed at least eighteen hundred dollars to Charles's *Journal of Morphology*.[71] He became a trustee of the MBL in 1897, when Charles was under fire for promoting expansion in the face of debts. L. L. offered to pay all expenses rather than allow the trustees to cancel the session of 1897.[72]

In the second great MBL crisis in 1902, L. L. was a principal in the

"Chicago plan," whereby a small group of businessmen and University of Chicago officials would guarantee operating funds and in turn manage MBL operations. Charles endorsed the plan, but when adoption of a similar proposal from the Carnegie Institution seemed imminent, he rallied the membership to choose independence over financial security. No other episode better illustrates his idealistic impracticality.[73] Writing to Emily in 1906, L. L. estimated that he had spent over twenty thousand dollars on Woods Hole, including property purchased for the family, "trying to obtain results for you there."[74]

L. L. was more interested in his sister's welfare than Charles's career or the advancement of science. He viewed Emily as a self-sacrificing person who had suffered much. Once he described her as having possessed since her youth "the disposition to disregard comfort and live on as little as possible—an attitude which goes hand in hand with an appetite for speculation resulting in bad business ventures and the loss in large sums of the miserly, unholy savings."[75] His special, protective affection for Emily and her sons led L. L. to honor her requests on behalf of Charles. Emily's 1894 letter to her brother is representative. "Did I tell you Carl [Charles] talked of giving up his insurance policies? He is so much in debt—he doesn't know how to get up the $500 due on June 2d.—could you let him have the am't—& be paid back $250 April & $250 August? It might be of great use if you could be in Woods Holl in June or later."[76]

Emily also had the power to turn L. L. against Charles. In one acrimonious letter Emily asked L. L. to send her Charles's IOU for her to collect. Charles owed his sister-in-law Ellen five hundred dollars for caring for the Whitman children. Emily had saved to pay the debt "cent by cent. It was literally taken out of me, in all manner of hard ways, & it sh[oul]d be left to C. to squander." Charles had just received fifteen hundred dollars from an insurance policy, and according to Emily, "this is the time to pay up."[77]

The relationship between the brothers-in-law deteriorated as they competed for control over Whitman family affairs. In 1904 Charles made L. L. pay five thousand dollars for a deed to Woods Hole property that had mistakenly been made out to Charles, though L. L. had purchased it for Emily.[78] At this time, the Whitman boys had begun to attend out-of-town schools, and Emily traveled extensively. Charles accused her of avoiding her sons, entrusting them to tutors while she went to "Woods Hole, Boston, Lowell, Salem, Lynn, New York, Philadelphia, Chicago, &c."[79]

The following year, Charles and L. L. exchanged bitter letters over the children's education, moral development, and loyalty. L. L. questioned the value of Charles's scientific work. Charles charged L. L. with unwisely giving cash to the boys, while refusing the only direct request for help he ever made.[80] Finally L. L. emancipated Emily from Charles's financial hardships by aiding her directly, rather than answering various distress calls on behalf of her husband. In 1906 he set up a bond fund to yield a monthly allowance of $150 for Emily. "If you will give up trying to do anything with Whitman, you may still be happy in your two boys."[81]

Emily and the boys spent an entire year on the Nunn family farm during parts of 1907 and 1908.[82] One detects both resignation and defiance in Charles's refusal to pay Carroll's tuition bill of $256. He referred the boy back to his mother and uncle: "I cannot allow myself to be a convenient football, or to be squeezed to settle affairs undertaken by others, in which I had no voice or part of any kind."[83]

In 1909 the Nunn family farm was occupied by tenants, and Emily worried about a trunk she had left there. "I have no home—no place to store a thing safely," she wrote to L. L., in spite of the Whitman home in Chicago. Deeply regretful, she reflected on the distance thirty years had placed between her and the treasures stored in the trunk—"manuscripts and letters—a book I wrote—and things I never looked at since Wellesley days. Letters from Huxley, Darwin, Tyndall and others." It upset her greatly that on a recent visit with the prominent Lyman Abbott family, once close friends, she could not even participate in the conversation. "My life is a thing of the past."[84]

Charles died on December 6, 1910, and Emily struggled to care for his birds in hopes that someone would complete his research. She obtained Carnegie Institution funds and authorized posthumous publication of the results. She soon regretted the agreement made with the Carnegie as providing too little compensation for use of the Whitman home, which the family had vacated. She convinced the Carnegie to remove the birds to its facilities so that she could sell her house.[85] Charles's work no longer took precedence over Emily's financial security.

Emily Nunn Whitman died in Ithaca, New York, on October 19, 1927. Emily can by no means be viewed as a professional scientist stymied only by gender discrimination or oppressive marital arrangements. Her unacceptable behavior ruined some opportunities, and recurring mental illness may have affected work and social relations. Strict codes of feminine behavior meant severe punishment for offenders, often for the very qualities admired in men. An elder sister among eleven children, Emily was strong-willed, inured to difficult circumstances, and convinced of her right to access on the same level as males. In her quest for education, Emily did not confine herself to women's institutions. She was not docile but defiantly independent, and self-supporting for much of her life.

Just as gender stereotyping exacerbated opinion against Emily, unequal opportunities in education and employment kept Emily and other women out of the mainstream of professional life in science. Despite her association, often as a "first" or "only" woman, with many of the most celebrated people and places of emerging biology, her handicaps were evident by the time she married. The weight of family responsibilities assured that a goal which had been retreating steadily would slip hopelessly out of reach.

Moreover, the commitments of her mate drained both personal and family resources, and damaged their relationship. Significantly, Emily's most explicit contributions to scientific work, preserving Charles's scientific legacy after his death and winning her brother's support for his career, fell well within the role of wife.

Initially, the similar experiences and passion for science that Charles and Emily shared must have figured in their mutual attraction. It seems more likely that they expected to strengthen each other in science than that Emily consented to give it up. Yet the success story was ultimately to be Charles's alone. The few surviving references to Emily are largely negative. Neither Charles nor his biographers mention Emily's scientific background. This remarkable omission mirrors Emily's feeling of lost identity. It was a casualty of the Whitmans' bitter failure to wed intimate and professional life.

JOHN STACHEL

Albert Einstein and Mileva Marić
A Collaboration That Failed to Develop

On December 28, 1901, twenty-one-year-old Albert Einstein assured his fiancée Mileva Marić:[1] "When you're my dear little wife, we'll diligently work on science together so we don't become old philistines, right? My sister seemed so crass to me. You'd better not get that way—it would be terrible."[2] Yet, in almost two decades together,[3] during which he became a leading theoretical physicist and published dozens of papers,[4] he never acknowledged her help in any of them, nor did she publish anything of her own. What went wrong?

It has been suggested that Marić actually made major contributions, perhaps even doing the preponderance of the work in some cases, to important papers published in Einstein's name, contributions that he simply failed to acknowledge.[5]

The available evidence does not support such claims, as I have argued elsewhere[6] and will argue here. A sketch of Marić's life up to her separation from Einstein,[7] with emphasis on a discussion of her work in physics and its relation to his,[8] leads to the conclusion that she played a small but significant supporting role in his early work, a role that later diminished to the point that she felt excluded from his career. Finally, I shall consider some possible reasons why a full collaboration between them never developed.

Marić's Student Years

Mileva Marić was born in 1875 to a mother of Montenegrin extraction, in Titel, a town in the Vojvodina, then part of the Austro-Hungarian Empire. Her Serb father was a middle-level official in the Hungarian bureaucracy, who saw to it that she received an education quite unusual for a young woman of that time and place, including two years as a private pupil at the Royal Gymnasium in Zagreb, where her father was then working. After receiving special permission to attend the otherwise all-male physics class, she got the highest grades in both physics and mathematics. She finished her secondary education in Zurich, graduating from a girls' school in 1896. After a semester of medical studies at the University of Zurich, she transferred to the Swiss Federal Polytechnical School, Poly for short, enrolling in Section VIA, which trained teachers of mathematics and physics.

Marić's move to Switzerland is not hard to understand. French universities were the first in Europe to admit women.[9] Switzerland was second with Zurich in 1865, and other Swiss universities soon followed. The Poly did so in 1876, and the first woman graduated from Section VIA in 1894.[10] Young women in search of a higher education, many of them Slavs, went to Paris if comfortable with French, like Marie Skłodowska, or Switzerland if more at home in German, like Rosa Luxemburg.[11] Russians and South Slavs from the Austro-Hungarian Empire flocked to Switzerland.[12]

Einstein and Marić were the only two physics students to enter Section VIA in 1896. Both took basically the same required courses, but rather different electives.[13] During her second year, she went to Heidelberg to attend mathematics and physics lectures, returning after one term.[14] As a result, she passed the Poly's intermediate examinations a year later than he did, using his physics lecture notes to help prepare.[15]

After her return, the two became very closely attached, spending most of their time together. In spite of the firm opposition of his parents to the liaison[16]—an opposition that led to dramatic clashes between Einstein and his parents—the two lovers resolved to live together after graduation, marrying as soon as economic circumstances permitted. Their relationship included more than romance; to supplement the meager offerings of the Poly in theoretical physics, they jointly studied many classic works.[17] They also spent a great deal of time working in the well-equipped laboratories of Heinrich Friedrich Weber, senior of the two professors of physics.

In 1900 both took the final examinations. Her physics grades were comparable to his, but she got a decidedly lower grade in mathematics; he passed with an average of 4.91 out of a possible 6, while she failed with an average of 4.0.[18] Still hopeful, she reregistered the next year to retake the final examinations.

Both saw Weber as their potential mentor in the process of gaining entry to the physics community. She continued to work in Weber's laboratory on her diploma thesis (see below), which she hoped to use as the basis for an eventual doctorate.[19] Einstein also expected to remain at the Poly as Weber's *Assistent* (the lowest rung on the European academic ladder) while working on a doctorate. But his failure to obtain this position, which he felt had been more or less promised to him by Weber, led to increasing friction between Weber and the young couple.[20] Einstein's efforts to get an assistantship in mathematics at the Poly and at numerous other universities also failed, and he and Marić tried to find other jobs, again without success.[21] For the next couple of years, he lived from hand to mouth, working at a series of temporary academic and tutoring jobs outside Zurich.

In the midst of this trying period, while studying to retake the final examinations, Marić became pregnant. After again failing,[22] she left for home vowing never again to work with Weber. It is quite likely that the friction over Einstein played a role in her estrangement from Weber. Cut off from the physics community, she was now entirely dependent on her relationship with Einstein for intellectual as well as emotional support. He, on the other hand, had found another mentor, Alfred Kleiner, professor of physics at Zurich University, and begun work on a doctoral thesis.[23]

During the remainder of her pregnancy, the couple were reunited only once. After his parents sent a letter to hers making their hostility painfully clear, Marić fled the ensuing crisis to be near Einstein, who was working as a substitute teacher. To preserve the proprieties, she stayed in a nearby town for a few weeks, meeting him only on weekends, and then returned home. Her letters to Albert during this period sound notes of real despair, while his reassure her of his devotion and depict a rosy future (see the opening quotation of this paper) once this difficult period in their lives has passed.

Their daughter, referred to as "Lieserl" in Einstein's letters, was born early in 1902.[24] The same year, Einstein moved to Bern to start work at the Swiss Patent Office, where he remained for seven years. Marić soon followed, but without Lieserl, and the couple married early in 1903.[25] It was not uncommon at the time to legitimize a birth by a subsequent marriage,[26] and Einstein had earlier resolved that the child would join them after theirs.[27] But Lieserl was never reunited with her parents, and, in spite of recent efforts to find more information, her ultimate fate remains unknown (see below).

The episode undoubtedly placed a great strain on their relationship, as their elder son, Hans Albert, seems to have later surmised. A biographer with unique access to information from him[28] reports:

> Friends had noticed a change in Mileva's attitude and thought the romance might be doomed. Something had happened between the two, but Mileva would only say that it was "intensely personal." Whatever it was, she brooded about it and Albert seemed to be in some way responsible. Friends encouraged Mileva to talk about her problem and get it out in the open. She insisted that it was too personal and kept it a secret all her life. . . . Mileva married Albert despite the incident. . . . She did not think of the shadow her "experience" would cast over their life together.[29]

Married Life

"All happy families resemble one another, but each unhappy family is unhappy in its own way," wrote Tolstoy,[30] and so it came to be for the Einsteins. At first things went well for the newly married pair—at least in the picture both painted for their friends. Einstein wrote to his friend Michele Besso: "Well, now I'm an honorably married man, and lead a very nice, comfortable life with my wife. She takes care of everything exceptionally well, cooks well, and is always cheerful."[31] Shortly after, Marić wrote to her friend Helene Savić in a similar vein: "I am, if possible, even more attached to my dear treasure than I already was in the Zurich days. He is my only companion and society and I am happiest when he is beside me." She also inquired about the possibility of teaching jobs for her and Einstein in Belgrade, her last known reference to the possibility of a career for herself.[32]

In September 1903, while she was visiting her parents, Einstein wrote: "Now come back to me soon. 3½ weeks have already passed and a good little wife shouldn't leave her husband alone any longer. Things don't look nearly as bad at home as you think. You'll be able to clean up in short order."[33] Marić presumably went to her parents to see to Lieserl's future, which Einstein discussed (the last known reference to her) in a way that suggests they had already decided not to keep her ("As what is the child registered? We must take precautions that problems don't arise for her later"). He mentioned a serious illness ("I'm very sorry about what has befallen Lieserl. It's so easy to suffer lasting effects from scarlet fever. If only this will pass"), and she may have died subsequently; but the reference to her future suggests that she survived. If so, she may have suffered permanent mental or physical damage and been placed in some institution. If she survived unharmed, she may have been adopted by a relative, or given up for a "normal" adoption.[34]

Einstein had apparently just learned about Marić's second pregnancy: "I'm not the least bit angry that poor Dollie [his nickname for Marić] is hatching a new chick. In fact, I'm happy about it and had already given some thought to whether I shouldn't see to it that you get a new Lieserl."[35] But there never was a "new Lieserl." The second child was a boy, Hans Albert, born in 1904; another son, Eduard, was born in 1910.

During his seven years as patent clerk, especially from 1905 on, Einstein produced a steady stream of scientific papers, which, by the end of the decade, gained him a reputation as one of the most promising young theoretical physicists. He left the Patent Office in 1909 to take his first full-time academic post, as an assistant professor at the University of Zurich. By this time the marriage was in trouble. Marić confided to Savić: "In mid-October on the 14th we leave Bern, where I have now spent 7 years, so many beautiful and, I must say, also bitter and difficult days."[36]

We have seen one source of her bitterness: the final decision about

Lieserl, taken early in the Bern years. Another source was quite recent: a marital crisis, involving Einstein's friend, Anna Meyer-Schmid.[37] Suspecting her of designs on Albert, Marić wrote Meyer-Schmid's husband. Enraged, Einstein wrote Herr Schmid attributing Marić's conduct to unmotivated jealousy.[38]

Marić's letter to Savić goes on to boast of Einstein's success: "He is now counted among the leading German-speaking physicists and is being frightfully courted. I am very happy about his success, which he has really earned; I only hope and wish that fame does not exert a detrimental influence on his human side."[39] A letter written to Savić soon after Einstein and Marić settled in Zurich further explains her fears:

> You see, with such fame, not much time remains for his wife. I read a certain maliciousness between the lines when you wrote that I must be jealous of science, but what can one do, the pearls are given to one, to the other the case. . . . I often ask myself . . . whether I am not rather a person who feels a great deal and passionately, fights a great deal and also suffers because of that; and out of pride or perhaps shyness puts on a haughty and superior air until he himself believes it to be genuine. And I must ask you, even if the latter were the case, and my innermost soul stood less proudly, even then could you love me? You see I am very starved for love and would be so overjoyed to hear a yes, that I almost believe wicked science is guilty, and I gladly accept the laughter over it.[40]

Einstein's academic star rose with dramatic speed: In 1911 he accepted a full professorship at the German University in Prague, and the next year was called back to a similar post at his alma mater in Zurich.[41] In 1914 he was named a member of the Prussian Academy of Sciences and head of the prestigious Kaiser Wilhelm Institute for Physics, moving to Berlin to take this full-time research job. A letter to Savić in 1911 gives further insight into Marić's feelings during this period:

> I . . . believe we women cling much longer to the memory of that remarkable period called youth, and involuntarily would like things always to remain that way. Don't you find that to be so; men always accommodate themselves better to the present moment. Things are going well for mine; he works very hard, gives courses that are very well liked and attended, as well as many lectures, which I never fail to attend. Since there are rather many musical occasions in our house, we really have very little time that we can pass together in privacy and tranquillity.[42]

These touching and remarkably frank letters depict a woman who feels she is losing her husband, not least because of his successful career in science. They convey a growing sense of exclusion from that career, but no sense of deprivation of credit for his scientific work. Her own earlier ambitions seem completely subsumed by ambitions for him, ambitions that go hand in hand with forebodings of what his success augurs for their relationship.

The toll on Marić became apparent to those around her. Referring to the period around 1912, Peter Michelmore gives us an insight into how things appeared to their son, Hans Albert:

> Close friends . . . worried because [Marić's] dark moods were becoming more frequent. She was far too introverted. She never talked about herself. Even alone with the family, she had little to say and her long periods of silence irritated Albert. If they ever discussed the root of the trouble, that mysterious pre-marital incident, nobody knew about it.
>
> Hans Albert, then an eight-year old with a distinct mind of his own, sensed the tension between his parents. But his fat'.er's personality assured him all would be well.[43]

In retrospect, Hans Albert evidently thought that the loss of Lieserl was at the root of the estrangement of his parents. At the time, he served as his mother's surrogate for the waning love of his father. In 1909 she wrote Savić: "[Hans Albert] should start school early next year, but unfortunately he entered the world a week too late and probably will not be accepted. Then he will stay with his mama for another year; we are actually inseparable and cling terribly to each other."[44]

By 1912, whether she knew it then or not, Marić was competing with more than science for Einstein's affections. During a visit to Berlin, he had started a romantic liaison with his cousin Elsa Löwenthal, a divorcée with two young daughters and literary aspirations, then living there with her parents.[45] His letters to her refer to Marić, often alluded to as "my cross," in increasingly bitter terms: "Miza [nickname for Marić] is the sourest sourpuss that has ever been. . . . I cannot be at ease at home . . . she herself is the most tormented one, and cannot understand that she herself creates the graveyard atmosphere. Miza is by nature unlovable and mistrustful. When one responds accordingly, she feels persecuted."[46]

By the end of 1913 Marić was aware to some extent of the situation, as he informed Löwenthal: "She [Marić] doesn't ask about you, but I believe she does not therefore underrate the significance that you have for me."[47]

Shortly after their move to Berlin in April 1914, Marić realized that one of its chief attractions for Einstein was cousin Elsa, and returned to Zurich with the two boys, never again to live with Einstein as husband and wife.[48]

The Nature of Their Intellectual Relationship

From the letters Einstein and Marić exchanged as students, a picture emerges of two young people very much enchanted with each other, not least because of their common love for physics.[49] However, the contrast between their comments on this subject is striking. Einstein's show a young man passionately engaged with his subject, constantly telling Marić about his readings of both the classics and new papers. Rather than giving bare reports, he critically evaluates his readings, often adding ideas of his own bearing on their subject matter. Without the benefit of hindsight, one cannot point to anything in these early letters giving proof positive of budding genius, but they do convey the distinct impression of an original and imaginative mind at work.

Marić's comments depict an eager, hardworking student, but without a spark of originality, or more precisely, of scientific originality, for she does display flashes of literary talent, catching fire in some descriptive passages rather than in comments on physics.[50]

What was the nature of their intellectual relationship during the student years? They studied physics together, which was very important to Einstein, who at first was quite dependent on her. During the summer break of 1899 he wrote: "When I read Helmholtz for the first time I could not—and still cannot—believe that I was doing so without you sitting next to me. I enjoy working together very much, and find it soothing and less boring."[51] Later that year, Marić requested his help in preparing for her intermediate examinations, which she took a year after he did (see the previous section).[52] Einstein's physics notes contain a correction in her hand, confirming that she read them carefully.[53]

In discussing his ideas, Einstein occasionally called upon her for help, such as finding data to corroborate them (see next section); but the letters suggest that the most important role she played in their intellectual relationship during these years was "that of a sounding board for Einstein's ideas," as the editors of the *Collected Papers* (myself included) put it. He had a strong need to clarify and develop his ideas in dialogue with others, a "role also played on occasion by his friends Michele Besso and Conrad Habicht" after his move to Bern.[54]

It is difficult to gauge the nature of her responses to his ideas, since many of his letters and even more of her replies are lost. But fortunately we have her reply to the letter containing his most important original ideas. Almost half of Einstein's letter is devoted to his earliest discussion of the electrodynamics of moving bodies.[55] Her reply comments on every topic Einstein discussed in this and his previous letter (which she had received at the same time): family matters, vacation, examination preparations, and so on, with the sole exception of the electrodynamics of moving bodies. None of her ten other extant letters comments on his scientific ideas, so this exchange may be typical.

While we can never know their private conversations, later reminiscences suggest her taciturnity in discussions. Philipp Frank, who knew Einstein and questioned him extensively for a biography, discussed their student years: "For Einstein it had always been pleasant to think in society, or better perhaps, to become aware of his thoughts by putting them into words. Even though Mileva Maritsch [Marić] was extremely taciturn and rather unresponsive, Einstein in his zeal for his studies hardly noticed it."[56] Just after their marriage, Einstein and two friends set up the mock "Olympia Academy" to discuss topics in the foundations of science, usually holding their sessions in the Einsteins' home. Maurice Solovine recalled that "Mileva, intelligent and reserved, listened attentively to us, but never intervened in our discussions."[57]

Marić's Work in Physics

Marić's only comments on her own work concern her studies at the Poly. The most interesting concerns her diploma thesis, prepared as part of the final examinations:[58] "Prof. Weber has accepted my proposal for the diploma thesis, and was even very satisfied with it. I am very happy about the investigations I'll have to do for it. E[instein] has also chosen a very interesting topic."[59] Both of them carried out experimental studies of heat conduction, one of Weber's pet research topics, under his supervision. Einstein also commented on her work in a letter to Marić: "For the investigation of the Thomson effect I have again resorted to a different technique, which is similar to your method for determining the dependence of K [the coefficient of thermal conductivity] on T and which also presupposes such an investigation."[60] Weber graded her work 4 (out of 6) and his 4.5.[61] In retrospect, Einstein characterized the topic of their work harshly, as "totally uninteresting to me";[62] neither thesis led to a publication and the Poly routinely discarded such student theses, so an independent judgment is impossible. At any rate, evidence that Marić devised an experimental technique has no bearing on the question of her talent in theoretical physics, the area in which Einstein made his name. Marić's other references to her work in physics are limited to discussions of preparations for examinations (see previous section), as are Einstein's other comments on it.

More relevant are two letters to Savić discussing Einstein's work in theoretical physics. The first states:

> Albert has written a paper on physics that will probably soon be published in the physical Annals.[63] You can imagine how proud I am of my dear treasure. It is really no ordinary work, but very significant, on the theory of fluids. We have sent a private copy to [Ludwig] Boltzmann, and would really like to know what he thinks of it, hopefully he will write to us.[64]

The work in question is a theory of molecular forces. Discussing this work, Einstein wrote Marić:

> The results on capillarity I recently obtained in Zurich seem to be entirely new despite their simplicity. When we're back in Zurich we'll try to get some empirical data on this subject from [Professor] Kleiner [of the University of Zurich]. If this yields a law of nature, we'll send the results to Wiedemann's Annalen [der Physik].[65]

Marić's second letter discusses the doctoral thesis based on the same theory that Einstein submitted to the University of Zurich in 1902 and then withdrew:[66]

> Albert has written a splendid work that he has submitted as a dissertation. In a couple of months he will probably receive the doctorate. I have read it with great pleasure and true admiration for my dear little treasure, who has such a clever head. When it is printed, I will send you a copy. It deals with the investigation of molecular forces in gases on the basis of various known phenomena. He is really a splendid fellow.[67]

In both letters, Marić states that the works were written by Einstein, claiming no role in the formulation of the theory; he also speaks of his results.[68] Nevertheless, in discussing this work both slip easily into the "we" mode, which should be kept in mind when evaluating similar uses of the first-person plural in his letters.

The most notable of these is a reference to "our work" on a problem of much greater significance than his theory of molecular forces (see below), one of the complex of problems that led to the special theory of relativity, and the passage has been cited to support claims that Marić was coauthor of that theory.[69] Leaving aside the fact that his letter was written in 1901, whereas the theory was not finished until 1905, it is important to put the passage into context.

Physics aroused emotions in Einstein that, during the early stage of their courtship, he felt impelled to share with Marić, come what may. For example, soon after she told him she was pregnant—surely a difficult time for both—he opened a letter as follows: "I have just read a wonderful paper by Lenard. . . . Under the influence of this beautiful piece I am filled with such happiness and such joy that I absolutely must share some of it with you. Be happy and don't fret, darling. I won't leave you and will bring everything to a happy conclusion."[70] It is striking how many of his few references to joint work were penned at difficult moments in their relationship, amid reassurances of his love and devotion.

For example, Einstein referred to "our work on relative motion" after he left Zurich to stay with his parents, whom she knew to be violently opposed to their engagement. Here is the context:

> You are and will remain a shrine for me to which no one has access; I also know that of all people, you love me the most, and understand me the best. I assure you that no one here would dare, or even want, to say anything bad about you. I'll be so happy and proud when we are together and can bring our work on relative motion to a successful conclusion! When I see other people I can really appreciate how special you are.[71]

His words here are moving in their emotional intensity, but provide no clue about her contribution to "our work." Elsewhere in his letters, he does mention specific ideas about "relative motion" and many other topics in physics, but he always refers to his own work. Here is an example: "I'm busily at work on an electrodynamics of moving bodies, which promises to be quite a capital piece of work. I wrote to you that I doubted the correctness of the ideas about relative motion. But my reservations were based on a simple calculational error. Now I believe in them more than ever."[72]

Naturally enough, their correspondence practically stops after Marić joins him in Bern. Their few letters from the crucial years 1903–1905 that led up to the final formulation of the theory of relativity contain nothing relevant, nor is there any other contemporary documentation. Later reminiscences suggest that she continued to play a modest role in his work. The one I find most significant comes indirectly from Hans Albert Einstein, presumably based on information he got from his parents (see note 28). Discussing Einstein's work on special relativity, Michelmore writes: "Mileva helped him solve certain mathematical problems, but nobody could assist with the creative work, the flow of fresh ideas. . . . [After he wrote up his work] Mileva checked the article again and again, then mailed it. 'It's a very beautiful piece of work,' she told her husband."[73] The mathematics involved does not go beyond elementary calculus, and it seems unlikely that Marić contributed unique mathematical expertise to the paper; one may speculate that she might have suggested methods of proving certain results and/or checked calculations.

Einstein indeed does thank someone "who stood faithfully at my side and to whom I owe many valuable suggestions" at the end of his paper,[74] but it is his "friend and colleague M[ichele] Besso."[75] Taken together with his silence about Marić, this is interesting—if negative—evidence of his attitude toward her role in his work.

Other documents suggest that Marić played the role of amanuensis on occasion after 1905. Einstein's notebook for his lecture course on mechanics, given during the winter semester of 1909–1910 at the University of Zurich, includes "seven pages of notes in Mileva Einstein-Marić's handwriting, containing material very closely corresponding to the introductory sections of the first notebook, followed by an eighth page with a drawing of three intersecting circles, also in Einstein-Marić's hand."[76] And a document entitled "Reply to Planck's Manuscript," dated to 1909 or 1910 and included in a letter of Einstein to Planck, is also in Marić's hand.[77]

Another Einstein lecture notebook from 1910–1911 testifies not only to her familiarity with the notes but to her continued affection. She inserted the words: "Here give a dear little kiss to his [word not deciphered]."[78]

Her letters to Savić from 1909 on, cited in the last section, indicate that she attended his public lectures but bear witness to her growing sense of isolation from his career, as does her only letter to Einstein from this period, written after the 1911 annual meeting of the Society of German

Scientists and Physicians in Karlsruhe, which he attended: "It must surely have been very interesting in Karlsruhe; I would have all too gladly also listened a little, and seen all these grand people [*diese feinen Leute*]."[79]

To sum up, Marić seems to have encouraged and helped Einstein in a number of ways during their years together, notably as the alter ego to whom he could express his ideas freely while developing them in isolation from the physics community. She also appears to have helped by looking up data, suggesting proofs, checking calculations, and copying some of his notes and manuscripts. He never publicly acknowledged this help, nor did a true collaboration ever develop. As he took an increasingly prominent place in the physics community after 1909, she felt increasingly isolated from his work and threatened by his success.

Why Did a Real Collaboration Never Develop?

Was it possible for a married couple to successfully and publicly collaborate in physics at the beginning of the century? Two well-known contemporary couples managed to do so: Marie Skłodowska and Pierre Curie,[80] and Paul Ehrenfest and Tatiana Afanasieva.[81] There are interesting similarities between them and the Einsteins. All three wives were Slavs with a higher education living in milieus not free of prejudice against educated women.[82] All three husbands came from secular backgrounds; Einstein and Ehrenfest were Jews, raised in South-German urban environments (Munich and Vienna respectively), who had yet to establish their careers when they married.[83]

There is also a striking contrast. In the case of the Curies and Ehrenfests, there is abundant contemporary evidence of the importance of the woman's role in their joint work, and each wife pursued a scientific career after her husband's death. Marić, of course, did not pursue a scientific career before or after her separation from Einstein, but we see it cannot have been because of the impossibility of such a career.

It also cannot have been because she lacked initial motivation and support. In the face of the many prejudices against and obstacles to women's higher education, particularly in the physical sciences, Marić possessed sufficient talent and drive, and got sufficient familial and institutional support, to successfully pursue an academic career that brought her to the brink of graduation from the Poly and pursuit of scientific research, alone or in collaboration with Einstein, or at the least a career as a teacher of science. What went wrong?

I suggest three interrelated factors may help to explain why Marić never pursued a scientific career, and in particular why a truly creative collaboration between Marić and Einstein never developed:

1. Her talents in physics were modest, so that she could not take advantage of the "exceptional" status sometimes granted even then to women

of the stature of Marie Curie and Lise Meitner in physics, or Sofia Kovalevskaia and Emmy Noether in mathematics.

2. She lost the inner self-confidence and drive necessary to pursue a
career in science in the face of the many obstacles that women face.

3. Despite his earlier-expressed intentions, after their marriage Einstein
failed to encourage her to pursue an independent career or to involve her
in serious collaboration.

I shall elaborate a little on these factors. As we have seen, in spite of
her early successes as a student, there is no evidence that Marić ever made
the crucial transition from student of physics to independent research
worker. There is no record of her original ideas, and her comments on
the ideas of others are uncritical. For example, she vastly overrated Einstein's theory of molecular forces (see above) when a critical judgment
might have helped him to discard it sooner than he did.[84]

There is nothing really surprising about this;[85] most physicists, male
or female, would have had to play a subordinate role in collaborating
with Einstein. The evidence suggests that she did so, even without
public acknowledgment, which might have been painful to her had she
not been so willing to acknowledge his superiority and subordinate her
own career goals to his. But she accepted this role without complaint,
and even accepted—but not without complaint—her growing exclusion
from this modest role in his work. Part of her resignation may be
ascribed to her great love and admiration for him. But I think there is
more to the story.

Her early letters evidence the good cheer, drive, and talent necessary to
enable a young woman of her generation to get from a remote region of
the Balkans to Section VIA of the Poly. But by the end of her student
days—at a crucial point in her intellectual development—she lost the inner
self-confidence so vital to overcoming the considerable obstacles on her
path to a career in physics, a self-confidence that Einstein possessed in
abundance and never lost, even at the most desperate moments in his life.
Two failures to pass the final examinations and the loss of Weber as a
mentor were undoubtedly contributing factors, but her relationship with
Einstein also played a role. The pressures on a woman to subordinate her
intellectual to her emotional life were even stronger then than they are
today. As her letters attest, she was painfully shy and fearful of criticism,
and his parents' opposition, of which she was well aware even though
she had never met them (rather than shield her, Einstein reported their
comments),[86] must have afflicted her. Above all, her pregnancy out of
wedlock and the fate of Lieserl seem to have contributed to an underlying
depression that grew as the years passed. Perhaps partly in reaction to the
loss, she became exceptionally devoted to her first son, born early in the
second year of their marriage; and she was unable to find a way to combine

her conception of the duties of motherhood with those of a career outside the home.

Given all these factors, she still might have played a more satisfying if subordinate collaborative role in his work, as did several male physicists during this period.[87] A married woman at that time was hardly likely to find another mentor than her husband, so her fate as a physicist was entirely dependent on Einstein.[88] But after their marriage, he failed to foster such a full collaboration. However modest her talents, he could have publicly acknowledged her contributions to his work, and helped her to enter the world of physics after he gained recognition. He is reported to have helped around the house,[89] but obviously he was not engulfed by household duties and could have done more to ensure that she was not. Instead, he seems to have been content to let her play the "philistine" role of hausfrau, involving her in his work as little more than occasional amanuensis, and never publicly acknowledging her contributions. Again, there is an obvious contrast with Pierre Curie and Paul Ehrenfest, who took pains to assure that their wives' contributions to joint work were publicly acknowledged,[90] so that success was shared. Far from bringing Einstein and Marić closer, the widespread recognition of Einstein's scientific activities became an important factor in their ultimate estrangement.

✦ 14 ✦

SUSAN HOECKER-DRYSDALE

Sociologists in the Vineyard
The Careers of Helen MacGill Hughes
and Everett Cherrington Hughes

The career possibilities are great. A young man may be fairly sure that he may choose from among a number of open places when he finishes his training, and that he may from there on move about from position to position to suit his talents and his special interests . . . no competently trained and talented sociologist need want for a choice of jobs.[1]   —*Everett Hughes, 1954*

In contrast to the male sociologist's well-defined track, the female counterpart, typically in the past and often in the present, has worked any unoccupied portion of the vineyard. There are exceptions, but sociology still awaits its Margaret Mead.[2]   —*Helen MacGill Hughes, 1977*

The case of Everett Cherrington Hughes, the son of an Ohio Liberal Methodist minister, and Helen MacGill Hughes, the daughter of Vancouver lawyers in a Canadian family of social liberals and feminists, is a particularly interesting instance of the problematic of collaboration for scientific couples. Everett and Helen played singular and significant roles in the development of sociology in North America, particularly at the University of Chicago, where they met, received their training, and eventually worked. Their influence was substantial through several generations of sociologists, not insignificantly because they *were* a couple.

Entering the field of sociology in the early days of its expansion in the leading American sociology department and becoming protégés of a major figure like Robert E. Park, both would have expected to "move from position to position" according to one's interests and talents. Indeed Everett did just that. In fact he became a central figure in the development of North American sociology over several decades.

Helen's training and qualifications were similar and equally strong. The formal resemblances in their training and interests were offset by the fact that as a woman Helen had to excel in motivation and performance to enter and then to compete successfully in such rigorous and increasingly patriarchal environments as the University of Chicago. In fact she had graduate scholarships, while Everett and many others did not. In spite of her high performance, Helen experienced a very different kind of career, wrought as it was within limits imposed by her gender and marital status.

Finding her professional place in an "unoccupied portion of the vineyard" was a solitary, lifelong task.

In 1892 Albion W. Small established at the University of Chicago the first graduate department of sociology in America and in 1895 sociology's first and for some time to come most important professional journal, *The American Journal of Sociology*. At first women were welcomed on the faculty and as students in the university and in the Department of Sociology. But soon after the turn of the century classes became officially sex-segregated, because women's high performance and overrepresentation among scholarship students were thought to be detrimental to male students. The decision was supported actively or passively by most of the male sociologists, including Small, George Vincent, W. I. Thomas, and Charles Henderson, but strongly opposed by the women faculty, Charles Zeublin, and George Herbert Mead, whose wife, Helen Castle, was an educated feminist.[3] Moreover, women with doctorates in sociology and other social sciences, who had played active roles in their departments, became increasingly marginalized into the "applied" and apparently less prestigious departments (Social Work, Extension, Applied Statistics) and into "special" (low) academic ranks.

While the pioneers of sociology at Chicago were people of broad humanistic and scientific perspective, the second generation beginning around 1920, led by Robert Park, were significantly more concerned with the professionalization and scientific credibility of the discipline, and wished to separate it from the applied field of social work. In his early years Park, trained in philosophical pragmatism (John Dewey and William James) and German sociology (Georg Simmel), had been a muckraking journalist and had worked with Booker T. Washington. A man of some contradictions, it seems, Park was involved in reform movements and matters of race relations but vigorously criticized "do-gooders," applied sociology, and feminism.[4] When Everett Hughes and Helen MacGill entered the graduate sociology program, in 1923 and 1925 respectively, the influence of the first generation was still present, but during their graduate studies Park's "professional" orientation became dominant.

Career Beginnings

Everett Cherrington Hughes was born in Beaver, Ohio, in 1897. His father, Reverend Charles A. Hughes, was a learned man from a family that stressed the importance of university education for the sons but not for the daughters. Everett's mother, Jessamine Roberts, like most women of that era, did not attend college.[5]

Everett's family imbued in him a concern for and curiosity about human beings, and a liberal and even skeptical view of the world.[6] Following the completion in 1918 of his undergraduate education at Ohio Wesleyan, Everett spent several years teaching English to immigrant workers in

Wisconsin and Chicago and working for the Chicago Park Service. For a time he lived in a Methodist settlement house in Pullman, Illinois, a model industrial town designed by George Pullman for the employees of his Pullman Car Company. The young Hughes was impressed; he described it as "that ideal town, that Utopia [which] is itself part of the history of industry and of social experiment and of the labour movement in America."[7]

As a result of these experiences, his interests in communities and immigrants, and some knowledge of Park gained from a friend, Hughes enrolled in graduate studies at the University of Chicago in 1923.[8] Under Park's tutelage Hughes decided on a career in sociology. Hughes also became a close friend of the anthropologist Robert Redfield, who was married to Park's daughter.[9]

In his memoirs Hughes recalled that some of the graduate students had formed a research and study fraternity called Zeta Phi, which he was asked to join. A serious group, which met regularly at dinner, it seems to have been concerned mostly with coursework and examinations.[10] "Our hoity-toity group did not have any women in it. . . . As a matter of fact, there hadn't been many women there until then."[11] Soon there arose a rival group, Sigma Theta, composed of students who had come in a year or so later and including some women. These student groups continued at least until the late 1930s.[12]

Born in Vancouver, British Columbia, in 1903, Helen Elizabeth Gregory MacGill was the daughter of Helen Gregory (B.A., M.A. and LL.D.), feminist lawyer, juvenile court judge, and writer from an avant-garde family of educated women, and James Henry MacGill, a lawyer interested in race and ethnic relations.[13] Helen Gregory had two sons from a previous marriage and as a widow had supported her family as a journalist before studying law. She and MacGill had two daughters, Helen and Elsie, born in 1903 and 1905. Mother and grandmother were important feminist role models for Helen and Elsie. Helen Gregory—who had regular contact and correspondence with Jane Addams, Charlotte Perkins Gilman, and Christabel Pankhurst and other British feminists—was "the best educated and wittiest woman in any circle, uncommonly venturesome and in many respects unconventional."[14] Her second husband, James, and her two sons were in sympathy with her goals and values, although at times James apparently resented his wife's economic independence.[15]

Young Helen MacGill received a bachelor of arts honours degree in German and economics from the University of British Columbia (UBC) in 1925.[16] Near the end of her final term Helen heard lectures by Robert Park's assistant Winifred Raushenbush[17] on Vancouver's Chinatown and by Park himself on the form and structure of cities, which enlivened her interest in sociology. Raushenbush and Park were on the west coast studying oriental immigration. Park persuaded Helen to apply to the graduate program in sociology at the University of Chicago, although she had already been offered a graduate fellowship in economics at Bryn Mawr.

Helen entered the graduate program at Chicago in 1925 as the only woman among six or seven Laura Spellman Rockefeller fellows. She quickly became the center of her graduate student circle. To the seminar on community studies, which convened once a week for students to report on their work, Helen added an afternoon tea, which became a regular event in the Social Science Building. When students gathered to study for their Ph.D. language examinations in German, Helen again was the leader. In the intimate sociology community at Chicago, professors and students worked closely together, gathering data from the community on weekends and studying and discussing matters in formal and informal seminars during the week. In that context Helen, who was enthusiastic about sociology and whose curiosity and energy no doubt touched her colleagues, was an important member of the sociology student group. By the end of her first year of study, she had become engaged to Everett Hughes.[18] From the beginning they shared common interests and approaches deriving from their common mentor, Robert Park.[19]

Helen completed her master's degree at the end of her second year with a thesis entitled "Land Values as an Ecological Factor in the Community of South Chicago." Using Park's theory of urban invasion and succession, she related land values to land use and proximity to the urban center. During the summer of 1927 Helen MacGill and Everett Hughes were married; they moved to Montreal, where Everett had accepted his first academic appointment, at McGill University.

Everett finished his dissertation, "The Growth of an Institution: The Chicago Real Estate Board," under Park's supervision during his first year at McGill. Soon thereafter Helen began her doctoral work by returning to Chicago, where she lived with the Parks,[20] for at least a quarter of each year to satisfy residence requirements. As a result of Park's influence her thesis focused on newspaper human-interest stories as a form of popular literature. Helen's interest in the topic was abetted by the fact that both Park and her mother, Helen Gregory MacGill, had been journalists: "Though at first I found it hard to see it as a thesis subject, the proposed topic was attractive . . . as time went on, [I] wrote articles based on three chapters of my thesis—but the main corpus itself materialized very slowly. I missed the exchanges of fellow students and was pained to discover how much a scholar's morale can depend upon like-minded associates. At home, of course, Everett and I, both steeped in Chicago sociology . . . used each other as sounding boards."[21]

In 1936 Park left Chicago to spend his final years at Fisk University. Much of his supervision of Helen's thesis occurred by letters, which Hel n said were "the equivalent of thesis conferences."[22] Park also encouraged Helen to present a paper at the 1935 American Sociological Society meetings on news as defined in twenty-eight daily newspapers published in

Berlin in 1932, the year she and Everett, who was on sabbatical leave, were in Germany. It became the basis for her first published article. Helen was granted her Ph.D. in 1937, and published her thesis in 1940 under the title *News and the Human Interest Story*, with an introduction by Park. This was one of the earliest studies in the sociology of journalism, specifically the human-interest story as a cultural form.[23]

The McGill Years

From 1927 to 1937 Everett was an assistant professor of sociology at McGill University in Montreal. Promotions were not quickly dispensed in this new department. Everett had been hired by Carl Dawson; the two built up the McGill Department of Sociology, which had separated from the School of Social Work in 1925.[24] Hughes and Dawson obtained a grant of one-hundred thousand dollars from the Rockefeller Foundation in 1930 for research on unemployment. They used the grant as well to support graduate students' investigations of ethnic groups in Montreal.[25]

Helen had limited opportunities at McGill. While working toward her doctorate at Chicago part of each year, she was also a teaching fellow at McGill from 1927 to 1930 and an assistant in sociology from 1929 to 1937, in which capacity she may have served as a teaching and/or research assistant.[26]

One of the projects that developed within the McGill research program was the well-known community study of Cantonville (Drummondville, Quebec), begun in 1930 and published by Everett Hughes in 1943 as *French Canada in Transition*.[27] It was a major study and the first significant collaboration of Helen and Everett. Although Everett was sole author of the book *French Canada in Transition*, he and Helen "jointly did the field work for the study." As he notes in the preface: "We jointly thank the Social Science Research Council for a grant-in-aid. We are grateful also to the citizens of Cantonville and especially to the *curé* of the mother-parish of the community."[28] Footnotes in the book contain notations such as "we got some indications in interviews," "we walked every inch of street and lane," and "our field work."[29] In 1977 Helen recalled: "In our last years at McGill Everett and I, financed by a small grant from the Social Science Research Council, . . . lived one summer in Cantonville, absorbed in the zestful enterprise of analyzing together the population of the market-town-turned-industrial-town. . . . we walked on both sides of every street in town, mapping every house and recording its occupants . . . [we enjoyed] the challenge and delight of doing it all in French."[30] Although there is no other evidence of this as a joint project, the collaboration indicated in footnotes and the preface would have suggested coauthorship.[31] Helen's role in the project remains unmentioned in the literature.[32]

Certainly the impact of this classic study of French Canada on Quebec,

Canadian, and American sociology, particularly in ethnic and industrial studies, was profound, and the influence of its author profound as well.[33] David Riesman comments: "Indeed, to this day, Everett is a kind of patron saint of Canadian social science. . . . He remained the abiding neutral, dispassionate but not uncritical, hostile to nationalisms and ethnocentrisms of all sorts, including those of peoples believing themselves oppressed minorities, but at the same time a comprehending and not self-righteous person."[34]

The Chicago Years

In 1938 the Hugheses returned to Chicago, where Everett had been given a faculty post as an assistant professor.[35] Helen stayed home for the next five years (1938–1943) to raise their two daughters: "I had an undiluted diet of housebound domesticity until the youngest entered nursery school."[36] She then entered part-time paid employment as the liaison between the University of Chicago Department of Sociology and *Encyclopaedia Britannica*.

In 1944 she was asked by editor Herbert Blumer, Everett's Chicago colleague, with one day's notice, to manage the editorial office of the *American Journal of Sociology*. Her work as editorial assistant (1944–1946), assistant editor (1946–1954), and managing editor (1954–1961) at the *Journal* was part-time and poorly paid.[37] But Helen enlarged the responsibilities of the job. Appalled by the state of the articles submitted, with the permission of Blumer and using the excuse of limited journal space during wartime, Helen became an active editor in terms of the content of the articles, although she always maintained that she did not in fact select the articles to be published.[38] At the *Journal* Helen was rewarded for effectiveness and skill with slightly more prestigious titles rather than salary increases. Although Everett was the editor in chief from 1952 to 1957 and again in 1959, he claimed to have nothing to do with Helen's original employment at the *Journal* nor with the conditions of her employment during his term.[39]

The pressures on married couples because of antinepotism rules and the general invisibility expected of faculty wives exercised considerable constraint even on a couple with high personal and professional credibility, whose research interests, ironically, focused on professions and occupations. Helen's sensitivity to the discrimination she suffered is revealed some years later in "Maid of All Work or Departmental Sister-in-Law? The Faculty Wife Employed on Campus," published in the *Journal* itself in 1975: "That this was an outcome of sexism is perfectly clear. The position of editorial assistant of the *American Journal of Sociology* would certainly never have been offered to a male Ph.D., or even to a male doctoral student (after 1942 all editorial assistants were women) whose mentor would be watching paternally for an opening in which he could set his disciple's feet on the path to a career like his own. But a female

Ph.D. in 1944 and perhaps even in 1972, would find herself in this position, although in 1972 she would probably be better able to negotiate."[40] Margaret W. Rossiter has shown that in certain fields it was not unusual for women doctorates to take jobs as editors and librarians. There are interesting parallels between Helen Hughes and Gladys Wrigley, Ph.D., who edited the *Geographical Review* for thirty years, and Yvette Edmondson, Ph.D., who edited *Limnology and Oceanography* for many years.[41] The experience at the *Journal* "led to nothing further . . . not even to a place on any of the by-then numerous sociological periodicals."[42] However, in the late 1960s, Helen was asked to edit the multivolume *Readings in Sociology Series* for the American Sociological Association project Sociological Resources for the Social Studies (1966–1972).

Helen was struck by the anomaly of being a faculty wife (and qualified professional sociologist) in a student's role and by the status contradictions and identity problems which that posed. "For who were my colleagues? Who set the model for my role? There were wives on several of the Press-owned journals. Occasionally two or three of us would compare notes, capping one another's complaints about authors, editors, and Press people. But our working lives (by which, as everyone knows, is not meant our more complicated and demanding domestic lives) never converged sufficiently for us to have opinions of each other's performance or any strong sentiment of common interests."[43]

After twenty-eight years at the University of Chicago, which included his promotion to full professor and chair of the department, Everett accepted a position at Brandeis (1961–1968). Leaving her work at the *Journal* for a vague future was very difficult for Helen. Following seven years at Brandeis, Everett moved to Boston College, where he continued to teach in the graduate program. At various times during his career he held visiting professorships at universities in the United States, Canada, and Germany. Helen continued in part-time research and editorial work and occasional short-term visiting professorships.

Collegiality, Collaboration, and Career

Everett and Helen, who was six years younger, shared strong intellectual family roots, and a knowledge and love of languages, particularly German and French. Filled with common intellectual and professional interests and a dedication to the life of the mind, the Hugheses' relationship and their family life no doubt comprised a meaningful marital existence. They read a great deal, often to each other. They discussed ideas and used each other as sounding boards as they proceeded with their work.[44] They learned and practiced languages, traveled, and did fieldwork together. They shared a love of music and cultivated a stimulating social life with colleagues and students wherever they were and developed enduring friendships with individuals and couples. They arranged small gatherings at home one afternoon a week in Montreal and later in Chicago for discussion and intimate conversation with their colleagues.[45]

In most respects they appeared as equals, and yet their careers followed quite different paths. Everett Hughes had a long and professionally active and influential career during which he played a significant role in the development and expansion of sociology in numerous universities and professional associations in both Canada and the United States. His research in race and ethnic relations, comparative sociology, and industrial and occupational sociology remain rather prototypical for the discipline in that period.

In contrast Helen's editorial work and temporary and part-time positions added up to an "auxiliary" career. Discounting neither her contributions and accomplishments nor the fact that actually she practiced her profession in a different way, Helen was disadvantaged as a woman in a male-dominated profession and, perhaps more significantly, as a woman married to a prominent sociologist. As she herself asked, was she "maid of all work or departmental sister-in-law?" Her utter devotion to Everett, her family, and the discipline, serving wherever she was needed, and her reputation as an intellectual and an editor determined her career path.

There is no doubt that Everett and Helen had a close personal relationship and happy family life. They constituted what might be termed a "modified dual-career family." Although their work and lives contained the structural characteristics of dual careers—"jobs which are highly salient personally, have a developmental sequence, and require a high degree of commitment, [while] establish[ing] a family life"—early in her career Helen dropped out of the sequence and found it impossible to return on the same terms.[46] At the point of completion of her doctorate and preparation of her dissertation for publication, Helen, at age thirty-five, had had her first child.[47] She took responsibility for childrearing and running the household, and looked up to her husband, whose professional desires and plans came first. While "she always had some job in the works" and was a woman of great energy, she had to "work around everyone else," that is, to define and structure her work around the needs and activities of other family members and of her own profession, as in the case of her work with the *Journal*.[48]

To what extent did the strength of the partnership compensate for these pressures?[49] While Helen experienced limitations in employment because of the Depression and university antinepotism rules, Everett's view was that those limitations were perfectly normal:

> I am trying to think of faculty wives of our time and [*sic*] McGill who went out to work. I can't think of any who did. Helen—who had no children for most of the time—worked on in a leisurely way at her thesis for ten years—doing a bit of assisting in student conferences. She mastered the literature on newspapers in Europe and America, and on the popular arts of communica-

tion in some folk countries (Mexican corredos, etc.) But she was not especially career-minded; of course, it was depression time. There weren't any jobs. She now says once in a while: What ever did I do with my time in those years?

Most women had no thought of outside work. Nor did the faculty wives at Chicago—most of them college women. They were active in the Women's Voters, the Settlement organization, Parents, etc., etc.—but professional careers, no.[50]

Everett's comments notwithstanding, Helen seems to have begun her career with seriousness, choosing to complete her doctorate before having a family (against Everett's wishes), publishing a major article based upon her dissertation and then the dissertation itself. At certain points she became passionately involved in projects of her own, as in the editing and publishing of *The Fantastic Lodge*, the story of a young woman drug addict whom Helen came to know.[51] Nevertheless, as her daughter observed, "If she had it to do again she would have wanted to teach in a regular full-time position."[52]

Clearly, Everett and Helen had collaborated on the famous study of a French Canadian community, although she was not recognized as a collaborator or coauthor. In the major instance of announced collaboration, *Where Peoples Meet*, a book of essays on race and ethnic relations, Helen's participation seems to have gotten lost after the title page. Some years later Everett made it clear that, as far as he was concerned, Helen's role in the preparation of that book had been quite limited. In a 1969 letter to Alex Morin (Aldine Publishers) and Howard Becker (Northwestern University) regarding the compilation of Everett's papers into a volume, Everett wrote: "Helen Hughes is co-author of *Where Peoples Meet*. If we use the chapter called 'What's in a Name?', I will have to note that too, in fact, she did nothing in this book, except edit it, after it was completely in typescript."[53]

Why then was she listed as collaborator for the book and not for *French Canada in Transition*, for which she did so much of the fieldwork? And why did she receive one-third of the royalties for *Where Peoples Meet* if the work was not really hers?[54] In fact Helen stated in 1977 that "Everett and I worked together on *Where Peoples Meet*." In addition, she coedited with Everett and Irwin Deutscher *Twenty Thousand Nurses Tell Their Story*, "a report of a huge program of research . . . on the division of labour in the care of hospitalized patients."[55]

Because of their family backgrounds and their interests in race, ethnic, and minority relations, the Hugheses must have been interested in, and affected by, the second wave of feminism. Helen, of course, had been exposed to feminism in her childhood, but seems to have suppressed it. She may have been wary, because the social and economic independence of her mother, the judge, had produced some marital tensions in that household. Helen's sister Elsie, on the other hand, had strong feminist convictions and lived a life of independence and intense career involvement.[56] In the 1970s Helen published at least four autobiographical articles and edited an important report on the status of women in sociology, all of which reveal an awakening of her feminism. They are candid assessments of her own life as a woman sociologist and of the consequences for her own career of marriage, motherhood, and status as a faculty wife.[57]

How did the Hugheses, and others, evaluate their own careers? Everett ended his career with a long bibliography of publications including eight books and a curriculum vitae of remarkable professional accomplishment. In 1968 he was honored with a Festschrift reflecting his major research areas. In 1971 he published a volume of his selected papers. He received numerous awards and eight honorary degrees; he held many professional offices and visiting appointments. He served as president of three professional associations (the Eastern Sociological Society, the American Sociological Society, and the Society for Applied Anthropology) and was a fellow in four others.

In addition to her seventeen years with the *American Journal of Sociology*, Helen edited eight books, authored one, coauthored two others, and published fourteen articles. She had at least five visiting professor appointments in later years, usually in connection with Everett's appointments, and an association with Brandeis as faculty research associate and with the Radcliffe Data Resource and Research Center, Radcliffe College, as scholar in residence. Just as Everett was the recipient of the Award of Merit of the Eastern Sociological Society in 1972, Helen received the award in 1973. In 1978 she was elected the Society's president (1979–1980) and in the same year vice president of the American Sociological Association. Obviously, the profession itself held both in high esteem.

In his collected papers and memoirs, Everett Hughes indicated his satisfaction with his career and academic production. He had confidence that his work had been important and well received. Helen's assessment of her career was less confident. Whatever had happened or had not happened in her career, she blamed herself: "Why was I content, all those years, to let accidental connections and friendly interventions determine the course of my life as a sociologist? . . . Had I been more realistic, I should have envied the small handful of women of my academic generations who began in lectureships in the departments and climbed up the ladder. . . . My many jobs, often overlapping, add up to a busy and gratifying life. But do they add up to a career?"[58]

The professional work and private lives of the Hugheses were imbued with their devotion and intellectual compatibility. As we have seen, they did collaborate on at least three projects. But in certain respects this is an example of a two-person career. From the very beginning Helen deferred to Everett and his career, serving as research assistant, editor, and coordinator of events. She had to set aside her own interests. This was in part because the luxury of doing research and writing at that time depended almost entirely upon having a proper university position. Although we may observe that Helen worked as a sociologist in a different way, the increasing professionalization of the discipline effectively eclipsed her career. Furthermore, Everett's high status in the discipline and his dominant personality, even among men, overpowered Helen's gentle style.[59]

There were pressures on married women in this generation to see their own careers as secondary and to give priority to their husbands' research, often doing significant work with no credit or compensation. And an insidious compulsion existed in Helen's generation of women sociologists (their intelligence, training, and achievements aside) to play "the good wife" role in relation to their husbands, even in profession-related contexts. Given the meager opportunities for women academics at that time, some women may have felt that being married to a prominent academic would earn a woman greater respect and better treatment than being a marginalized single woman in the field.[60] Some of those who married, like Alice Rossi and Jessie Bernard, became radicalized in the 1960s, led the second wave in sociology, *and* managed to establish independent careers as sociologists. Others engaged in collaborative work with their husbands without sacrificing their own careers. But for still others, like Helen Hughes and Carolyn Rose, careers had already been defined and circumscribed. The 1960s did not entirely liberate any of these women, many of whom previously had hired household workers and suddenly no longer felt justified in doing so. Their husbands did not help with the household, so the domestic work of these professional women became even more energy- and time-consuming. And their sociologist mates accepted that.[61]

The academic system itself demanded that one's research appear to be one's own singular effort (with no collaboration or assistance), particularly for purposes of hiring, tenure, and promotion,[62] matters in which men (husbands) rather than women (wives) were given primary consideration. Those who have examined Everett Hughes's career in some detail report that Helen does not appear to have a major presence.[63] We know, of course, that in fact she did. Everett might have encouraged or helped to facilitate Helen's entrance into a regular academic position or expanded their collaborations, but he seems not to have considered her to be "career-oriented." In another time Helen's career might have been defined quite differently by Everett and by Helen, for that matter.[64]

The Hugheses' case must be seen therefore as one of defaulted scientific collaboration. Just as Helen never objected openly to Everett's professional opportunities, no matter how difficult their consequences for her, so she apparently did not make family issues over credit for collaborative work, the possibility of continuing their collaboration, or her desire to pursue her own career more vigorously. Pleading her own case would have seemed selfish and was out of character for Helen, so she suffered to some degree a self-victimization, blaming herself for her career situation.

In that period of professionalization and consolidation of male power and values in sociology, women's contributions in collaborative work, even or perhaps especially with their husbands, went largely unrecognized. Men's advantage in the structural context of the discipline, as in society, vitiated genuine collaboration with women, and its recognition, and generally ensured discrepancy and domination.

✦ PART V ✦

Comparative Study of Couples along Disciplinary and Transdisciplinary Lines

NANCY G. SLACK

Botanical and Ecological Couples
A Continuum of Relationships

"The more I work on these plants, the more the fascination of them grows on me; I only wish I could devote all my time to their study and collection." So wrote Violetta White, a woman who by age twenty-seven had published two important monographs on fungi. That was in 1902. Shortly thereafter she married a prominent New York lawyer and no more was heard of her in the botanical world.[1] This pattern was the prevalent one for American women botanists who married, starting with Jane Colden, the only eighteenth-century American woman botanist of note, a plant collector and correspondent of Linnaeus. In 1757 she produced an illustrated botanical manuscript, worthy enough to be printed two centuries later. By 1759, however she had married and left her botanical studies. These two women, and doubtless a great many of the 1,185 women discovered to be "actively interested in botany" in the nineteenth century,[2] followed the pattern of marriage followed by motherhood and the great unlikelihood of continued botanical research. Positions were lacking, it is true, but society's expectations were even more important.

Marriage itself in nineteenth- and early-twentieth-century America was a highly asymmetrical relationship. Women had constrained roles as wives and mothers; the separate spheres doctrine was proclaimed even by prominent women, including Almira Lincoln Phelps, the very successful author of the most popular nineteenth-century botanical textbook. She herself had earlier given up her intellectual pursuits for marriage and motherhood, or, as she wrote, she gave up "literature for receipt books . . . and the Young Housekeeper's Companion."[3] She also had three children in three years. Only the sudden death of her husband from yellow fever propelled her back into education and eventually into teaching hands-on botany and writing science textbooks.

The major achievers among women botanists and ecologists, both in research and in college teaching in the nineteenth and first third of the twentieth centuries remained single. Positions opened up as professors in women's colleges starting in the 1880s. Those women who found such employment usually were not allowed to retain their positions if they married. Others who worked for the U.S. Department of Agriculture (USDA) or in the new state agriculture experiment stations also remained single or, like the best-known researcher among them, Agnes Chase, were early widows. Widows, as I have discussed elsewhere,[4] were also among the prime achievers among nineteenth-century women botanists. What about the married women who did participate in continued botanical and ecological research? It appears that in the nineteenth and early twentieth centuries the best strategy was to marry a man in the same field.

This chapter considers briefly the Sullivants and the Farrs, and more fully two botanical couples, the Brittons and the Brandegees, and one ecological couple, the Clementses. Additional ecological couples, the Shreves, G. Evelyn Hutchinson and his first wife, Grace Pickford, and a few contemporary couples are compared and contrasted.

Several questions can be asked about the American botanical and ecological couples now to be considered, those of the nineteenth century and first part of the twentieth century. In this period I found a continuum from the earlier husband-creator/wife-executor mode to women who did important research of their own within a botanical marriage. What accounts for these differences?

Factors to be considered are the age and status as botanists of both the man and the woman at the time of marriage, and the social class to which they belonged. Related factors such as race, religion, and national origin, which could also be important, are in fact inapplicable, since all the botanical and ecological couples were white Protestants, and excepting G. Evelyn Hutchinson and Grace Pickford, were American-born. Immigrant women or even the daughters of immigrants, although important in other sciences (see, for example, Mildred Cohn's chapter on the Coris) do not appear among the married botanists.

The relationships, both personal and professional, between the members of a couple are important to consider. Did the couple enjoy a companionable marriage? Were there children? Did either member of the couple do published research before marriage? If they collaborated in their research, were they equal partners? Did each do her or his own separate work? Did one or the other or both have professional positions?

Societal factors may be influential, too, particularly changes in the opportunities provided for women during this period and the expectations of married women (including mothers) by society. The other side of that coin is the perception of these expectations by the women themselves and by their husbands.

Collaborative marriages were common in botany even in the first half of the nineteenth century in England, where Ann Shteir[5] has documented the little-known work of Mary Turner, Maria Turner Hooker, and Frances Henslow Hooker, wives of prominent botanists. They did botanical drawings, engravings, and translations for their husbands or, as Shteir put it, "the family firm." During these women's lifetimes there was no possibility of university education for women or of professional positions; marriage and family mentors gave them the opportunity to use their talents and to have their work published, whether or not they received the credit. As Marilyn Ogilvie wrote, "collaboration with a male was a back door through which women could enter science unobtrusively." But in may cases collaboration took the form of the "husband-creator, wife-executor mold."[6] That went on for a long time on both sides of the Atlantic.

A collaborative marriage of this type occurred during the same period in the United States. Eliza Wheeler married an early Ohio settler and established botanist, William Starling Sullivant, in 1834. She, like her British contemporaries, was married to a dominant male with prior expertise. He referred to her as his "co-laborer." She had no previous experience with botanical techniques but learned to do microscope work and to prepare specimens; she drew outstanding botanical illustrations. She also was the mother of five children, the rule not the exception for nineteenth-century women. Asa Gray and the outside botanical world did not acknowledge Eliza Wheeler Sullivant's contributions. Her husband, William Sullivant, did. He wrote to Asa Gray, "in [botanical studies] she was my constant companion and helpmate for the last fourteen years; and I can say with the strictest truth that more than half of whatever I have done in these pursuits is due to her. . . . In the most difficult and delicate dissections and microscopical examination she was astonishingly successful. . . . No female I am sure ever equalled her and few of the other sex excelled her." But that was after her death in 1850 of cholera. He later wrote to Gray that "her attainments amply entitle her to a short notice among the obituaries of those devoted to the Natural Sciences" and asked him to write one.[7] But Gray did not. Hers remained a derivative career as the largely unknown, talented helpmate of her well-known botanist husband. The case of English ornithologist John Gould and his artist wife, Elizabeth, discussed by Janet Bell Garber in this volume is very similar.

The dominant-male-of-prior-expertise model persisted in botany and other fields of science until very recently. Botanical wives received financial support and encouragement from their better-known, well-positioned botanist husbands. Wanda Kirkbride, born in 1895, married a botany professor, Clifford Farr, and worked with him on plant cells in his laboratory at both Washington University and the University of Iowa. She did not have a university position herself until her husband's early death, when she received his academic post. Farr had a child to support and thus the approbation of society to work, and she went on to do her own research on plant fibers in other positions, at the University of Maine, the USDA, and the Boyce-Thompson Institute, that is, in academia, government, and industry. She is the only botanist included in Edna Yost's biographical collection of American women scientists. She is another of those achieving widows. One wonders whether, had her husband lived, this collaborative marriage would have kept her as perpetual "wife-executor" in her husband's laboratory, albeit with his financial support and encouragement?

The collaborative marriage of Elizabeth G. Knight to Nathaniel Lord Britton was one in which the husband provided the financial support, encouragement, and indeed a position, though unpaid, for the wife.

N. L. Britton was born in 1859 to a family that had settled on Staten Island, New York, in 1664. He was originally trained in geology at the Columbia School of Mines under John Strong Newberry, who also gave some botany lectures, but Britton was largely self-taught in this field. He coauthored a flora of Richmond County with a classmate while still an undergraduate. He received a doctorate from Columbia College in 1881 and was appointed professor of botany there in 1891. In 1896 he became director in chief of the newly established New York Botanical Garden, and Emeritus professor at thirty-seven. He held that position for the rest of his professional career, one of the most prestigious and powerful botanical positions of his time. He published, with Judge Addison Brown, the *Illustrated Flora of the Northern United States and Canada* in 1896–1898, edited the *Bulletin of the Torrey Botanical Club*, the earliest American professional botanical journal, and wrote many books and professional papers. He made thirty botanical expeditions to the West Indies and elsewhere, many of them self-financed.

In addition to his research, he found time for social duties, political contacts, architects, fund drives, and running the Botanical Garden. The establishment and development of the New York Botanical Garden was his greatest accomplishment. He married Elizabeth Gertrude Knight, a Hunter College graduate who had spent much of her childhood in Cuba, in 1885, and in 1888 they went to England and visited the Royal Botanical Gardens at Kew. She asked, "Why couldn't we have something like this in New York?" By 1891 such a garden had been chartered by the New York State Legislature with a board including Cornelius Vanderbilt, Andrew Carnegie, J. Pierpont Morgan, and N. L. Britton. A group of women, led by Elizabeth Britton, collected $250,000, an enormous sum at that time, as an endowment.[8]

The Britton marriage proved different in several respects from those discussed above. First, Elizabeth Knight was a published botanist before she married in 1885; she in fact met her husband in the Torrey Botanical Society, to which they both belonged. In this she was unlike Eliza Wheeler, who became interested in botany because of her marriage to William Sullivant and was trained by him.

Secondly, Nathaniel Britton's financial support freed her from the role of several of her contemporaries, that of a teacher in a women's college.

Elizabeth taught at Hunter College before her marriage. Her position at the Botanical Garden enabled her to devote her time to research and the publication of many botanical papers. Her contemporaries, botany professors Susan Hallowell and Emily Gregory, although better educated than Britton, who had no advanced degrees, taught much but published little. They were not invited to be charter members of the newly formed research-oriented American Botanical Society. Elizabeth was one of twenty-five charter members.[9]

Thirdly, Nathaniel Britton's position enabled him to appoint his wife to a curatorial position at the Botanical Garden. Although "honorary" and thus unpaid, this was a position entailing resources and authority. It is very unlikely that she would have had any such position had she not been his wife; he even signed her appointment letter![10]

Fourthly, as an upper-class woman she was already in a position of authority in her own sphere and for better or worse transferred some of these managerial skills and authoritarian attitudes to her employees and even her graduate student at the Botanical Garden. Her upper-class status enabled her to have considerable household help and several domiciles, but she preferred her work, albeit unpaid, to leisure. She wrote in 1897, when about to move into New York City for the winter: "I feel as if I ought to make the most of it . . . as I seem to be buried in the country most of my life . . . ever since I married. I must confess, I do not like it, and cannot work to advantage so far away from books and specimens, nor can I keep energetic with so little to amuse me. So you see I am looking forward to the change with pleasure."[11]

The relationship between the Brittons was certainly companionable and supportive. As noted above, they were the prime movers and among the primary fund-raisers in the establishment of the New York Botanical Garden. They traveled together to European herbaria and spent their winters in Cuba, Jamaica, and other of the West Indian islands, where they both did research studying floras little known previously, but the plants were different ones. His research and publications were on higher or vascular plants; hers were on lower plants or cryptogams, largely mosses, of which she was curator at the Botanical Garden. She published 346 papers and reviews, edited two journals, and supervised a Ph.D. student, A. J. Grout, who went on to write the definitive books in Elizabeth's field, the mosses. She carried on an immense correspondence with literally hundreds of American and European botanists, including negotiations for the purchase of collections for the Botanical Garden. But there was also time to entertain both the elite of New York and European botanical visitors in the city and at their country home. It is unlikely that Elizabeth Britton ever had to do the cooking.

They each did some of their botanical traveling separately and kept in touch with delightful letters. At least Nathaniel's are delightful; he did not save hers apparently. From the labels on her specimens and her letters to

Charles Horton Peck, the state botanist, [12] it is clear that she traveled on her own to some out-of-the-way mountain passes in the Adirondacks to collect rare plants. She hiked in her long dress and button-up boots to places that are still hard to reach. There is a fascinating unpublished account of one of her botanical climbs up Whiteface Mountain. As an upper-class late-nineteenth-century Adirondack traveler, however, she had someone to carry her picnic basket and set out her lunch. She did make many new discoveries and published them in the then relatively new botanical journals.

The Brittons thus enjoyed a personal and botanical partnership. While she was not entirely an equal partner, since he had the funds and the more powerful position, she wielded considerable power and influence through ability, force of character, and—well, yes—nepotism.

While the Brittons were part of the eastern botanical establishment, the American West was the scene at the turn of the century of much botanical exploration, which included both men and women. [13] The botanical and ecological marriages I would especially like to contrast are those of two western couples, the Clementses and the Brandegees.

Edith Schwartz married Frederic E. Clements in 1899; Mary Katharine (Kate) Curran married Townshend S. Brandegee in 1889. Although the Brandegees were older, the research work of both couples overlapped in time, the Brandegees largely in California and adjacent Mexico, the Clements in Nebraska, Colorado, Wyoming, and elsewhere. Kate and Townshend Brandegee traveled extensively together, but each made botanical expeditions separately as well. Edith and Frederic Clements traveled five hundred thousand miles—always together; they were separated for more than a few hours only once in forty-six years of marriage. Kate Brandegee had an important and paid position at the California Academy of Sciences. Edith Clements was a "field assistant" without salary at the Carnegie Institution, of which Frederic Clements was a staff member. She also had a two-year replacement position teaching botany at the University of Minnesota, but only while her husband was department chairman. Yet it was Edith Clements who had the Ph.D. in Ecology; Kate Brandegee's only advanced degree was an M.D.

Edith Schwartz Clements was the first woman to earn a Ph.D. at the University of Nebraska. Her doctoral thesis, which was later published, was quite substantial, in a then-modern field now called physiological ecology. But it was her first and last scientific paper.

In terms of her husband's ecological research she became the most extraordinary helpmate. She typed Frederic's letters and manuscripts and took his dictation of field notes "sitting on the running-board of the current automobile"; she made the illustrations for his published papers and books, she drove the automobiles, starting with a 1915 Buick, and she even became an "expert diagnostician of car-trouble" and of "remedying the trouble." All of this, as she said, "left her husband free to devote his entire attention to the study of vegetation." Later her duties included photography; she took thousands of photographs of subjects concerned in his researches and later slides to be used at conferences as well. She also did translation and typed extracts from foreign authors, not to mention her summer work at the Colorado field station, the collection, pressing, and packaging of plants. These botanical helpmate activities were in addition to what Arlie Hochschild has recently called the "second shift" of modern working wives, in Edith Clements's words, "running of the house, entertaining the faculty and friends, sewing, marketing, etc. etc.," the usual wifely role of the middle-class woman without servants, even though Frederic and Edith had no children. Later in their married life she even added wifely medical duties, acting as "dietitian and special nurse in all his illnesses even in the hospital." They were never apart even when he was hospitalized.[14]

Frederic Clements spent his college summers on expeditions commissioned by the Department of Agriculture and Forestry, work for which he was paid, an opportunity not offered to women students. Frederic had already earned a Ph.D. when Edith entered the University of Nebraska. He was an instructor in botany at the University of Nebraska and had written an important ecological book by the time they were married in 1899.

It is difficult to know Frederic Clements as a person and even harder to assess his view of his wife. Their lack of separation resulted in an absence of letters after their marriage. According to an account Edith wrote of him, he grew up "in grinding poverty" on the Nebraska prairie. It was an abstaining family, and Frederic did not smoke or drink all his life. A collaborator later complained that Clements had "no redeeming vices." In spite of poverty Frederic was able to attend the University of Nebraska, where he did extremely well not only in science but in literature, according to Edith.[15]

Although much of Frederic Clements's theoretical work has been replaced by newer concepts, he was without question an innovator in his own time, laying the theoretical groundwork for plant ecology that lasted into the 1940s. There is no evidence that his wife collaborated on any of these theoretical works.[16]

They did collaborate in other ways. They collected and sold sets of plants including fungi and mosses as well as flowering plants, and she helped to found the Alpine Laboratory on Pike's Peak. She was also his unpaid but recognized field assistant when he became a full-time researcher for the Carnegie Institution. On her own, she wrote popular guides to western wildflowers and *National Geographic* articles illustrated with her own paintings. But her attitude toward her botanical career seems to be summed up in her advice to a young daughter of a friend, who was considering a scientific career. Edith Clements's advice to her was, "It's much easier to marry someone who knows it all and if you know some yourself . . . can draw and paint and take photographs and drive the car and typewrite . . . you can go along the way I do."[17] She also wrote, "Each brought individual talents to [this partnership] that were complementary to a complete whole. He furnished the brains and I the manual dexterity. His dislike of using his hands and extraordinary mental gifts and training led to a division of labor."[18]

It is perplexing but possible to understand Edith's satisfaction with this role. She would have had to terminate her studies and go home to help support her family if she had not married Frederic. Thus he, already a salaried instructor, provided financial security—and the opportunity to continue her studies. Previous to her marriage these were in Germanic languages; she had taught German at the University of Nebraska. She says in her biography of her husband, "After marriage I immediately turned my attention to this subject [ecology] under my husband's stimulation and enthusiasm with the result that I received my Doctor's degree in 1907." She was thus in part his protégée and professionally dependent on him in the early years of their marriage; she perhaps never became an independent researcher.

But there are still unanswered questions both about their relationship from his point of view and about the positions available to her. She was quite proud of her teaching success at the University of Minnesota. She wrote that her teaching attracted many students to advanced courses in botany—though even here she was using her husband's theories of education by which the student is trained to think instead of simply taking notes on lectures or making drawings. However well these theories may have worked, her opportunity to teach appears to have ended after two years, when the regular faculty member returned. Not long after, Frederic Clements left academia for full-time research with the Carnegie Institution.

While it is difficult to get any picture of Frederic's view of his wife, there are glimpses of his own character through her eyes. In her biographical sketch, she characterized his personality as "not demonstrative, often repressing feelings that should have been expressed . . . kind, generous to a fault, an earnest seeker after perfection . . . impatient of human frailties and especially of loose thinking and prejudice." There is much more about his brilliant conversation and his moral character—but it is all quite impersonal, making it difficult to see their relationship through his eyes. He occasionally mentioned her in his letters to other botanists, such as Elizabeth Britton. One entry from Edith's journal quotes a comment she overheard of Frederic's to John Philips, visiting from South Africa. "Mrs. Clements would hold that position today [like that of Dr. Philips 'near the top on the list of world ecologists'] . . . had she not devoted herself to furthering my career instead of winning recognition as ecologist in her own right."[19]

Edith proclaimed Frederic "a staunch feminist" in her short biography of him. Perhaps he was, in relation to women other than his wife. Women did work as researchers in the early days of his Colorado laboratory and later in Tucson; for example, Frances L. Long, who held a Ph.D. and was a paid Carnegie researcher, worked and published with him.[20] Edith Clements's last book, *Adventures in Ecology, Half a Million Miles from Mud to Macadam*, published long after Frederic's death and half a century after some of the events, is a curious mixture of anecdotes of their travels and ecology minilectures—of her own and quoted from Frederic. In the introduction she describes herself and Frederic as "two plant ecologists who lived and worked together for four decades, traveled hundreds of thousands of miles in the study of natural vegetation and grew hundreds of species of plants in greenhouse and garden, and then cooperated with conservationists, foresters, agriculturalists and cattlemen in putting the knowledge so gained to practical use." However, the frontispiece, showing her husband among the cacti, is captioned "Frederic, Director-in-Chief."

Only rarely in her book does she come close to complaining about Frederic: "I expressed my feelings in a poem" (about the beauty of the prairie), "although Frederic, who wrote beautiful poetry in the Browning manner, called my efforts doggerel." And perhaps more to the point, when he received his Carnegie research appointment, she remarked that he would also be allowed a (paid) assistant, but a "mere wife" would work just as hard for nothing.

The "mere wife" did work hard, long, and under difficult conditions in the pursuit of his research. On one trip from California to Utah she drove between ten and twelve hours for five successive days at an average speed of twelve miles per hour! She did nearly all the technical drawings for their joint book, a popular one called *Flower Families and Ancestors*, as well as all the ecological photographs for his many books. Frederic's father was a photographer, and Frederic took the early pictures. Later he taught her photography, told her what to photograph, and left her all the problems. One quote, from the Montana badlands research, perhaps sums it all up:

Not only were the slopes steep and slippery, but the deep soft shale made climbing them almost impossible. . . . Of course, some of the tiny pioneer plants there were to be photographed seemed to be in inaccessible situations. Once in a more or less favorable position, however, the tripod must be kept from slipping or blowing over by my frantic grabs as the wind blows my hair into my eyes and my hat from my head. The camera cloth flaps and flutters: it is a struggle to keep it in position to hold the tripod, insert the film pack, pull the slide, snap the shutter. Perspiration pours into my eyes and down my nose and every fresh gust threatens to send the entire outfit including myself skittering down the steep slope. By this time I am a wreck. . . . I tell Frederic that I am offering myself [as] a burnt sacrifice on the alter of his love of work. He looks at my peeled nose and red neck and assents as calmly as though a mere wife were a small price to pay for the rewards of scientific research. But Frederic can't very well write a book about badlands unless we study them.[21]

Perhaps Edith Clements's contributions to the works of Frederic Clements have not been sufficiently recognized. Even the self-effacing Edith wrote, "unless we study them."

The contrasting lifestyle of Kate Brandegee is quite striking. Mary Katharine Layne Curran Brandegee, born in 1844, is certainly a highly unusual character among the nineteenth-century women botanists and, in view of both her life and work, most worthy of a biography. She was raised in mining towns in the West, married a constable named Curran, was widowed at thirty, and went to San Francisco, where she received her M.D. in 1878. Young female doctors were not overrun with patients, and with an interest in and knowledge of plants gained from her materia medica course, she began to collect plants and to work in the California Academy of Sciences herbarium. The Academy, in San Francisco, was probably the most important scientific institution in the western United States; it had been started by a number of prominent citizens, largely physicians, several of them competent botanists.[22] They offered the paid curatorship of botany to Katharine Curran in 1883. The Academy was flourishing at this time; California botany was in the process of becoming independent of the eastern establishment.[23] Kate Curran had what was at that time the most important paid botanical position of any woman in America. She gave up her small medical practice and devoted herself full-time to botany. As she wrote, "My botanical trips are too numerous to catalogue. By the aid of the railway companies I enjoyed a general pass . . . which allowed me to ride on anything from Pullman to engine."[24] Kate Curran had been left a widow without children at an early age. Her education and her botanical career were well established before her second marriage, to Townshend S. (T. S.) Brandegee in 1889.

T. S. Brandegee was also an established botanist at the time of their marriage. He was born in 1843 in Connecticut. He served two years in the Civil War and later attended Sheffield Scientific School at Yale, receiving a degree in civil engineering in 1870. He also studied botany at Yale with Daniel Cady Eaton. From there he went to Colorado, taught school, served as a county surveyor, and collected plants. He was appointed assistant topographer to Hayden's "Exploring Expedition of S.W. Colorado," on which, in addition to his topography, he was to collect plants. Later he worked as an engineer on a number of railway surveys, transferring from one survey to another and thus to new floras. "Continually, depending on work and the season, I collected plants." He wrote that he was put in charge of a division in the Wasatch Mountains, which extended from seven thousand to eleven thousand feet in altitude, and which "gave me a fine opportunity to collect the Alpine flora . . . especially on Sunday." Later he worked in New York on a state forest survey under Governor

Cleveland and in Washington State on a transcontinental railway survey as a botanical collector and surveyor. None of these occupations were open to women during this period, nor was the Sheffield Scientific School training in engineering and science that T. S. Brandegee received.[25]

In 1886 he came to California to collect tree trunks for the American Museum of Natural History in New York. Henceforth California became his home and botany his sole occupation. He became a member of the California Academy of Sciences and an associate of "the scientific men [*sic*] of the day . . . Dr. H. W. Harkness, Dr. Albert Kellogg, Professor Edward Lee Greene and Dr. Mary Katharine Curran."[26]

He married the last of these associates three years later in San Diego. They are supposed to have walked the five hundred miles from San Diego to San Francisco on their honeymoon, collecting plants all the way. So began a long companionable botanical marriage. Townshend Brandegee had an independent income. He was able to provide financial support to the Academy journal, *Zoe*, which Kate founded and edited after her marriage, and for which she wrote many botanical papers and reviews. Later she turned the curatorship over to her protégée, Alice Eastwood, a noted botanist who later heroically rescued important Academy plant specimens from the San Francisco fire. The Brandegees left the Academy and the Bay Area and made San Diego their home and botanical headquarters.

Both the Brandegees worked in the same field, plant taxonomy; both had discovered species new to science before they married. Townshend's specimens had usually been sent east to George Englemann or Asa Gray for identification and publication, but after he moved to California all his publications of new plants were in California journals, as Kate Brandegee's had been from the start. T. S. Brandegee's interests were largely in plant exploration and taxonomy with an emphasis on Mexico; he collaborated for some time with C. A. Purpus on Mexican plants.

Kate Brandegee had wider interests. Although also a taxonomist and plant explorer in a rich California flora in which there was still much to discover, she was interested in more fundamental questions having to do with the evolution of plants, such as the nature and extent of variation within a species, which she studied in evening primroses and lupines among other plants. There are extensive notes of hers with comments, some negative, on lectures that Hugo deVries gave on his mutation theory at Berkeley. His opinion in these lectures was apparently that there were only two kinds of variation, that which was mutational and smaller variations that were merely environmental. With this she disagreed. There is evidence that she was an evolutionist; she corresponded with Asa Gray and his followers at Harvard on this subject.[27] Townshend Brandegee was also said to be an evolutionist, though I have not seen direct evidence of this. Their colleague Edward Lee Greene, another major California plant collector and early Berkeley professor, certainly was not. Kate Brandegee wrote very critical reviews of Greene's work in *Zoe*. She also criticized other botanists including Nathaniel Britton, probably well-deserved criticism, though not written to make her popular with these eminent men and their friends.

What was Kate Brandegee like? She and her husband had both experienced the rough life of California mining camps, although she had done so at a much younger age. This upbringing perhaps helped make her an independent person, not so conscious of or at least beholden to society's role for women. According to Marcus Jones, "she was always a rebel against convention. Her sister Susan used to chide her for the way she kept her house." Albert Herre mentioned her poor housekeeping and little attention to cooking. The Brandegees went to their favorite French restaurant in San Francisco once or twice a week "for a French dinner such as the city was famous for," and Townshend "could endure the sketchy food in between."[28]

Jones has written a description of her, undated but at some time after 1906, when they returned to Berkeley: "As I knew her she was a person rather angular and unconventional, with a strong face, compelling consideration and respect without an effort on her part. . . . The atmosphere she carried was that of Englemann, Rose, Vasey, Parish, Setchell [all male botanists]. . . . An air of comradeship." On her professional qualities, Jones wrote: "Her mind was masculine [sic] in its grasp, philosophical, discriminative to the last degree, and the keenness of observation and memory of things, and capacity to correlate was marvelous. I was always impressed by the mental grasp she had on any subject she tackled. She would at once, out of rubbish of descriptions, select the crucial characters and reveal them."[29]

Another portrait, by Robert Brandegee, Townshend's artist brother, written to Townshend from Connecticut in 1913 when Kate was visiting Harvard and his family, said: "You will be glad to hear that your 'better half' has arrived in Farmington. We were all quite impressed that she is quite original and unusually handsome."[30]

In addition to these contemporary accounts there is a much later account by Frank S. and Carol D. Crosswhite, who stated that eastern botanists classed Kate Brandegee as a virago, "a manlike woman; a bold, impudent, turbulent woman." The Crosswhites, however refer to the Latin meaning of virago, "a heroic maiden: a heroine: a female warrior"—a warrior "of science and of women's rights as professionals in science." There is some contemporary evidence, again recounted by Jones, that Kate Brandegee did feel undervalued as a professional botanist. "I said to her, 'You owe it to the world to let us have the story of your life,' and she savagely replied 'What does the world care for me?' . . . . the cry of . . . one longing for recognition long over-due."[31]

An account of "an odd expedition" the Brandegees made together to Lower [Baja] California in 1894 as reported in the *San Francisco Examiner* gives a picture of scientific expeditions of the time and of the general

public's view both of such expeditions and of the women like Kate Brandegee who took part in them. As reported:

> the two men engaged in this queer line of business [buying rattlesnakes,
> lizards, centipedes, and such from the natives] were Dr. Gustav Eisen and
> Townshend Brandegee. . . . and the woman Mrs. Brandegee. All are well-
> known members of the Academy of Sciences and they were on an expedition
> in the interest of that body. After a week's stay at the village on the Cape a
> mule train was organized and with a couple of guides and servants and some
> boys to help chase bugs and butterflies the scientists started for the mountains
> where Mr. and Mrs. Brandegee were to begin their real botanical work. . . .
> Mrs. Brandegee on this trip rode astride of her mule man-fashion in the
> pantalooned suit that she took with her for the purpose."[32]

Kate returned from this expedition earlier than the two men and was
shipwrecked on the *Newbern* but rescued in a lifeboat. She seems to have
carried their plants, which, according to the newspaper, included fifteen
to twenty undescribed species, safely back to San Francisco.

They each made separate botanical expeditions as well and wrote to
each other constantly while separated, detailing everything. For example,
in July 1897 from a forestry expedition near Idaho Falls T. S. wrote Kate
of having to carry the body of one of their horsemen who had been killed
by a horse falling on him. He added that it was "cold as Greenland," that
the wash water was frozen solid, and that he appreciated his fur sleeping
bag.[33]

She wrote him from one of her expeditions: "I am in the oak belt. . . .
tomorrow I leave on the stage for the Giant Forest [giant sequoias] where
. . . I will be for nearly a week. Then I try to get higher—with a packer
guide and finally fetch up across country to Mineral King. It may be 3 or
four weeks before I reach there. Of course you know I am trying by study
of their variation [oaks] to get the bounds of the species." And four days
later from the Giant Forest, "There are lots of plants and I am by way of
settling many things [botanically]. Monday I will get sent to Alta Meadow
which is a great deal higher than this and may stay a week." The letter
ended, "Write to me often to this address until I give you some other,—
for I am comfortless and lonely—I wish I were with you tonight . . . but
I have put my hand to this plough and will not turn back. Lovingly,
K.B." They were always writing "Arrived safely" from different parts of
the country. They wrote about the difficulties of travel: "couldn't get a
sleeper," she from Laramie east to Boston, he another year from New
York south because of the 1893 presidential inauguration. Each scolded
the other for lack of mail, as (October 1915): "Dear Townie. I think you
are very unkind. It is now four days since I have heard from you and
there can have been no good reason for such neglect." Meanwhile he was
writing "Haven't had a letter from you for a week."[34]

In 1917 she wrote to him that everyone at Harvard was very kind but that there was no hotel in Cambridge, that she had to take the underground from Boston, and that she missed their good San Francisco French restaurant. "The food is of course poor compared to San Francisco, a dinner like the St Germain would hardly be possible to get." She is longing for a salad.[35]

The couple left San Diego in 1906 and moved to Berkeley, giving their eighty thousand joint specimens to the university and working there the rest of their lives. Townshend was an honorary curator—unpaid like the contemporary Elizabeth Britton in New York. At one point Kate was ill in the hospital at Berkeley after minor surgery, and he was in San Diego. Kate, unlike Edith Clements, wrote, "Dearest . . . It will not be necessary to telephone or come to see me there being no prospect of fatality." But she wrote him the bus stop just in case he did come.[36]

Clearly the Brandegees had a very close and interdependent relationship, a truly companionable marriage, but each was free to be an independent botanist—to travel both in the field and to other herbaria, alone as well as together, in order to work on chosen research projects. Albert Herre was probably reflecting their own and others' views when he wrote that "they were completely in love and entirely devoted to each other."[37]

In the collaborative marriages of these three couples, we see many of the factors suggested above at work. Independent research by the woman was favored by training and publication before marriage, exemplified by Elizabeth Britton and Kate Brandegee but not Edith Clements. She had turned to plant ecology only under her husband's tutelage. Upper-class status and financial support both helped Elizabeth Britton; the California Brandegees were not upper-class in the same sense, but they had an independent income, and Kate Brandegee, against the prevailing views of society and even of her own family, avoided the "second shift" duties of cooking and housekeeping. Middle-class Edith Clements, however, seems most of all to have internalized society's views, and her marriage was in some respects at least one that Hochschild would currently characterize as "traditional," based on "gender strategy" and "gender ideology." In these marriages the woman wants to identify with her home and to hold less power than her husband, and wants her husband to identify with his work. The husband wants the same. For Edith Clements, however, her "home" was often on the road, but the same principles applied. Elizabeth Britton's and, even more so, Katharine Brandegee's marriage would be characterized as "egalitarian," in which the woman identifies with the same work areas as her husband and wants equal power.

Although it is not possible in this chapter to examine their lives in as great depth as those of the three couples above, it is revealing to compare and contrast the creative lives and work of other twentieth-century couples in ecology.

Forrest and Edith Bellamy Shreve, both born in 1878 and living until midcentury, were contemporaries of the Clementses. Forrest completed

Marie and Pierre Curie, 1904. Courtesy of Archives Curie et Joliot-
Curie.

Pierre Curie—"Marie Curie's favorite portrait of her husband," 1906.
Courtesy of Archives Curie et Joliot-Curie.

The Joliot-Curie family: Irène Joliot-Curie, Frédéric Joliot-Curie, Hélène, and Pierre. Courtesy of Archives Curie et Joliot-Curie.

Carl and Gerty Cori. Courtesy of Washington University School of
Medicine Library.

Left, John Gould, at the age of 45. Lithograph by T. H. Maguire, 1849.
Courtesy of Gordon Sauer, M.D. Right, Elizabeth Gould. Courtesy of
Brigadier David Edelsten, great-great-grandson of Elizabeth and John
Gould.

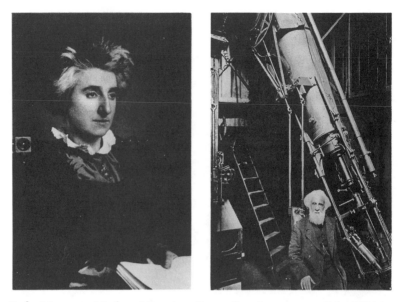

Left, Margaret Lindsay Huggins. Reproduced courtesy of the Whitin Observatory, Wellesley College, Wellesley, Massachusetts. Right, William Huggins at the Grubb star-spectroscope in the Tulse Hill Observatory. Reproduced from William Huggins and Margaret Huggins, *The Scientific Papers of Sir William Huggins* (London: William Wesley and Son, 1909).

Left, *Pseudohazis hera* from "Moonlight Sonata," drawn by Anna Botsford Comstock, in John Henry and Anna Botsford Comstock, *Manual for the Study of Insects* (Ithaca, N.Y., 1895), plate IV. Right, John Henry and Anna Botsford Comstock. The Charles P. Alexander Papers, Smithsonian Institution Archives.

Grace Chisholm Young and her son, Frank. Courtesy of Sylvia Wiegand and Laurence C. Young.

Will Young, shortly before the First World War. Courtesy of Sylvia Wiegand and Laurence C. Young.

Edith and Cyril Berkeley, 1902. Courtesy of Dr. Mary Needler Arai.

Helen and Frank Hogg and family, late 1930s. Courtesy of Dr. Helen S. Hogg.

Left, Frank Nelson Blanchard studying salamanders while wading in a swamp. Photograph by Frieda Cobb Blanchard. Courtesy of Dorothy Blanchard. Right, Frieda Cobb (Blanchard) studying *Oenotheria* at the Botanical Gardens of the University of Michigan. Photograph by Harley H. Bartlett. Courtesy of Dorothy Blanchard.

Thomas Lonsdale, unknown, Maureen Julian, Kathleen Lonsdale, and unknown (left to right). Photograph by Carl Julian after a Quaker meeting in 1968.

Left, Mary Corinna Putnam (Jacobi), Paris, 1866. Courtesy of the Schlesinger Library, Radcliffe College. Right, Abraham Jacobi. Courtesy of the Wellcome Institute Library, London.

Left, Emily A. Nunn, 1881. Photograph by Pach. Courtesy of Wellesley College Archives. Right, Charles Otis Whitman with Carroll (left) and Frank (right), Worcester, Massachusetts, 1890. Courtesy of the Marine Biological Laboratory Archives, Woods Hole, Massachusetts.

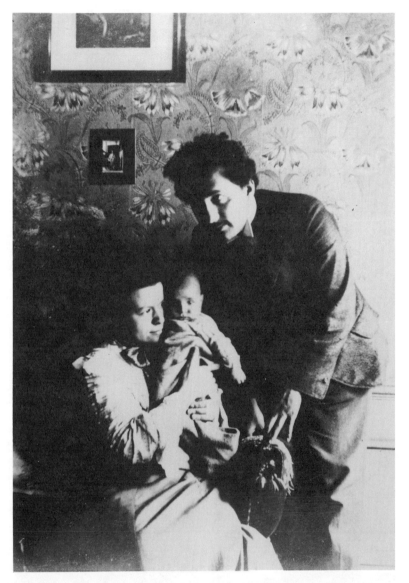

Mileva and Albert Einstein with their son Hans Albert in 1904. Courtesy of The Albert Einstein Archives, The Hebrew University of Jerusalem, Israel.

Helen MacGill Hughes and Everett Cherrington Hughes. Courtesy of Elizabeth Hughes Schneewind.

Left, Elizabeth G. Britton. Courtesy of the Library of the New York Botanical Garden, Bronx, New York. Right, Nathaniel Lord Britton. Courtesy of the Library of the New York Botanical Garden, Bronx, New York.

Frederick E. Clements and Edith S. Clements, c. 1899?: "The same skirt I had in 1898." Courtesy of the Joseph Ewan Archives, Missouri Botanical Garden.

Katharine and Townshend Brandegee. Courtesy of Special Collections, California Academy of Sciences.

Dr. Grace Pickford (center) with Dr. Penelope Jenkins, limnologist (left), International Limnological Congress, Zurich, 1948. Courtesy of the George Evelyn Hutchinson Papers, Manuscripts and Archives, Yale University Library.

G. Evelyn Hutchinson in the field. Courtesy of the George Evelyn Hutchinson Papers, Manuscripts and Archives, Yale University Library.

Elizabeth Campbell riding on a turtle, Flint Island, 1908. Courtesy of the Mary Lea Shane Archives of the Lick Observatory.

William Wallace Campbell, 1914. Courtesy of the Mary Lea Shane Archives of the Lick Observatory.

Annie Scott Dill Russell (Maunder)—standing at far right, holding a chair—in her matriculation photograph, Girton College, Cambridge, 1886. Courtesy of the Mistress and Fellows, Girton College, Cambridge.

Walter Maunder. Courtesy of the Royal Astronomical Society.

Bertrand and Dora Russell at Beacon Hill School in the early 1930s.
Reproduced from Katharine Tait, *My Father: Bertrand Russell* (Boston,
1975), 231. Courtesy of Katharine Tait.

Alva and Gunnar Myrdal at work in their model house office, c. 1936. Courtesy of the Archives of the Swedish Labor Movement.

Margaret Mead takes notes while Gregory Bateson films a children's play group in Bali, c. 1938. Reproduced from Margaret Mead, *Letters from the Field, 1925–1975* (Harper & Row, 1977). Courtesy of the New York Academy of Sciences.

a Ph.D. in plant anatomy at Johns Hopkins in 1905 but chose ecology as his research field, doing pioneering research in Jamaica, followed by many years at the Carnegie Institute's Desert Research Laboratory in Arizona. Edith studied chemistry and physics at the University of Chicago, where she started a Ph.D. but did not complete it. She taught physics at the Women's College (later Goucher) in Baltimore, where she met Forrest Shreve, who arrived to teach there in 1906. He obtained a research position with the Carnegie Institution in 1908 and spent a year at their Tucson laboratory. He and Edith married in 1909 and set off together for Jamaica, also under Carnegie auspices. In Jamaica their roles appear to have been traditional ones. Photos depict Forrest in field clothes and Edith in "floor-length gowns. . . . she sits stiffly at a table set for tea." Back in Arizona, however, she accompanied him everywhere in his field studies of desert plant distribution.[38]

Shreve often disagreed in print with Frederic Clements's ideas on desert vegetation. They were both supported by the Carnegie Institution, were cordial but not friends.[39] Edith Shreve's contribution to this research is not clear; they never published together. Forrest Shreve went on to become an internationally known desert ecologist, interested in vegetation and the factors involved in its distribution, and in plant demography.

In 1911 Edith began, with her husband's support and encouragement, her own research in what would now be called physiological ecology, utilizing her physical science background. She undertook a major research project on the water relations of a desert plant, paloverde. The Shreves moved to Baltimore for several months in 1912 so that Edith could work out some of the techniques with the help of Burton Livingston at Johns Hopkins. Between 1914 and 1940 she published original research in the top botanical and ecological journals.

Unlike the Brittons, Brandegees, and Clementses, the Shreves did have a child, Margaret, born in 1918. Edith Shreve not only continued her research following her daughter's birth but was helped in this by what appears to have been some role reversal in the Shreves' marriage. They worked in the laboratory together, often into and beyond the dinner hour. On one such occasion Forrest wrote: "We ate dinner on the hill for four consecutive nights. . . . Mrs. Shreve was trying out the balanced pot method of determining the water movement in her cacti, and I was drawing the vegetational map toward an early finish,—and doing the cooking."[40] That was in 1915, an arrangement envied by many women scientists eighty years later. Forrest Shreve's biographer suggested that his egalitarian attitude sprang from a Quaker upbringing. His own mother was college educated, and he did not feel threatened by an independently achieving wife. Less is known directly about Edith Shreve's motivation; she was clearly willing to set off on her own course, instead of following the prescribed one for wives and mothers of her era. And Forrest Shreve seemed to be able to carry out his field and laboratory research without as many semiprofessional services as Edith Clements provided for her plant ecologist husband.

Nevertheless, as we look at women ecologists of the next era, although many successfully obtained Ph.D.s, particularly at the University of Illinois before 1930, most taught in high schools or at women's colleges, and at the latter usually retained their positions only if they remained single. Some of the best known did remain single, for example, E. Lucy Braun (Ph.D., University of Cincinnati, 1914), who was the only woman to serve as president of the Ecological Society of America (1950) over a more than sixty-year period.[41] She lived and worked with her entomologist sister, Annette. They did extensive fieldwork together throughout the eastern deciduous forest. Plant ecologist Harriet George Barclay, however, who earned her Ph.D. in 1928 with famous University of Chicago ecologist Henry Cowles, did marry. Her husband was a fellow University of Chicago botany student. They both taught at the University of Tulsa. She was known as a gifted teacher and "inspired students through her infectious enthusiasm for field ecological studies of plants at university field stations," particularly in Colorado. In this case the wife was able to continue her work, although Tulsa was not a major research institution.

Ecologist couples who have worked together have not been uncommon throughout this century, especially in plant ecology. Murray and Helen Buell both received their Ph.D.s in the midthirties and were an important ecological team. In this case Murray had the academic position, at Rutgers, and received most of the honors. The graduate students were officially his, although Helen perceived her role as important in the training of generations of Rutgers graduate students in plant ecology.[42]

More recently trained couples have continued this pattern. Jerry and Carol Baskin both received their Ph.D.s at Vanderbilt University under ecologist Elsie Quarterman in the sixties. They went to the University of Kentucky, where they have both done research and published extensively in plant ecology. Throughout this period, however, Jerry has had the university position, and Carol has had no official appointment. In 1985 she was made an adjunct associate professor. Hazel and Paul Delcourt are another such couple, both with Ph.D.s, at the University of Tennessee. Their research is in paleoecology, but until 1993 Hazel Delcourt was a nontenure-track research assistant professor while her husband had long been a tenured professor. Nevertheless, Hazel became the first woman secretary of the Ecological Society of America in 1986. Relatively few women have been officers of ESA since its founding in 1915. Jean Langenheim notes that some institutions have looked upon married women with Ph.D.s as "captive spouses" and that these women scientists have been forced into "weedy opportunist" niches in order to continue their research.[43]

The highest honor of the Ecological Society of America is the Eminent Ecologist Award. Started in 1953, it has only twice been given to women.

In 1972 it was awarded to Ruth Patrick, a limnologist who has had an extremely important and well-honored research career, still continuing after sixty years. No other woman was selected as Eminent Ecologist until 1986, when the award was given to E. C. Pielou, a mathematical ecologist who never went to graduate school but eventually received a Ph.D. from the University of London for her published work. Neither of these women belong to ecological couples; their husbands were in other fields, but each was very supportive of his wife's work.

One other pattern that has not been discussed previously is that of the couple who, having worked creatively together during the time of their marriage, later divorce but then go on to separate research careers, in some cases each to considerable eminence. With the increase in the divorce rate particularly within the past twenty or thirty years, this has become a fairly common pattern for biologist couples. Divorce sometimes releases women from strictures on their professional careers, such as nepotism rules and the lack of mobility—and often from the housewife role as well. The most remarkable couple to have followed this pattern were G. Evelyn Hutchinson (1903–1991), the most prominent American ecologist of the post-Clementian era, and his first wife, Grace Pickford (1902–1986), both born in England These two were friends and collaborators on three continents, in England, South Africa, and New Haven.[44] Grace Pickford, who like Ruth Patrick retained her maiden name after marriage, was educated at Newnham College, Cambridge. She and G. Evelyn Hutchinson were zoology students together at Cambridge from 1921 to 1925. She was awarded a Mary Ewart Travelling Scholarship to South Africa, enabling her to spend two years there. Evelyn was awarded a fellowship to Naples to do marine biology, after which he also went to South Africa, obtaining a senior lecturer position at the University of Witwatersrand. He and Grace were married in South Africa. While there they worked together on the ecology of the shallow lakes or *vleis* near Capetown, including both the geochemistry and the organisms of these lakes, remarkable in having no fish but a variety of submicroscopic crustacea, both predators and prey. Grace and Evelyn did the fieldwork together with a South African woman biologist and published two papers, including one titled "The Inland Waters of South Africa," in *Nature*.[45]

Evelyn Hutchinson went on studying these closed arid lakes in other parts of the world including Indian Tibet (Ladak), where he was the biologist on the Yale North India Expedition in 1932. There is a whole series of letters from Evelyn to Grace during that expedition, relating the exact nature of his lake studies and asking her advice. These letters are collegial, but not at all like the loving letters of the Brandegees; he addresses her as "Dear old thing."[46] In 1932 they published another joint paper on Mountain Lake in Virginia. In this case the fieldwork was all hers; Evelyn was never there. He appears to have done some of the analysis, though it is not clear why he is the first author. He had the higher status, since by this time he was an instructor in the Zoology Department at Yale while she was doing research without an actual position. On the other hand she had by then earned a Ph.D. at Yale, something that Evelyn never did, although he was given an honorary Cambridge University doctoral degree late in his life.

G. Evelyn Hutchinson went on, with his many graduate students, to become the major force in modern ecology, in his over sixty years at Yale. Mathematical ecology, population ecology, and systems ecology, in addition to his original field, limnology (the ecological study of lakes), owe much to him. His four-volume *Treatise on Limnology*, the fourth volume finished after his death in 1991, is the standard work in this field. His theory of the niche is still the subject of new research He was elected both to the National Academy of Sciences and to the Royal Society as an overseas fellow. Twice awarded the President's Medal of Honor, he turned it down the first time during an administration he disapproved of. Grace Pickford remained at Yale and at nearby Albertus Magnus College for most of her career. She did research at Yale's Bingham Oceanographic Laboratory with titles like assistant in research and research fellow from 1931 until 1957, when she was appointed lecturer in zoology at Yale. She became the first woman full professor of biology at Yale only in 1969, the year before her retirement. She retired to Hiram College in Ohio as distinguished scientist in residence, at the invitation of a former student, James Barrow, and did further research there.

Divorce seems to have sent Pickford in several directions of her own; she no longer worked with Evelyn Hutchinson nor in ecology per se. She had already done doctoral research on the systematics of earthworms, work she had begun in South Africa. She worked out microanalytical methods for studying enzymes, including digestive enzymes of spiders. Later she did research on cephalopods, becoming the world authority on the octopus and relatives. At a time when research travel on American naval vessels was closed to women scientists, she was invited on a Danish expedition to study deep-sea cephalopods. In 1947 she began pioneering research in fish endocrinology, an indirect result of some World War II research carried out at the Bingham Laboratory. All her hormone work on the killifish (*Fundulus*) was supported by the National Science Foundation. She also worked on blue shark and stingray hormones. She invented ingenious microanalytical apparatus and techniques; comparative endocrinologists came from all over the world to visit her laboratory. Some stayed on as graduate students and collaborators. When she moved to Hiram College at sixty-eight, she started new work on the control of the pituitary by the hypothalamus, on which she published until she was over eighty, producing over 135 publications altogether.[47]

Her colleague, J. N. Ball, cited her "tremendous enthusiasm, . . . awesome dedication to research, . . . and her great talent for living and inspiring loyal friendship," also noted by her Cambridge associates. She usually wore trousers or shorts in a period when women wore skirts, and was sometimes mistaken for a young boy. Ball wrote that her unconventionality was "rooted in her sense that convention had discriminated against her in her early career simply because she was a woman."[48]

We do not have Grace's own word on this; she destroyed most of her papers. Women were not allowed Cambridge University degrees until after World War II, certainly not in 1925. Marriage was frowned upon at Cambridge in that era for women academics, but another of her well-known zoology classmates, Sydnie Manton, also continued her research after marriage. Otherwise there seems to have been surprisingly little discrimination; Cambridge women scientists of the twenties went far and had good professional positions at universities at a time when this was rare in the United States. At Yale, Grace Pickford did not have a real position in spite of better credentials than her husband in the early years. We know very little about how she was treated in South Africa, but she had English funding to carry out her research there.

In any case, Grace's professional career blossomed after her divorce. Interestingly, as was noted in a letter written by a more senior Yale professor, J. S. Nicholas,[49] Evelyn's professional accomplishments increased after the divorce and his marriage to a woman who was not interested in science and had no career. Nicholas viewed this in part as a release from competition with Grace. Nicholas's views may or may not be correct, but that kind of release from competition with a scientist spouse more often has happened to women scientists after a divorce. In any case we have here a possible negative affect of collaboration, particularly when both halves of the couple are working in very similar fields. A woman scientist in particular, if she can find the position and funds that make research possible in botany, ecology, or any other field of science, may strike out in new directions on her own after a divorce in a rather similar way to that in which a number of nineteenth-century women scientists were able to focus on their own professional accomplishments after becoming widows.[50]

✦ 16 ✦

MARILYN BAILEY OGILVIE

Patterns of Collaboration in Turn-of-the-Century Astronomy
The Campbells and the Maunders

Rapid changes during the last decade of the nineteenth century produced a "new astronomy" of astrophysics. Sophisticated spectroscopic techniques began to supplant eye and camera as prime sources of information about the stars and planets. Two solar eclipse expeditions mounted by Lick Observatory illustrate the change in instrumentation in the few years between 1889 and 1898. In the earlier expedition of 1889, astronomers S. W. Burnham and John Schaeberle used the older instruments to probe the Sun's secrets. Just a few years later, William Wallace Campbell (1862–1938, known as Wallace), with his scientific team including his wife, Elizabeth Ballard Campbell (1868–1961), teased apart the Sun's light using spectroscopic techniques.[1] Not only could astronomers determine the chemical composition of Sun, stars, and planets; they could use this information as a source to speculate on theoretical problems beyond the ken of eye and lens. Spectroscopy could speak to questions about life on other planets, the existence of galaxies beyond the Milky Way, and cosmological conclusions about the very nature of the universe.

For the Campbells, science was a masculine activity. Wallace believed in wresting the objective "truth" from nature, an activity he considered better accomplished by males. Elizabeth agreed that the male mind was better equipped to consider scientific questions, but was confident that her "feminine" abilities were vital to her husband's scientific success.[2] In England, another scientific couple, Edwin Walter Maunder (1851–1928, known as Walter) and Annie Russell Maunder (1868–1947), also effectively exploited the new techniques to add new data and interpretations. The science of both Maunders was less gender-dependent. Although they agreed that science consisted of positive knowledge, they both thought that objective truth could be gleaned by men and women equally well.

The synergistic effect of the interaction between Wallace and Elizabeth and between Walter and Annie produced a creative product more important than either member of the partnership could have accomplished alone. Moreover, the collaboration of these creative couples took distinctive forms because of differences in the couples' personalities, backgrounds, and conceptions of the nature of science.

Elizabeth and Wallace Campbell

Wallace Campbell and Walter Maunder had similar scientific interests. Both studied radial velocities of stars and planets, were interested in solar astronomy, and were fascinated by the "life on Mars" controversy. Walter Maunder was also interested in questions of astronomical history, astronomy and religion, and popularization, interests not shared by Campbell. Annie Maunder's interests were almost identical with those of her husband, but Elizabeth Campbell's scientific pursuits were confined to her work on eclipse expeditions. Her influence on science, however, far exceeded her obvious contributions. It would not be an exaggeration to say that Wallace Campbell's achievements in science depended upon Elizabeth.

Elizabeth Ballard Thompson and William Wallace Campbell were both born in the midwestern United States. Elizabeth, the daughter of Henry E. Thompson and Elizabeth Whitney Ballard Thompson, was born in Grand Rapids, Michigan, as were her two brothers. Wallace, the youngest of six children of Robert Wilson Campbell and Harriet Welsh, was born on a farm in Ohio. When Wallace was four years old his father died, leaving his mother with six children to rear.[3]

Encouraged to attend a large university by his high school principal, Wallace taught for a year at Fostoria, Ohio, and then entered Ohio State University. Financial problems forced him to leave school after a semester. He again taught and saved money, this time to enter the University of Michigan in 1882 to study civil engineering. During the summer between his junior and senior years, he chanced upon a copy of Simon Newcomb's *Popular Astronomy*. Fascinated by this book, he decided on a career in astronomy. Although such a career was difficult to launch, Campbell was fortunate in locating a sympathetic mentor. John M. Schaeberle, professor of astronomy at Michigan, was in charge of the University Observatory and was willing and able to teach the young man the intricacies of nineteenth-century observational and positional astronomy.[4]

After he graduated from Michigan in 1886 with a B.S. degree in civil engineering, Campbell accepted a position as professor of mathematics at the University of Colorado, where he met his future wife, Elizabeth Ballard Thompson, a student at the university and a sophomore in one of Wallace's mathematics classes.[5]

Wallace remained at Colorado for two years only, for in 1888 a position in astronomy opened at the University of Michigan when his former mentor, Schaeberle, left Michigan to join the staff at the newly established Lick Observatory on Mount Hamilton in central California. Wallace's tenure at Michigan was also short, for a vacancy occurred at Lick Observatory in 1891 when astronomer James Keeler left to become director of the Allegheny Observatory in Pittsburgh. Lick director Edward S. Holden "secured Mr. Campbell's appointment as an astronomer in the Lick Observatory in succession to Keeler with principal duty to take charge of the spectroscopic Department."[6]

In the meantime, Wallace had kept in contact with his former student, Elizabeth. She graduated from Colorado as one of only three students in 1890 and wrote and read the "Class Poem" at graduation. After graduation, she took a position teaching mathematics at Rockford College in Illinois (1890–1891). Once Wallace was assured of his future in astronomy, he proposed to Elizabeth (Bessie), she accepted, and they married in 1892.[7]

Elizabeth Campbell, by her own assessment, was "anything but scientifically inclined." Her people-oriented diaries are subjective. Her prose flowed smoothly as she described the appearance and feelings of the people she met during expeditions. For example, when she and Wallace were in India for the 1898 eclipse, she wrote of the "rows of natives sleeping [in the square], . . . stretched out on the grass wrapped in their white robes and looking like 'The sheeted dead.' " She described her outrage at the Brahman headman who had told the local people that "Campbell Sahib was digging the pit by order of the British government and on the twenty-second of January the pit would be filled with Hindus to be sacrificed to the god of the eclipse."[8] Subjective descriptions were of less interest to Wallace than attending to the details that would assure a successful objective scientific experience. He conducted repeated tests, equipment examinations, and drills of volunteers.[9]

A common love of travel led to Elizabeth and Wallace's cooperation in astronomy. Before marriage, neither had much opportunity to indulge this interest. Elizabeth explained that during the San Francisco part of their honeymoon, they gazed upon the ocean liners leaving for various parts of the world and dreamed of setting "out into the Pacific and keep[ing] on going west around the world until we got back to San Francisco." They entertained themselves by poring over an atlas and planning mock trips.[10]

The trips soon became more than imaginary and developed into the basis of Elizabeth and Wallace's scientific collaboration. From 1889 it was the policy of the Lick Observatory to send out parties to observe solar eclipses. In the early years of their marriage, Wallace had an opportunity to join the Lick party in Akashi, Japan, to observe the eclipse of August 9, 1896. However, Wallace declined because Elizabeth was pregnant with their second son, Douglas, and he did not want to be far away from home.[11] Having missed this opportunity, Wallace was delighted to be appointed chief of the expedition to India to observe the eclipse of January

22, 1898. The Lick policy was to save money by sending as few astronomers as possible and hiring laborers at the site. Schaeberle had found well-educated volunteers to help with the actual observations when he had gone to French Guiana and Chile, so it was assumed that Wallace could do the same in India. Wallace was to go as the lone scientist but wanted Elizabeth to go with him, as the couple put it, "at our own expense." Before they left, Elizabeth did not consider helping Wallace with the actual observations, but she knew she could be "of considerable assistance as secretary and in managing the camp where he would have to live for six weeks."[12]

In deciding to go with Wallace, Elizabeth faced a conflict that ambitious women and wives who are mothers must consider—mixed emotions about leaving young children behind. "At night when I put them [two-and-a-half-year-old Wallace and sixteen-month-old Douglas] to bed I would say to myself, 'I never can leave these babies for eight months. In the morning I'll tell Wallace I can't go.' " San Jose and Berkeley matrons with their opinions—"many taking sides for and against my going with my husband and leaving two babies behind"—complicated the issue. After deciding that she should not leave the children, she would wake up "in the night . . . and imagine Wallace alone in camp in India, threatened by fever, cholera, tigers, cobras, plague, pestilence and famine, battle, murder and sudden death."[13] Her fear for Wallace's well-being and, undoubtedly, her own longing to see unknown parts of the world overrode her reluctance to leave the boys. She accompanied Wallace on the eclipse expedition to India.

Elizabeth's first decision to leave the children and become Wallace's research companion was her most difficult one. When she realized her own importance to the success of the Indian expedition, she hardly considered staying home on subsequent occasions. She found she was needed to shepherd an emotionally fragile Wallace through frustrating setbacks. When confronted by last-minute changes, incompetent bureaucrats, and inefficient help, the outwardly controlled Wallace came close to breaking down. The problems that they had to face on the Indian and subsequent expeditions would have discouraged the most stable individuals. When they arrived in India, they found an epidemic of plague would force a change in the observation site. The substitute site was less than satisfactory, and Wallace began to brood: "the time was short. He wouldn't be ready—and so."[14] His fear that his first eclipse expedition would be a failure climaxed when he was unable to get accurate time signals to determine their exact longitude. Elizabeth reported an incident after this final frustration when, after going to bed "with a big dose of scotch and soda" and sleeping for a short time, "he kept waking up and talking as I never heard him talk before." She reported that "he seemed to see all at once every difficulty he was likely to meet or might possibly encounter, and his confidence in himself was shaken." Hearing Wallace doubt his own ability "that no one ever questions" and express his fear that he would not be "a credit to those dear little boys [their sons]" frightened Elizabeth. She talked to him and calmed him so that, as she noted, after "a while he went to sleep again, and then *I* lay awake and wondered if he was coming down with a fever, or had a sunstroke."[15]

The success of subsequent expeditions also depended on Elizabeth's ability to soothe Wallace when outside events seemed overwhelming. The eclipse expedition of August 30, 1905, to Spain was fraught with frustrations for Wallace. In spite of Elizabeth's care he became ill, and the eclipse preparations were delayed while he recovered. Inclement weather, equipment delays, absence of accurate time signals, signs torn down by children, and inadequate help were obstacles that Wallace could not have overcome by himself.[16]

In addition to assuring Wallace's effectiveness by keeping him calm, Elizabeth managed the supplies, assured healthful conditions, and arranged for tents and other equipment. For the eclipse of August 21, 1914, viewed from Russia, she no longer had to worry about leaving family members behind, for this time the party of eight included six Campbells: Wallace; Elizabeth; the couple's three sons, Wallace, Douglas, and Kenneth; and Elizabeth's mother.[17]

Elizabeth's contribution to the scientific success of the expeditions went beyond her administrative and personal skills. Originally because of necessity, she learned to operate the various instruments, interpret the observations, and record them accurately. The only observers to accompany Wallace to India in 1898 were "Mrs. Campbell and Miss Rowena Beans of San Jose," both of whom were volunteer observers. Wallace's care in choosing and training his volunteers was one of the reasons his expeditions were successful. He went through innumerable drills so that everybody knew exactly what was expected. As it was for Wallace, this eclipse of 1898 was Elizabeth's first chance to view a total solar eclipse. Unlike him, she was unfamiliar with the instruments. After he instructed her in the use of the spectrographs, she was stationed at one "of the two spectrographs on the equatorial mounting," and Wallace was at the second."[18]

Shortly before the Campbells left to observe the eclipse of 1898 in India, Lick director Holden resigned under a cloud and was replaced by James E. Keeler. Keeler's tenure was terminated by his early death in 1900, and Wallace Campbell was designated director on January 1, 1901.[19] Administrative requirements cut into his research time, although he still found time to head subsequent major solar eclipse expeditions. With the exception of the eclipse of 1900, viewed by the Lick team in Georgia, Elizabeth accompanied Wallace on all the expeditions: Spain, August 30, 1905; Flint Island in the central Pacific Ocean, January 3, 1908; Russia, August 21, 1914; Goldendale, Washington, June 8, 1918; and Australia, September 21, 1922. With each experience, her observational skills improved. The data collected by both Campbells and others on the expeditions led to important astronomical information about solar eclipses. In the earlier expeditions, some astronomers had explained a perturbation in Mercury's orbit by postulating an intramercurial planet. Wallace hoped to be able to solve the problem definitively through eclipse photographic plates. The eclipse of 1905 offered an excellent opportunity to settle the question, because of its long duration (a maximum of three and three-fourths minutes), its occurrence in August, when the weather was likely to be favorable, and the possibility of widely separated observation stations on three continents. The eclipse was to occur in Spain one and one-half hours later than in Labrador and in Egypt one hour later than in Spain. William H. Crocker, the chief financier for the Lick expeditions, recognized the scientific importance of this eclipse and supported three different expeditions.[20] A slight miscalculation of time by one of the astronomers (not Campbell) and a thick cloak of clouds, which did not completely obscure the event, nevertheless made it impossible to confirm or deny the existence of the intramercurial planet.

Crocker agreed to defray the expenses for the ninth Crocker Eclipse Expedition so that Lick Observatory astronomers could view the total eclipse of 1908 in hopes of settling the problem. The chosen site was Flint Island in the central Pacific Ocean. Because of the isolated location, plans had to be made very carefully, for once the researchers arrived on the island, resupplying would be almost impossible. Elizabeth, always violently seasick on the long ocean voyages, had not planned to go until the last minute. Less than two weeks before they were to sail, "like the warhorse of scriptures [that] sniffed the battle afar off and longed to be in the thick of it," she decided to go.[21] From an astronomical point of view this expedition was highly successful, although sporadic clouds kept the Campbells in suspense until the moment of totality. The rain that was falling about a minute before totality stopped, the skies cleared, and the team was able to put into practice the skills they had perfected over a period of a month before the eclipse. The photographs were excellent, and the results denied the existence of an intramercurial planet.[22]

Another project involving eclipses replaced the search for the discredited intramercurial planet. Campbell became involved in testing Einstein's prediction that massive bodies like the Sun could bend light. He attempted to test the hypothesis by comparing the position of a specific star relative to the Sun at two different times of the year. Only by measuring the deflection of light rays during a solar eclipse could such a comparison be established. The eclipse of August 21, 1914, was suitable for such a test. However, the luck that had prevailed in the Flint Island Expedition failed them in Russia: "Just as one minute of totality was called, a great black cloud roled [sic] up from the East, completely obscuring the sun. There was absolutely no observation possible." Thwarted in their scientific efforts, the Campbell family now faced even greater problems. World War I had been declared on July 30, and most travelers had left for home. Obviously, the Lick party, which had come halfway around the world to observe an eclipse, had remained. Although the return of people and instruments was difficult, arrangements were finally made.[23]

The problems of the Russian expedition spilled over to the next eclipse. This time the eclipse of June 8, 1918, was visible in the United States, and the Lick party chose Goldendale, Washington, as its observation site.[24] The instruments that the Lick party had been forced to leave behind in Russia were so delayed that Campbell had to borrow equipment from various observatories. The borrowed equipment did not produce plates sufficiently accurate to solve the deflection-of-light problem. The problem itself was deflected to the next eclipse, on September 21, 1922, when the Campbells went to Wallal in northwestern Australia. The observations made in Wallal eliminated the ambiguity left by the 1919 Eddington expedition and left little doubt that Einstein's theory of general relativity was correct. Elizabeth undertook her usual duties on the trip. She reported that the eclipse day was perfect and that they spent the following night— all night—developing film. These photographs supplied the additional evidence to support Einstein's theory.[25]

When the Campbells returned from Australia, they were met by a delegation from the regents of the University of California, offering Wallace the presidency of the university. He reluctantly accepted, retaining the directorship of the observatory without remuneration.[26]

During his tenure at Lick, Wallace Campbell trained himself in spectroscopic techniques and established an important program to determine the radial velocities of stars. He also became involved in the "Mars question." Campbell opposed the accepted view that Mars's spectrum contained small amounts of water vapor. He correctly concluded from observations made at high, dry sites—during an expedition of 1909 to Mount Whitney, for which Elizabeth made the provision list even though she did not accompany him— that the water vapor was actually in the atmosphere of Earth, not Mars.[27]

In 1930 Campbell retired from both the university presidency and the observatory directorship. His health had begun to fail, and he lost the sight of one eye. However, on July 1, 1931, he accepted the presidency of the National Academy of Sciences, and he and Elizabeth moved to Washington, D.C., but came back to California each summer. When his term of office was over in 1935, they returned to California and lived in an apartment in San Francisco.[28]

On June 14, 1938, the San Francisco *Call-Bulletin* screamed Wallace Campbell's farewell to his lifelong partner: "Good bye, dearest Elizabeth. Be of great courage. My aphasia convinces me that you will never make of me but a broken reed." Wallace Campbell left several notes to his beloved Elizabeth before he took his own life. Realizing that he was failing both physically and mentally and fearing that his life would "as long as it lasts, be a burden to Elizabeth," he committed suicide by leaping from a window of their apartment house in San Francisco.[29]

Annie and Walter Maunder

During the first decades of the twentieth century, astronomy became increasingly professionalized. One of the first effects of professionalization was the exclusion of amateurs from a discipline in which they had formerly been active and productive. Women, especially, were often locked into an amateur role and were victims of this exclusionary policy.[30] One astronomer, Walter Maunder, ardently defended the importance of amateurs to astronomy.

Walter Maunder's background may explain his support of the amateur tradition. He was one of seven children of a Wesleyan Society minister, Reverend George Maunder, and his wife, Mary Ann Frid Maunder.[31] The family grew up in a working-class neighborhood in London, and family members were, no doubt, influenced by the egalitarian social-gospel policies of the Wesleyans. Wesleyans believed that faith must be lived out in the world of the factories and shops as well as in the parlors of the wealthy. Walter attended classes at King's College but did not achieve a degree or its equivalent. In 1875 he married Edith Hannah Bustin in a Wesleyan Methodist ceremony conducted by his father.[32] The couple had six children who lived past infancy, two daughters and four sons. Family impressions indicate that Walter was more interested in his work than in his family.[33] The apparent ignoring of the children may have occurred after the death of Edith Maunder of tuberculosis when she was thirty-five years old.[34]

After working briefly in a London bank, Walter took the examination for the post of photographic and spectroscopic assistant at the Royal Observatory at Greenwich, the first such civil service examination. While he was a young employee at the observatory, he became incensed by the treatment of the underpaid computers. The older men who were the Greenwich computers in the early nineteenth century had been replaced by teenage boys. "The unfortunate boys who carried out the computations of the great lunar reductions were kept at their desks from eight in the morning till eight at night without the slightest intermission, except an hour at midday."[35] Although the personnel situation at Greenwich improved after W. E. Christie replaced G. B. Airy as Astronomer Royal, the pay remained pitifully low, particularly for computers. From 1891 to 1896, the Royal Greenwich Observatory experimented with hiring women computers. Since only women who had completed the equivalent of a bachelor's degree were hired, the women computers were better educated than most of the young boys.[36]

One young woman computer, Annie Scott Dill Russell, particularly caught Walter Maunder's eye. Born in Strabane, County Tyrone, in Northern Ireland, Annie was the daughter of the Reverend William Andrew Russell and Hester Nesbitt Dill. Her early education was at home and at the Victoria School and College, Belfast.[37] Annie entered Girton College, Cambridge, in 1886 but was ill prepared for rigorous university courses.[38] By working very hard, she overcame her educational deficiencies and scored well enough on the mathematical tripos to obtain the place of a high senior optime.[39]

When Russell was at Girton, Cambridge University itself had been undergoing some radical curricular changes, particularly in the sciences, but the impact of these changes on Girton was minimal. Both because it lacked the money to endow research and because its relationship with the university was weak, Girton was only mildly influenced by the new spirit. Girton alumnae, aware of its weakness, established several prizes (one of which Annie Russell later received).[40]

At that time astronomy was not an independent field, and a would-be English astronomer studied mathematics and mathematical physics (usually at Cambridge) before undertaking a kind of apprenticeship at a major observatory. As a woman, Russell had no opportunity for such an apprenticeship, so she became a mathematics teacher earning a salary of eighty pounds per year with residence included.[41]

During the short period of time that the Greenwich Observatory hired women computers, young women who wanted to be astronomers applied in droves for these few jobs. Annie, who had never wanted to teach school, joined the multitudes and wrote the observatory secretary for information, enclosing recommendations from "Miss Welsh, Mistress of Girton and from my Coach." Welsh and her tutor, William Henry Young, praised her "diligence, intelligence, and conscientiousness."[42] To strengthen her application, she enlisted the prestige of Robert Ball, a friend of her father and a fellow of the Royal Society. Ball wrote Christie that his friend "the Rev. A. A. Russell . . . asked me to write to you and solicit your influence to procure for his daughter Miss A. S. D. Russell a position in your ladies' staff at Greenwich."[43]

When Russell was offered the position, she found that the salary of four pounds per month—a sum "so small that I could scarcely live on it"—was much lower than the six pounds per month she had expected. Nevertheless, after Russell "carefully considered the subject," she decided "to accept the post at Greenwich Observatory as I believe I shall like the work."[44] Annie Russell went to work as a lady computer in September 1891 and met the crusading widower Walter Maunder.

Walter Maunder had become so unhappy at the observatory that he applied, unsuccessfully, for other positions.[45] After being passed over for several promotions, he compensated for his unsatisfactory job and discontent with the exclusionary policies of the Royal Astronomical Society by forming an inclusive astronomical organization. Even though he was now a fellow of the Astronomical Society and had served both as a council member and secretary, he was not tolerant of what he perceived to be injustices in this organization. As editor of *Observatory Magazine*, a magazine concerned with popular astronomy, Maunder processed correspondence from numerous amateur astronomers, including disaffected members of the Liverpool Astronomical Society. Many of its members felt that the Liverpool society had not been faithful to its ideals and demanded a new organization that would reflect its original goals. The Liverpool society had welcomed amateurs, including women, and was divided into "Sections," each headed by an experienced observer. This plan was used by Maunder as a model for a new organization, the British Astronomical Association (BAA). The BAA was established in 1890 "to meet the wishes and needs of those who find the subscription of the R. A. S. too high or its papers too advanced, or who are, in the case of ladies, practically excluded from becoming Fellows."[46]

One year after Walter formed the BAA, Annie was assigned the task of examining his sunspot photographs. In 1895, after working together on many astronomical projects, Walter and Annie married. They were almost inseparable until Walter's death in 1928, just after his seventy-seventh birthday.[47]

The Royal Greenwich Observatory's appointment of Walter Maunder had signified a commitment to the new astronomy of astrophysics—a hollow promise, for the observatory provided only inferior spectroscopic equipment. Consequently, Walter's observations of the radial motion of stars and his determination of the spectra of planets and comets were not noteworthy. His achievements, and Annie's, were in the areas where they worked together: the behavior of sunspots and of the eclipsed Sun, observations of Mars, popular astronomy, religion and astronomy, and historical astronomy.

From the time of his original assignment at Greenwich, Walter was involved in observing and interpreting data regarding the dark splotches on the Sun known as sunspots. He agreed with reports indicating that about eleven years passed between sunspot maxima (times of the greatest number of sunspots) and sunspot minima (times of the least number of sunspots). By analyzing historical records, Maunder reported that hardly any sunspots were seen in the sixty-year period from 1645 to 1705 (the so-called Maunder Minimum). Maunder reported that not only do sunspot numbers vary during a cycle but the positions of the sunspots change as well. Annie joined him in the sunspot project in 1891, hypothesizing that the decrease in sunspot frequency as spots pass from east to west of the Sun's disk as viewed from Earth is "apparently" caused by a terrestrial effect.[48]

Sunspots fascinated the Maunders, but they, like the Campbells, were also intrigued by what the eclipsed Sun could tell them about that body. Before astrophysical considerations became paramount, amateurs under the direction of professional astronomers collected a wealth of observations. Although Walter Maunder was a member of the Joint Permanent Eclipse Committee of the Royal Society and the Royal Astronomical Society, the Royal Observatory at Greenwich gave little financial or moral support to his work on solar eclipses. He was forced to rely upon telescope, camera, and inferior spectrographic equipment. Sometimes he was sent, on the lamest of excuses, to inferior observing sites. In 1900 Annie wrote to their mutual astronomer-friend, Captain Molesworth, that the Astronomer Royal "proposes to send my husband to Mauritius on the ground that he is experienced in eclipses and the totality is shorter there, and Mr. Dyson [in favor] who wants to try long exposure photographs is to go to Sumatra because he is new to the work. I don't quite follow that logic." The group that went to Sumatra was large—as Annie put it, "in fact everyone but ourselves."[49] By organizing expeditions with the BAA, Walter was able to circumvent potential problems with the "establishment" groups. He and Annie were a part of eclipse parties of August 9, 1896 (Vadso Island, Norway), January 22, 1898 (India), May 28, 1900 (Algiers), May 1901 (Mauritius), and August 30, 1905 (Labrador).

Both Maunders became involved in the controversy sparked by the highly regarded Giovanni Schiaparelli's announcement in 1877 that he had discovered a system of "canals" on Mars. The Maunders, like Wallace Campbell, did not believe that the canals were human creations. As individuals and as collaborators, Walter and Annie published numerous articles and books in which they argued that most of Mars's "canals" were merely optical illusions. They thereby played key roles in helping to discredit Schiaparelli's canal observations, which (according to Michael Crowe) were rejected by the majority of astronomers by 1912.[50]

Annie and Walter maintained that significant contributions could be made in astronomy without resorting to huge telescopes with tremendous light-gathering capabilities, for it seemed ridiculous to "use a cannon to kill a fly."[51] The people's astronomy, in which they firmly believed, could be pursued by all. It could both yield important information and serve as an enjoyable hobby. In her book *The Heavens and Their Story* Annie explained how hobbyists could gain fascinating information as the Sun tells "his" story.

Annie and Walter were equally interested in religion and astronomy and in historical subjects. In Walter's *Astronomy of the Bible*—dedicated "To my wife. My Helper in this book and in all things"—he located and analyzed scriptural references to astronomical phenomena. This topic emerged in many of his popular books as well as in his papers.[52] Annie not only quoted the Bible in her scientific works but also wrote several papers discussing allusions in Eastern religions to astronomical phenomena.[53] Both Walter and Annie wrote on historical subjects. Walter's *Royal Observatory, Greenwich: A Glance at Its History and Work* (1902) and *Sir William Huggins and Spectroscopic Astronomy* (1913) recorded recent astronomical history. However, both were interested in the more remote past—the origin of the ideas and the names of the constellations.[54]

Conclusion

The Maunders and Campbells were separated by a continent and ocean, yet both couples confronted similar professional problems and were involved in the "new astronomy" of astrophysics, so they were working on a common set of astronomical questions. It was their personal situations that determined the differences in the nature of their collaboration. Elizabeth Campbell's interest in astronomy came after her marriage to Wallace, whereas Annie Maunder was involved in astronomy before she met Walter. Because Elizabeth had no academic training in science, Elizabeth and Wallace's relationship was that of student and mentor. Although her role was circumscribed (she basically followed Wallace's instructions), she went beyond his tutelage in creative planning and in recording scientific events. She observed, photographed, recorded, trained others, and generally assured that each of his scientific projects was successful. For his part, Wallace was dependent on Elizabeth to assure that his physical and emotional needs were met. As for the Maunders, by the time they married Walter was established professionally at Greenwich, but Annie was locked into the only position in which she, as a woman, could be paid for astronomical work. Since the observatory discontinued the policy of hiring lady computers in 1896, she soon would have been out of any job. Annie and Walter alternated editorships, wrote prolifically, and were generous about sharing credit.

Children were a consideration for Elizabeth but much less so for Annie. Isolated on Lick's Mount Hamilton, Elizabeth raised and supervised the education of the couple's three sons. Annie and Walter had no children together, and by the time they married some of his children were grown and the others were old enough not to need constant attention. Although she was involved with the care of his two younger children and for a time her own young nephew, she did not have to endure the constant pressure of raising a young family.

The Maunders' career took a turn that the Campbells' did not—an emphasis on people's astronomy. Perhaps because of his background, Walter was disappointed with the snobbishness of professional astronomy. His disillusionment led to the formation of the BAA, which in turn supported the couple's collaborative science. The intimacy of long years of close association enhanced the creativity of both couples. They shared success and disappointment, and each partner accepted the other's strengths and foibles. Long-term collaboration affected how each thought, wrote, and felt. Elizabeth Campbell's support was not enough to keep the severely debilitated Wallace from taking his own life, and the end of Wallace Campbell's life was a bitter closure to their intimacy. Although Elizabeth understood that the Wallace who committed suicide was not the same person with whom she had shared forty-six years of her life, it remained a bitter blow. She continued to correspond with their mutual friends, but her contributions to astronomy ended when Wallace died. After Walter died, Annie gamely continued to publish and remained active in the society to which they had contributed so much, the BAA.

The lives of the two couples illustrate two patterns adopted by husbands and wives in scientific collaboration. In the Campbell pattern, the wife assisted her husband while gradually becoming more involved herself. In the Maunder pattern, the partners' contributions were as equal as the times allowed.

PNINA G. ABIR-AM

Collaborative Couples Who Wanted to Change the World

The Social Policies and Personal Tensions of the Russells, the Myrdals, and the Mead-Batesons

The interface of social policy and social science scholarship is a particularly interesting arena for observing collaborative couples. Here, the partners in a collaborative couple have a built-in edge over the individual actor precisely because they are a couple. Their relationship *as a couple* inevitably bears upon the social policies they advocate, because, as a couple, they exemplify the most basic unit of social structure and function, the nuclear family.

In other domains, most notably the natural sciences, it is possible to consider the collaborative couple chiefly in terms of its ability to combine skills and interests that society has traditionally segregated by gender.[1] In the realm of social policy, however, the collaborative couple's ability to transcend the gender polarity of society, without appearing to undermine that polarity, is equally crucial. This contradiction has both enhanced and constrained the social impact of couples who strove to combine social science scholarship and social activism in a joint package. Inevitably, it has forced them to act out a complex and delicate balancing act between the worlds of private and public life, thought and action, basic and applied research, their sexual identities as females or males, and their sociocultural roles as mothers and fathers.[2]

The lives and works of three well-known couples who wanted to change the world—the Russells, the Myrdals, and the Mead-Batesons—shed light on these interlocking dichotomies of social order. Their stories are individually fascinating, but the ensemble is even more so. Taken together, these case studies suggest new ways of looking at the history of twentieth-century science, at the nature and uses of the social sciences, and at the spectrum of relationships possible in a marriage.

The three couples were jointly active over three successive decades, from 1920 to 1950, a crucial period of social change around the world.

They sought to influence policy at every level, from the household to the nation to international politics, and in many parts of the world. Whether in a Britain facing the end of its empire, a Sweden on the verge of becoming the model democratic welfare state, a United States recognizing its racism, or a Polynesia still epitomizing a prehistoric paradise—the policies reflected in the public careers of the Russells, the Myrdals, and the Mead-Batesons had major consequences for their own contemporaries and for generations to come.

But their influence *as couples* was equally potent. Their highly visible experiments in combining intellectual creativity, social activism, and non-conformist marriages attracted the kind of intense press attention that is now beamed on contemporary political leaders who have challenged the still-prevailing gender hierarchies. The careers and unconventional marriages of, for example, Golda Meir, Indira Gandhi, Margaret Thatcher, Benazir Bhutto, the Roosevelts, and the Clintons have critical precedents in the lives of the Russells, the Myrdals, or the Mead-Batesons.[3]

Dora Black Russell and Bertrand Russell: Diverging Critiques of Socialism, Marriage, and Education

Through the 1920s Bertrand Russell (1872–1970) and Dora Black Russell (1894–1987) collaborated on a variety of social issues, ranging from British parliamentary elections to the consequences of the Soviet Revolution inside and outside the Soviet Union, sexual freedom, legalized birth control, open marriage, pacifism, and progressive educational practice. Over the course of the decade, they came to disagree on many of the subjects that originally united them. The greatest strain on their marriage and collaborative work, however, came from the everyday realities of running the school they had founded for their friends' children and their own. A relationship that began in shared intellectual enthusiasm and ideals ended in vindictiveness, alienation, and victimization. Indeed, the Russells' separation is a paradigm of the worst that can happen when a collaborative couple's marriage goes sour.

Bertrand Russell and Dora Winifred Black were introduced by a common friend, Dorothy Maud Wrinch (1894–1976), a fellow Girtonian and student of Russell in mathematical logic, in 1916.[4] Black, like Wrinch, had just graduated from Girton College, where she had studied French and German, and looked forward to an academic career. Russell's own brilliant career at Cambridge University—he had been best known as coauthor, with Alfred North Whitehead, of *Principia Mathematica* (1913) and as a member of the Apostles—had just ended in scandal: Trinity College had dismissed him from his lectureship in mathematics and logic for his objection to conscription and to World War I.[5]

Their serious relationship began two years later in the midst of postwar London's heady political and cultural movements. They were brought together by a fascination with the Bolshevik Revolution and joined a group of British intellectuals who were eager to see for themselves the mammoth social reconstruction underway in Lenin's Soviet Union.

The first rifts between them were revealed by their trip to the Soviet Union in 1920. Dora Black showed her staunch sense of independence by refusing to give up the trip, as Russell wanted her to do for safety's sake, and instead met Russell and the group of British intellectuals that he was part of after traveling there on her own through Finland. Although initially they had shared enthusiasm for the great Soviet experiment, their opinions diverged sharply soon after the trip. In his book *The Practice and Theory of Bolshevism* (1920), one chapter, Dora's, continued to support the new political system, while the rest of the book, all Bertrand's, was a devastating critique of its totalitarian tendencies. Their daughter, Katharine Tait, summed up the conflict oversuccinctly: "They did not seem ever to have agreed on anything to do with Russia—in fact, they quarreled bitterly about going to Russia, they quarreled about Russia itself, they quarreled in retrospect of the circumstances of the trip. They went on quarreling about Russia for years."[6]

Nonetheless, the year after, Dora gave up her research fellowship at Girton and the prospects of an academic career to accompany Bertrand on a lengthy and stimulating trip to China, where the Chinese addressed her as the "very intellectual Miss Black" and regarded her and Bertrand as collaborators. While in China, Bertrand fell seriously ill, and Dora nursed him devotedly. When they returned to England, unmarried, Dora was visibly pregnant. They eventually married with the expectation, on Dora's side certainly, that they would be equal partners in the work of raising a family, writing, and promoting radical social change. According to Tait, "It was the beginning of a remarkable partnership. He was an urbane man of the world, much traveled and highly sophisticated; an aristocrat, a brilliant philosopher and a famous man. . . . My mother, still young, loved and admired him deeply, but she was never a docile worshipper. She expected to be a partner in his work, not an auxiliary to her husband's career."[7]

The birth of their son John Conrad, in 1921, and daughter, Katharine, in 1923, thwarted those plans. Their only source of income was their writing, and publishers saw Bertrand's fame (or notoriety) as the selling point for anything either of them wrote, jointly or separately. Although they talked out their next book together—*The Prospects of Industrial Civilization* (1923)—Bertrand again wrote all but one chapter. The demands of two small children and Dora's political activities (she stood for Parliament and led the Workers Birth Control Group, a lobby in the Labour Party) left her little time and energy to collaborate fully on the writing. Bertrand protested only weakly when the publisher, Stanley Unwin, proposed to credit Dora as "Mrs. Bertrand Russell": "What I have put on the title page, 'by Bertrand Russell, in collaboration with Dora Russell' represents the facts as exactly as I can; but I don't think it is fair to leave out her name, so the only short formula is 'by Bertrand and Dora Russell.' I will not however insist upon this if you have strong objections."[8]

Dora too was ambivalent in the face of the economic consequences of asserting her right to be listed in her own name, even on the books she wrote entirely by herself. As she explained in an interview:

> The publisher persuaded me to allow him to put "Mrs. Bertrand Russell" in parentheses under my name on the title page. He said it would add two thousands to the sales, and the thing I cared most about was to get the book read, to make my ideas known, and besides we needed the money. Our only income is from our writing. So I agreed. You see, I used to write a good deal, but I could not get any paper to print what I wrote when I was Dora Black. . . . Symbols are vastly important, and certainly this taking your husband's name is one of the most devastating symbols of "subjection" that remain. And yet I don't think keeping your own name is the ultimate test of a Feminist.[9]

Ultimately, the book, *Hypathia*, was promoted entirely on the basis of Dora's derivative status as the wife of a famous man, for even plain "Dora Russell" was not included on the title page. Although Dora eventually managed to write several more books of her own,[10] they were inevitably overshadowed by her husband's best-sellers (he would receive the Nobel Prize for literature in 1950).

In contemplating marriage to Bertrand, Dora had foreseen some of the ambiguity of their situation: "I am really very unhappy because I do not want to get married and yet I am given no peace. And I am too much in love to do my work the better for it. . . . If I want a life of my own I shall have to do without him. And I don't WANT to give all my life to love."[11]

But the full, unhappy consequences of the contradictions between their socially progressive, but not feminist, ideals and the sheer weight of patriarchal tradition did not emerge until after the Russells embarked on their joint project of Beacon Hill School in 1927.

The school was born out of their hopes of putting their social and educational philosophy into practice. They saw it as an endeavor that would nourish their ideal of a partnership of equals, stabilize their precarious income, and permit their own children to learn in an atmosphere free of competition, regimentation, and sexual repression. As with many other would-be collaborative couples, their good intentions were undone by the economic efficiency of sticking to the traditional sexual division of labor and by the husband's inability to escape the deep-seated, "old-fashioned" belief that the division was normal and right.

> "Bertie," Dora wrote, "was an old fashioned husband. His first need was for a mother to his child, his pride to support both; he accepted as normal, and was very dependent upon his wife's management of his household and

domestic affairs. Beyond this he did not much explore the inner workings of the female psyche, but would minister to a mind if it bore resemblance to that of the superior male. This male attitude has made of wives, especially wives of intelligent men, no more than an adjunct to their husbands' lives and not an integral part of them."[12]

Running their school proved to be quite an ordeal. Many students came not for the bold educational experiment but because other schools had turned them down. In particular the costs were higher than expected, forcing both but especially Bertrand to spend most of his time churning out popularizations or lecturing in America for fund-raising purposes. Dora also became increasingly frustrated by the burden of running the school, although she persisted until 1943, when war-related difficulties brought the venture to an end.

Acting on their belief in and former practice of open marriage, both turned to other partners, a situation that when coupled with the accumulating strain from running the school led first to a separation in 1932, with Bertrand leaving the school altogether. Having begun an affair with their children's nanny, Patricia Spence, a student from Oxford, Bertrand married her shortly after the divorce from Dora in 1937. Dora had two children by an American journalist, Griffin Barry, whom she did not marry (Bertrand recognized the first child but not the second, possibly because the second child was a boy who might have had a claim to Bertrand's hereditary title). The family was destroyed. As Tait recalls, "[Dora and Bertrand] came from Cornwall full of joy and hope, to start a school in which their children would blossom into the finest flowers of mankind. At the end of seven years they had lost each other, their children's confidences, their money and much of their hope. Those years shattered the crystal of our happiness and left us like jagged splinters, unable to touch one another without wounding."[13]

The pioneering collaboration came to a pathetic close in a bitter custody battle over the children. To win custody, Bertrand resorted to invoking all the laws and social mores that he and Dora had previously, deliberately, flouted together. He used all the influence of his title and fame to discredit Dora as a fit mother to his children (though he did not dispute her fitness as a mother of someone else's children).

The ugliness in the Russells' divorce stands in a bitter contrast to the idealism that drove their collaboration at its start. Their relationship was founded, they asserted, on the assumption of equality between the partners. Yet, it is clear that wishing did not make it so: the partnership was, in fact, never one of true equals. Dora came into the marriage with all the disadvantages: Although a promising scholar in a prestigious college at an elite university, being twenty-one years younger than her husband and still in her twenties at the time of marriage, she could hardly have had an established reputation of her own or prospects for an income of her own. And social disapproval of their joint nonconformity fell heavier on her.

Both she and Bertrand had given up their academic careers, but he could claim higher principles as the ground of his sacrifice. During and especially after the divorce, she was even more vulnerable to the asymmetries of the gender hierarchy. She lived on in genteel poverty, severed from her children with Bertrand during their adolescence and unable to carry out any of the projects in social change that she and Bertrand had dreamed of working on together. Though Bertrand too encountered various social difficulties and job insecurity (on account of his socially and sexually radical views he was dismissed from teaching jobs in both New York and Pennsylvania during World War II), he not only retained his world fame as perhaps the most famous philosopher of the twentieth century but emerged in the 1950s as a leader of the antinuclear movement. He also retained better access to his children by both Dora and the younger, former student nanny, whom he married after her, while having two subsequent considerably younger wives. (Shortly after his 1950 Nobel Prize for literature, Bertrand, by then close to eighty, married for the fourth time. Edith Finch, his fourth wife, was yet another woman who abandoned her professional plans upon marriage to him.)

This asymmetric impact of the separation upon the former partners suggests that even the Golem, or the metaphor for a work or product that turns to destroy its creator, is sexist, as it turned so selectively upon the female partner in this joint creation of a progressive school.

Alva Reimer Myrdal and Gunnar Myrdal: From the Idea of a Welfare State to Incipient Solutions of the American Race Dilemma

Two of the twentieth century's most influential tracts in social policy, *Crisis in the Population Question* (1934), which defined the main parameters of the Swedish model of a welfare state, and *An American Dilemma: The Negro Problem and Modern Democracy* (1944), which pinpointed the central tension of American society, were produced in the context of a long-term, ongoing pattern of collaborative interaction between two Swedish social scientists, Alva Reimer Myrdal (1902–1986) and Gunnar Myrdal (1898–1987). The Myrdals, and the public, were convinced that the collaborative relationship of the authors was critical to the inception and impact of these books.[14]

But the Myrdals also worked in other modes equally well, pursuing separate projects in international policy, especially in the 1950s and 1960s. In fact, they are the only married couple to have won Nobel Prizes for their achievements in different fields: while Gunnar shared the 1974 Nobel Prize in economics, Alva was the sole recipient of the 1982 Nobel Peace Prize, for her diplomatic work on nuclear disarmament. Why then was their collaboration on these two books so significant?

The Myrdals married in 1924, both relatively young, both without clear objectives beyond a strong desire to leave their mark on the world, a world they agreed could use a great deal of reform. From the start, the Myrdals rejected the notion of separate spheres for husband and wife. They relished each other's company to such a degree that they spent much of their time discussing ideas, traveling together, and doing joint research. This pattern of collaboration and companionship continued after their three children were born: Jan in 1927, Sissela in 1934, and Kaj in 1936.

In 1936 they began living in a model house that Alva had deliberately designed around the complementary needs of family and work. The ground floor had rooms for the children and their caregivers, and the standard "living quarters." Upstairs, Alva and Gunnar shared a huge office, with a large, double, face-to-face desk, a room for their private "Archive," and a bedroom whose sliding partition permitted both solitude and companionship. Thanks to publicity about the house's novel features, the house became a pilgrimage site for Swedish couples eager to see the dwelling that encouraged spouses to produce both joint books and children.

The house's novel architecture and functional use of space seemed to epitomize the radical ideas of its residents, ideas that had already attracted a great deal of publicity especially when packaged as a joint book, *Crisis in the Population Question* (1934), while further illustrating the book's message with the joint example of an additional child. For complex reasons of innovative contents and disciplinary range, politically smart timing, paramount relevance to key problems of Swedish society, and the Myrdals' own flair for public relations, this book had had a profound effect on social debate, legislation, and cultural discourse in the country. It was instrumental in Sweden's becoming a pioneering welfare democracy that served as model and inspiration for other nations around the world.

The Myrdals argued dramatically that Swedish national survival required attention to such sensitive issues as fertility, family planning, sexuality, housing, wages, childcare, and social benefits. Unless social and political reforms encouraged and enabled Swedes to have more children, they declared, the ongoing decline in population would lead to a population crisis possibly amounting to the country's self-annihilation.

The book's unusual disciplinary range, from the "hard" data of macroeconomics and demography to the "soft," impressionistic ideas drawn from early childhood education or labor relations, and its integration with so many disparate issues, was made possible only through the Myrdals' collaboration. Together they commanded fields and points of view that would have been lost to one or the other if they had acceded to the traditional segregation of fields by gender.

Moreover, the Myrdals' collaborative habits of mind were reinforced by parallel collaborative habits of body. They sought and sustained an intimacy that was psychological, emotional, and sexual as well as intellectual. The multidimensional richness of that intimacy amply revealed itself in their work and lives. No single author, male or female, could have written the book in a way that demonstrated at once such real-life experience and such scholarship. The Myrdals practiced in their own lives what they preached in *Crisis in the Population Question*. Although it was probably only a coincidence that their second child was born just a month after the book's publication, the press regarded the timing as emblematic, at first by castigating Alva for her extensive lecture tour so soon after the baby's birth but later by praising the Myrdals for doing their personal bit to reverse the Swedish population decline.

Through their marriage (virtually the only framework available to legitimize the high degree of intimacy that they needed to forge a long-term collaboration), Alva in particular gained access to resources that would likely have been denied to her on her own. Her formal credentials were not only weaker than Gunnar's, who had been steadily rising as a brilliant professor of economics at the University of Stockholm, but remained incomplete, for she left unfinished her doctoral thesis in psychology. Yet, she compensated for this lack of formal advanced degrees by gaining practical experience in new educational institutions where her degree in sociology apparently sufficed.

But it was the pattern of joint travel abroad, initially designed to satisfy Gunnar's needs for convenient, dual, sexual and research companionship, that more than anything else transformed Alva into a confident, fully fledged collaborator. The trip to the United States in 1930–1931, with both of them receiving fellowships from the Rockefeller Foundation's Social Sciences Division,[15] was particularly influential, for it brought them in contact with other collaborator couples, most notably the sociologists Dorothy Swaine Thomas and William J. Thomas, who became both friends and an inspiration; as well as with accomplished women researchers in the social sciences, most notably the émigrée child psychologist Charlotte Buhler. At the same time, this trip led to a rift between the Myrdals and the two-year-old son they left behind with relatives, a rift that never healed and eventually spilled into a public scandal with Jan Myrdal's bitter autobiography, *Childhood*, which had also been serialized in wide-circulation newspapers around the time of Alva's award of the Nobel Prize.

Their year-long sojourn in the United States was crucial to their collaboration as well as to their future individual careers. The intense experience of traveling together in a foreign, albeit friendly, society helped clinch their techniques of collaboration. Their continual exposure to American society provided both husband and wife with a more critical view of their own Swedish society, thus laying the ground for the combined radical voice of *Crisis in the Population Question*.

A second, longer trip to the United States, between 1938 and 1943, resulted in *An American Dilemma*. This time all three children accompanied them. The existence of a large team of research assistants including a capable partner in difficult data collection, Ralph Bunche, supplied by the Carnegie Foundation, which underwrote the entire project including the recruitment of Gunnar Myrdal as a non-American and hence more objective social-scientist project director; the additional family responsibilities; and Alva's preoccupation with her own book *Nation and Family* (1941) kept both from becoming so deeply involved in a collaboration as before. Though both books were not coauthored, the pattern of intellectual sharing continued.

This sharing was evident in references to each other's work, but of special interest is the debt to Alva of *The American Dilemma*'s key insight, namely, that "the Negro problem" was not so much a problem inherent in African Americans, who were blamed as victims, but a problem ingrained in American society's practice of racism. That insight, which led to this book's being cited in the landmark Supreme Court decision on desegregation in education, *Brown* v. *Board of Education of Topeka* (May 1954), strongly paralleled Alva's discussion of the "woman's problem" in *Nation and Family*. She had argued there, while developing an earlier argument from Harriet Martineau, that the problem did not reside in women themselves but in a patriarchal society that construed women as inferior to men.[16]

The analogy between race and gender as key categories in an inequitable social order that blamed its problems on its victims was to be articulated in public policy in the decades after the publication of *An American Dilemma*. By then, interestingly, the order of influence was reversed: it was the civil rights movement of the 1950s that inspired the women's liberation movement of the 1960s. But it is impossible to imagine either without the pioneering contributions of the Myrdals in the 1930s and 1940s.

Margaret Mead and Gregory Bateson: A Collaborative Couple in Polynesia and America

Margaret Mead (1901–1978) and Gregory Bateson (1904–1980) collaborated chiefly in the 1930s, from their initial meeting in 1932 until the outbreak of World War II in 1939, when their different nationalities took them to separate places and tasks in the war effort. During a decade or so of collaborative interaction, their work included Mead's *Sex and Temperament in Three Primitive Societies* (1935), Bateson's *Naven*, and their jointly authored book, *The Balinese Character* (1942).[17]

In the post–World War II period and especially in the 1950s, Mead became America's most visible anthropologist and a veritable cultural guru, always willing to explore the relevance of findings from fieldwork with small-scale Pacific societies to urgent problems of American and other Western societies, whether those problems were in wartime Britain, the postwar United States, or the newly independent state of Israel absorbing immigrants from diverse cultural backgrounds, to name just a handful of places where she served as a consultant on social policy. Her pronouncements on topics ranging from the cultural determination of sexual roles and personality to childrearing, nutritional habits, decolonization, and race and ethnic relations had enormous influence among social scientists concerned with applications of their research, within the circle of professional anthropologists, and indeed within American society at large.

Bateson, too, became in his own way an influential interdisciplinary social scientist studying patterns of communication in humans and animals, as well as a leading figure in the counterculture of the late 1960s, whose ideas touched the lives of many individuals, especially students. Although after their divorce in 1950 Mead and Bateson followed diverging disciplinary paths, paths that brought them to their distinct and separate forms of fame, the period of their collaboration in the 1930s, though relatively short, still remains crucial for understanding their emergence later as celebrities of social thought.

Mead and Bateson had the remarkable opportunity during the 1930s of studying small-scale societies, virtually untouched by Western civilization, simultaneously from the viewpoints of the two different schools of anthropology they came from: Mead's American cultural anthropology and Bateson's British social anthropology. Their two books on New Guinea tribes reflect both the different intellectual traditions *and* the continuing stimulation of endless discussions together.

Their collaboration has to be seen, though, as far more complex than an intellectual, cross-cultural (Mead had been presumably fascinated by Bateson's background in the British intellectual aristocracy) admixed with sexual give-and-take between two extraordinarily gifted observers of other and their own cultures. When Mead and Bateson met in 1932, Mead was doing fieldwork in New Guinea with her second husband, the New Zealander anthropologist Reo Fortune. The three of them argued anthropology for the good part of a year, in the claustrophobic conditions of a small tent, swathed in mosquito netting, while undergoing occasional malaria bouts.

To complicate these relationships still further, a fourth voice was very much present in the trio's universe: Ruth Benedict's. Benedict had been Mead's mentor and lover, and a copy of Benedict's *Patterns of Culture* (1934) reached Mead, Fortune, and Bateson in the field at the height of their own fieldwork.[18] Stimulated by the dramatic circumstances of the intense quadruple interaction, the three anthropologists emerged from the jungle with three very different manuscripts in hand. Mead also emerged with a new, third husband as she switched her interest from Fortune to Bateson; both Mead and Bateson later described to their respective mothers their subsequent marriage as being in the service of anthropology, as both planned to do extensive fieldwork in Bali. Marriage to another anthropologist was not only highly congruent with such a goal, but, as it turned out, it facilitated the innovative approach of photographic ethnography.

Mead's and Bateson's joint book, *The Balinese Character*, pioneered the innovation of photographic anthropology, an innovation made possible by their joint fieldwork, or a form of utmost professional intimacy in anthropology, for they "shared" the same society. Indeed, their projected collaboration provided for both the very justification for their marriage, as well as the inkling that the marriage might not last beyond the fieldwork. Yet, the conditions of the collaboration also pooled their respective complementary skills, with Margaret excelling in interventionist or active, extensive note taking and holistic interpretation of the observable details and Gregory focusing on more passive recording of a multitude of fragmented sociocultural aspects (such as the hands of Balinese men watching a cockfight).

The emphasis on visualisation was a new, neutral ground on which they could meet and share new insights without having to part with one's own ingrained professional traditions. It was also an innovative medium that enabled each to express one's own strength in interacting with people, or with (photographic) instruments, in a complementary, noncompetitive manner, while capitalizing on Mead's strong sense of professional mission and Bateson's looser view of his career.

The question thus persists as to the precise professional and personal conditions that sustained their collaboration in the 1930s but dissolved it in the late 1940s. Companionship in fieldwork was their joint rationale for getting married, or, as they put it, their marriage was in the service of anthropology. Away from their respective societies, they shared the common predicament of combating the isolation and danger to one's sense of identity that fieldwork in another society brings. They were also free from gender roles obtaining in their own societies. In contrast, upon returning to Mead's America or visiting Bateson's England, the delicate balance as to whose cultural code, and hence persona, was in charge was disturbed to the effect that the relationship that they had established in the context of noncompetitive fieldwork could no longer obtain.

The issue of Margaret being already a celebrity by the time she met Bateson, following the reception of her work on Samoa in the late 1920s,[19] may also have created alienation between them in the post–World War II era, the last era of unchallenged male chauvinism in American society. At a time when the feminine ideal was pursued by most women in suburban, mythical versions of domesticity, Mead became one of the few women to enjoy celebrity status as a scholar turned cultural icon. The disparity between their careers was further evident in Bateson's subsequent marriages to younger women without professional aspirations, while pursuing an erratic career occasionally rescued by Mead's letters of recommendation.

Perhaps of even greater public interest is the Mead-Batesons' collaboration in the innovative rearing of their only child, Mary Catherine Bateson, born in 1939, whose childhood was the most documented one in America, with both anthropologist parents extensively taking notes and photographing her "non-Balinese character." Mead, in particular, applied to her own daughter childrearing insights she had gained from other cultures, such as (breast-) feeding upon demand rather than on schedule or picking up crying children, insights that became widely diffused once her pediatrician, Dr. Benjamin Spock, codified them as a centerpiece of postwar American childrearing practices. Mead's own elaborate solutions to the problem of childcare, while building extended families with various friends and colleagues, anticipated the now mainstream solutions to childrearing by professional single parents.

The Mead-Bateson brief but intense collaboration raises the questions of disciplinary determinants such as fieldwork, cross-cultural marriages, war-induced separation, professional responsibilities and markets in sustaining or undermining certain collaborative patterns, and hence the innovative work that is predicated upon such collaborative arrangements. Mead's and Bateson's pioneering experience with collaborative intimate relationships remains an ongoing source of lessons for those of us whose professional and personal lives have been so dramatically transformed by the shift to greater cultural flexibility in performing gender roles since the 1970s.

Conclusions: The Synergy of Creativity and Intimacy in the Social Policy Arena—Why So Fleeting?

In parallel with the above-mentioned diversity of topical coverage for comparative purposes, inherent in the ensemble of these three collaborative couples, the Russells, the Myrdals, and the Mead-Batesons, a number of common features become noticeable as a plausible indication of some deeper structures embedded in the social context of interlocking creativity and intimacy by collaborative couples in the domain of social policy, shaped as it was by the historical context of the interwar and immediate postwar period.

First, all these couples enjoyed relatively early in life, especially in the historical context of the interwar period, *a certain degree of celebrity status, often mixed with an equal degree of notoriety*, eventually acquiring the permanence of social icons. For example, following the controversial reception of the Myrdals' jointly authored book *Crisis in the Population Question* (1934) and the tremendous impact it had on Swedish society and politics, their name came to be associated with some of the meanings of their work or the circumstances of its production; "Myrdal couches" came to signify intimacy and love, "Myrdal buildings" came to connote buildings inhabited by families that chose to have many children, while "Myrdal cactis" meant objects with many offshoots.

If in the case of the Myrdals both partners or rather the couple became the object of social iconization, in the other two cases the social iconization applied to the more famous partner, Bertrand Russell and Margaret Mead, respectively. Some spillover occurred vis-à-vis the less famous partner, as when Dora Black Russell was requested to sign her books as Mrs. Bertrand Russell, or when cultural guru Margaret Mead wrote letters of recommendation for her former, job-insecure (third) husband, Gregory Bateson.

The aura of an early celebrity status, whether jointly shared or chiefly bestowed upon one partner, not only increased these couples' social impact but also accelerated the transformation of their private lives into public ones, with major consequences for the couples' lives, and for those who looked upon them as role models. Therefore, one lesson from these historical case studies pertains to the historical context in which creative couples become celebrities and to the influence of celebrity status on propagating or defusing their activities in the domain of social policy. The interwar period in general (the 1930s in particular) thus emerges as a historical period conductive to the rise of cross-gender collaboration in the arena of social policy.[20]

Yet, another remarkable common feature of these otherwise diverse collaborative couples pertains to *the centrality of socialist, communal, radical political philosophies in propelling the couple toward seeking a greater measure of gender equality*, companionship, and partnership in their own lives than that prevailing in (bourgeois) patriarchal society at large. In their turn, the private lives these couples sought to create for themselves served as a resource in inspiring social policies such as those of the welfare state in the case of the Myrdals, or those of the birth-control movement and the Labour Party in the case of the Russells, or those of multiculturalism in the case of Mead and Bateson. Political radicalism both underlay and reinforced the activities of collaborative couples, as a social unit defying patriarchy by conferring upon the female partner a measure of apparent, if not always real, power.

Indeed, a related form of radicalism, vis-à-vis the dominant patriarchal regime, shared by these couples pertained to their *avant-garde conduct in sexual matters*, thus suggesting that the resonance of social, political, and sexual radicalism reinforced the challenge that these couples mounted against the patriarchal social order in the aftermath of the great watershed of World War I. The Russells, as befit Bertrand Russell's earlier membership in the Edwardian Bloomsbury group of culturally hegemonic writers, artists, and intellectuals, preached and practiced open marriage. They even displayed the same Bloomsbury-inspired tolerance of recognizing children conceived with other mates, though eventually this tolerance extended only to the first such child. The precariousness of alternative sexual patterns, or patterns other than those prevailing within the institution of marriage, surfaced again and again, with all but Mead reverting to more or less traditional patterns of sexual, albeit often sequential, exclusivity.

The sexual radicalism of the Myrdals pertained to their advocacy of family planning and extension of welfare benefits to single parents but also to their private lives when a period of intense companionship in the first twenty-five years of their married life was followed by a commuting marriage or by extended periods of actual geographical and emotional separation. In the case of Mead and Bateson, sexual radicalism involved their triangle with her second husband during joint fieldwork in Polynesia, her homosexual relationship with mentor Ruth Benedict, and his expecting a child from a future wife prior to his separation from Margaret.

Another related arena of social radicalism pertained to these collaborative couples' *innovative approaches to parenting, especially mothering and the integration of one's own child(ren) into research projects.* Thus, the Russells began a progressive school in 1927 not only in order to implement their different educational philosophy but also in order to provide a different rearing environment for their own children. The Myrdals' children were widely viewed as exemplifying their social outlook in their 1934 book on the population crisis, with the birth of a second child in 1934 being announced as an ironic commentary on the controversy surrounding the parents' book, including the scolding of Alva for lecture touring for the book at a time she should have presumably focused her total energies on their newborn. Mead and Bateson's preoccupation with both innovative parenting practices and with the role of children and of their education in reshaping society by peaceful means further suggests that collaborative couples possessed yet another edge for effectiveness in social policy, namely insights into social reproduction as a vehicle for social change.

In this connection, it is perhaps of special interest to mention that each of these three couples active in social policy received one or more book-length biographical treatments by one or more of their offspring, with the effect that their own parenting theories and practices are judged by the products of those activities. The existence of these biographical narratives by the children of all these couples further encourages an analysis of their parental innovations and educational theories and practices, an analysis that can then be juxtaposed with the reflections of the very products of those practices. No comparative social analyst could hope for more enticing source material than such a brilliant ensemble of biographer-offspring.

Last but not least, a key feature of utmost importance displayed by these couples is the fact that *collaborative couples serve as a vehicle for expanding the options for transdisciplinary innovation, a form of innovation typical of twentieth-century science,*[21] in the domain of social policy. For example, the collaborative venture of Dora and Bertrand Russell in establishing a progressive school is an instance of combining educational philosophy with pedagogical practice, a combination made possible only by the pooling of their complementary resources as a couple, namely his background in social and educational philosophy and her capacity to implement those ideals in the form of managing a real school.

In a similar manner, the Myrdals' seminal work on the population crisis, which contains the main ideas later to be incorporated into welfare-state social programs, reflects a new transdisciplinarity resulting from integrating his expertise in economic theory and macroeconomics with hers in sociology, social work, pedagogy, and psychology. In a similar manner, the analogy between race and gender developed in *The American Dilemma* as a powerful explanatory device built on Alva's better acquaintance with the gender issues and Gunnar's intensive study of race relations.

In contrast, Mead's and Bateson's collaborative work did not synthesize American cultural with British social anthropology as one might naively have expected, but it did stimulate innovative joint work on the pioneering use of photography in anthropological fieldwork. Though their pioneering of photography may be viewed as a "merely" technical innovation, it was made possible by Mead's and Bateson's willingness to experiment together, by their marriage-induced copresence in the field, and by their search for a form of joint work that could transcend their unbridgeable respective training via the "neutral" vehicle of the visual image. The professional and political consequences of this "merely" technical innovation were tremendous for post–World War II anthropology.

However, the issue of how collaborative relationships in general and a relationship as a collaborative couple in particular can induce new forms of scientific and social transdisciplinary authority, as preconditions for new forms of knowledge, both basic and applied, through their unique capacity to convert intimacy into creativity and vice versa, as two of the most basic needs of human nature, remains to be explored further for other historical and cultural contexts.

The historical and cultural specificity of these three collaborative couples exerted such a unique impact on social policy and hence on social change at the multiple, interlocking levels of educational institutions, national social and political discourse, international multiculturalism, and moral world order that only further studies of other collaborative couples can render a wider validity to the interpretations offered here.

PNINA G. ABIR-AM, HELENA M. PYCIOR,
AND NANCY G. SLACK

Appendix

Additional Collaborative Couples and Other Cross-Gender Collaborators

In the course of research for this book, the editors and contributors discovered many other cross-gender collaborators in the sciences. This appendix, the bulk of which provides the names of more couples who are noted for their innovative scientific work, may serve as a starting point for further research on issues of cross-gender collaboration.

The first section lists only married couples (alphabetized by the woman's last name). This list suggests some of the more obvious couples who deserve scholarly attention in the future, and many other couples would certainly qualify for inclusion. These forty-five couples represent many countries and a great variety of scientific disciplines. All the women and men were born before 1940; many are still living and continue to do noteworthy work in science.

The second section lists women and men who were *not* married to one another but collaborated on research that contributed to a Nobel Prize for at least one of them. In four of these six pairs of collaborators, the Nobel Prize was awarded only to the male collaborator; in the other two, both partners shared the prize.

Selected bibliographical sources follow the two lists. In preparing this appendix, the editors acknowledge the help of Bonnie Brady and Susan Pae.

Collaborative Couples

Name	Discipline	Country	Dates	Source
Bailey, Florence M.	zoology	US	1863–1948	AMS
Bailey, Vernon	zoology	US	1864–1942	AMS
Donnay, Gabrielle	X-ray crystallography	US, Canada	1920–1992	AMWS
Donnay, Jose	X-ray crystallography	Belgium, US, Canada	1902–1987	AMWS
Ehrenfest, Tatiana A.	physics	Russia, Holland	1876–1964	
Ehrenfest, Paul	physics	Holland, Russia	1880–1933	AAE
Eigenmann, Rosa S.	zoology	US	1858–1947	Oglv
Eigenmann, Carl H.	zoology	US	1863–1927	NASBM
Farquhar, Marilyn	cell biology	US	1928–	AMWS
Palade, George	cell biology (Nobelist, 1974)	US	1912–	AMWS
Gaige, Helen T.	zoology	US	1886–1976	AMS
Gaige, Frederick	zoology	US	1890–1976	AMS
Geiringer von Mises, Hilda	mathematics	Austria, Germany, US	1893–1973	WM
von Mises, Richard M.E.	mathematics	Austria, Germany, US	1883–1953	WM
Goeppert-Mayer, Maria	physics (Nobelist, 1963)	Germany, US	1906–1972	NASBM
Mayer, Joseph	chemistry	US	1904–	AMWS
Goldhaber, Gertrude	particle physics	US	1911–	AMWS
Goldhaber, Maurice	particle physics	US	1911–	AMWS
Harvey, Ethel Browne	embryology	US	1885–1965	NAWM
Harvey, Edmund N.	embryology	US	1887–1959	NASBM
Herzberg, Luise	spectroscopy	Germany, Canada	1906–1971	
Herzberg, Gerhard	spectroscopy (Nobelist, 1971)	Germany, Canada	1904–	AAE
Herzenberg, Leonore	immunology	US	1931–	
Herzenberg, Leonard	immunology	US	1931–	AMWS
Hollingworth, Leta	psychology	US	1886–1939	AMS
Hollingworth, Harry	psychology	US	1880–1960	AMS

Hubbard, Ruth	biochemistry	US	1924–	AMWS
Wald, George	biochemistry (Nobelist, 1967)	US	1906–	AMWS
Jeffreys, Bertha	mathematical physics	UK	1903–	WWBS
Jeffreys, Harold	mathematical physics	UK	1891–1989	BMFRS
Karle, Isabella	X-ray crystallography	US	1921–	AMWS
Karle, Jerome	X-ray crystallography (Nobelist, 1985)	US	1918–	AMWS
Klein, Eva	immunology	Sweden	1925–	WWSE
Klein, George	immunology	Sweden	1925–	WWSE
Polubarinova-Kochina, Pelageya Y.	mathematics	Soviet Union	1899–	WM
Kochin, N.E.	mathematics	Soviet Union	1901–1944	WM
Koprowska, Irena	cancer research	Poland, Brazil, US	1917–	AMWS
Koprowski, Hilary	virology	Poland, Brazil, US	1916–	AMWS
Koshland, Marian	immunology	US	1921–	AMWS
Koshland, Daniel	biochemistry	US	1920–	AMWS
Leakey, Mary Douglas	anthropology	UK	1913–	AAE
Leakey, Louis S.	anthropology	UK	1903–1972	AAE
Lederberg, Esther	microbial genetics	US	1922–	AMWS
Lederberg, Joshua	microbial genetics (Nobelist, 1958)	US	1925–	AMWS
Lewis, Margaret R.	tissue culture	US	1881–1970	AMS
Lewis, Warren	tissue culture	US	1870–1964	AMS
Lwoff, Marguerite	microbiology	France	1905–1979	PIA
Lwoff, André	microbiology, zoology (Nobelist, 1965)	France	1902–1994	IFADS
Lynd, Helen Merrill	sociology	US	1896–1982	DABS
Lynd, Robert S.	sociology	US	1892–1970	DABS
Macintyre, Sheila S.	mathematics	Scotland, US	1910–1960	WM
Macintyre, Archibald J.	mathematics	Scotland, US	1908–1967	WM

Name	Field	Country	Dates	Source
Metchnikoff, Olga	immunology	Russia, France	1862–1944	PIA
Metchnikoff, Elie	immunology (Nobelist, 1908)	Russia, France	1845–1916	AAE
Migeon, Barbara	embryology	US	1931–	AMWS
Migeon, Claude	embryology	US	1923–	AMWS
Morgan, Lilian V.	zoology	US	1870–1952	Rstr
Morgan, Thomas H.	zoology (Nobelist, 1933)	US	1866–1945	NASBM
Needham, Dorothy	biochemistry	UK	1896–1987	WWoBS
Needham, Joseph	biochemistry	UK	1900–1995	WWoBS
Neumann, Hanna	mathematics	Germany, UK, Australia	1914–1971	WM
Neumann, Bernhard H.	mathematics	Germany, UK, Australia	1909–	WM
Onslow, Muriel Wheldale	physiological genetics	UK	1880–1932	Fruton
Onslow, Huia	physiological genetics	UK	1890–1922	Fruton
Parkes, Ruth D.	physiology	UK	1903–	
Parkes, Sir Alan	physiology	UK	1900–	WWoBS
Payne-Gaposchkin, Cecilia	astronomy	US	1900–1979	UCIL
Gaposchkin, Sergei	astronomy	Russia, Germany, US	1898–1984	UCIL
Pirie, Antoinette	biochemistry	UK	1905–	Fruton
Pirie, Norman W.	biochemistry	UK	1907–	WWoBS
Rakic, Patricia G.	neurobiology	US	1935–	
Rakic, Pasco	developmental biology	US	1933–	AMWS
Robinson, Julia Bowman	mathematics	US	1919–1985	WM
Robinson, Raphael M.	mathematics	US	1911–	WM
Schrader, Sally Hughes	zoology	US	1895–	Rstr
Schrader, Franz	zoology	US	1891–1962	Rstr
Stadtman, Thressa	biochemistry	US	1920–	AMWS
Stadtman, Earl R.	biochemistry	US	1919–	AMWS
Stanier, Germaine	molecular biology	France, US, France	1920–	PIA

Stanier, Roger	microbiology	Canada, US, France	1916–1982	BMFRS
Sussman, Raquel R.	molecular biology	Chile, US	1921–	AMWS
Sussman, Maurice	cell biology	US	1922–	AMWS
Tréfouel, Thérèse	pharmacology	France	1894–1967	PIA
Tréfouel, Jacques	pharmacology	France	1897–1977	SCFA
Vogt, Cécile	neurobiology	France, Germany	1875–1962	FN
Vogt, Oskar	neurobiology	Germany	1870–1959	WWWS
Weiss, Mary Catherine B.	mathematics	US	1930–1966	WM
Weiss, Guido L.	mathematics	Italy, US	1928–	WM
Wollman, Elisabeth	microbiology	Belgium, France	1890–1944	PIA
Wollman, Eugene	microbiology	Belgium, France	1890–1944	PIA

Cross-Gender Collaborations That Contributed to Nobel Prizes

Name	Discipline	Country	Dates	Source	Nobel Prize Year
Chase, Martha	microbiology	US	1925–		—
Hershey, Alfred D.	microbiology	US	1908–	ABET	1969
Elion, Gertrude B.	biochemistry	US	1918–	AMWS	1988
Hitchings, George H.	biochemistry	US	1905–	AMWS	1988
Franklin, Rosalind	molecular biology	UK	1920–1958	ABET	—
Klug, Aaron	molecular biology	UK	1926–	WWoBS	1982
Levi-Montalcini, Rita	embryology	Italy, US	1909–	AMWS	1986
Cohen, Stanley	biochemistry	US	1922–	AMWS	1986
Meitner, Lise	nuclear physics	Austria, Germany	1878–1968	ABET	—
Hahn, Otto	nuclear physics	Germany	1879–1968	ABET	1944
Vogt, Marguerite	animal virology	US	1910–		—
Dulbecco, Renato	animal virology	Italy, US, UK	1914–	AMWS	1975

Selected Bibliographical Sources

AAE   *Academic American Encyclopedia.* Danbury, Conn., 1992 edn.

ABET   Asimov, Isaac. *Asimov's Biographical Encyclopedia of Science and Technology.* New York, 1982.

AMS   *American Men of Science.* New York, 1929.

AMWS   *American Men & Women of Science.* New York, 1989.

BES   *Biographical Encyclopedia of Scientists.* New York, 1981.

BMFRS   *Biographical Memoirs of the Fellows of the Royal Society.* London, 1955–1982.

DABS   Garrity, John, and Mark Carnes, eds. *Dictionary of American Biography: Supplement 3, 4, 5, 8.* New York, 1966–1970.

FN   Haymaker, Webb, and Francis Schiller, eds. *The Founders of Neurology.* 2d ed. Springfield, Ill., 1970.

Fruton Fruton, Joseph S. *A Bio-Bibliography for the History of the Biochemical Sciences since 1800.* Philadelphia, 1982.

IFADS   *Institut de France Académie des Sciences Annuaire pour 1992.* Paris, 1992.

NASBM   *National Academy of Science Biographical Memoirs.* New York, 1959.

NAWM   *Notable American Women: The Modern Period.* Cambridge, Mass., 1980.

Oglv   Ogilvie, Marilyn Bailey. *Women in Science, Antiquity through the Nineteenth Century: A Biographical Dictionary with Annotated Bibliography.* Cambridge, Mass., 1986.

PIA   Pasteur Institute Archives, Paris.

Rstr   Rossiter, Margaret W. *Women Scientists in America: Struggles and Strategies to 1940.* Baltimore, 1982.

SCFA   *Sociéte Chimique Française Annuaire.* Paris, 1975–1976.

UCIL   Abir-Am, Pnina, and Dorinda Outram, eds. *Uneasy Careers and Intimate Lives: Women in Science, 1789–1979.* New Brunswick, N.J., 1987.

WM   Grinstein, Louise S., and Paul J. Campbell, eds. *Women of Mathematics: A Biobibliographic Sourcebook.* Westport, Conn., 1987.

WWBS   *Who's Who in British Science.* London, 1953.

WWoBS   *Who's Who of British Scientists.* Columbus, Ohio, 1971.

WWSE   *Who's Who in Science in Europe.* London, 1989.

WWWS   Debus, Allen G., ed. *World Who's Who in Science.* Hannibal, Mo., 1968.

Notes

Series Foreword

1. Notable examples of this effort include Margaret W. Rossiter, *Women Scientists in America, Strategies and Struggles to 1940* (Baltimore, 1982); Rossiter, *Women Scientists in America: The Limits of Opportunity, 1940–1972* (Baltimore, 1995); and Pnina Abir-Am and Dorinda Outram, ed., *Uneasy Careers and Intimate Lives: Women in Science, 1789–1979* (New Brunswick, N.J., 1987). See also Billie Melman, "Gender, History, and Memory: The Invention of Women's Past in the 19th and 20th Centuries," *History and Memory* 4 (1993): 5–41.

2. See in particular Renée Riese Hubert's penetrating analysis in *Magnifying Mirrors: Women, Surrealism, and Partnership* (Lincoln, 1994). See also the many references in note 14 to the Introduction of this volume. For this collection's relevance to couples in politics, see my comments quoted in Elizabeth Cody, "The Buck Stops with Him . . . and Her" (The sidebar reads: "Bill and Hillary Clinton are breaking down the old taboos against couples working in tandem, high-powered jobs. The country doesn't know what to make of it."), *Atlanta Constitution*, February 21, 1993, F-1, F-3; see also Mary Matalin and James Carville, *All's Fair: Love, War, and Running for President* (New York, 1994); Doris Kearns Goodwin, *No Ordinary Time: Franklin and Eleanor Roosevelt—The Home Front in World War Two* (New York, 1994).

3. The Einsteins are a particularly telling and well-documented example of this asymmetry in a scientific couple; see John Stachel's piece in this collection and *Albert Einstein and Mileva Maric: Lettres d'amour et de science* (Paris, 1993).

4. For a critique of the myth of Marie Curie the chemist and Pierre the physicist, see Helena M. Pycior, "Reaping the Benefits of Collaboration While Avoiding Its Pitfalls: Marie Curie's Rise to Scientific Prominence," *Social Studies of Science* 23 (1993): 301–323.

5. Added in proof: For the widest and most up-to-date perspective on these topics, see Sandra Harding and Elizabeth McGregor, "The Gender Dimension of Science and Technology." *World Science Report* (Paris: UNESCO, 1995), 1–47; see also Pnina G. Abir-Am, "Women in Science: A Historical Overview," ibid., 48–56.

Introduction

1. On the exponential growth in scientific collaboration during the first half of the twentieth century to the point where 50% of all abstracted scientific research in biology, chemistry, and physics was collaborative by 1950 (and 60% by 1960) and on limitations to women scientists' access to collaborators, see Mary Frank Fox, "Gender, Environmental Milieu, and Productivity in Science," in *The Outer Circle: Women in the Scientific Community*, ed. Harriet Zuckerman, Jonathan R. Cole, and John T. Bruer (New York, 1991), 197–203.

2. The study of research schools provides a good example of the fruitful broadening of history of science beyond heroic scientists to other units of analysis. If J. B. Morrell's article, "The Chemist Breeders: The Research Schools of Liebig and Thomson" (*Ambix* 19 [1972]: 1–46), pioneered the study of research schools, a recent volume of *Osiris* attests to an increasing acceptance of the research school as a unit of analysis and an appreciation of the "new directions in the historiography of science" that it opens (Gerald L. Geison and Frederic L. Holmes, eds., *Research Schools: Historical Reappraisals*, published as *Osiris*, 2d ser., 8 [1993]: 238 [quote]). See also Pnina G. Abir-Am, "Women in Research Schools: Approaching an Analytical Lacuna in the History of Chemistry and Allied Sciences," in *Chemical Sciences in the Modern World*, ed. Seymour H. Mauskopf (Philadelphia, 1993), 375–391.

3. On early "family firms," see Ann B. Shteir, "Botany in the Breakfast Room: Women and Early Nineteenth-Century British Plant Study," in *Uneasy Careers and Intimate Lives: Women in Science, 1789–1979*, ed. Pnina G. Abir-Am and Dorinda Outram (New Brunswick, N.J., 1987), 31–43, 34 (term "family firm"). See also Dorinda Outram, "Before Objectivity: Wives, Patronage, and Cultural Reproduction in Early Nineteenth-Century French Science," ibid., 19–30, and Sally Gregory Kohlstedt, "Parlors, Primers, and Public Schooling: Education for Science in Nineteenth-Century America," *Isis* 81 (1990): 424–445.

4. Ruth Hubbard, Isabella Karle, and Mari Krogh, for example, were coinvestigators with their Nobel laureate husbands. On this point, see Patricia Farnes, "Women in Medical Science," in *Women of Science: Righting the Record*, ed. G. Kass-Simon and Patricia Farnes (Bloomington and Indianapolis, Ind., 1990), 288–289.

5. Marilyn Bailey Ogilvie, "Marital Collaboration: An Approach to Science," in *Uneasy Careers and Intimate Lives*, ed. Abir-Am and Outram, 104.

6. In the 1920s and 1930s there were twenty-seven new women "stars" in the *AMS*, eleven of whom (40.7%) were married; of the eleven, nine (81.8%) were married to scientists. See Margaret W. Rossiter, *Women Scientists in America: Struggles and Strategies to 1940* (Baltimore, 1982), 143, 292.

7. Ibid., 393.

8. See, e.g., chapters 5 (by Nancy G. Slack), 6 (Marilyn Bailey Ogilvie), 10 (Helena M. Pycior), 11 (Peggy A. Kidwell), and 12 (Pnina G. Abir-Am) of *Uneasy Careers and Intimate Lives*, ed. Abir-Am and Outram. Earlier, Ivor Grattan-Guinness had discussed Grace Chisholm Young and William Henry Young, a mathematical couple who because of their consciously intertwined personal and professional lives (documented in their rich surviving letters) perhaps defied reduction to the then-current mold of solitary genius (I. Grattan-Guinness, "A Mathematical Union: William Henry and Grace Chisholm Young," *Annals of Science* 29 [1972]: 105–186).

9. Marianne Gosztonyi Ainley, "Last in the Field? Canadian Women Natural Scientists, 1815–1965," in *Despite the Odds: Essays on Canadian Women and Science*, ed. Marianne Gosztonyi Ainley (Montreal, 1990), 55–58; Joan Pinner Scott, "Disadvantagement of Women by the Ordinary Processes of Science: The Case of Informal Collaborations," ibid., 316–328. A "two-person single career" refers to one career (traditionally a hus-

band's) supported by the activities of two persons (husband and wife), whereas a "dual-career marriage" indicates that both spouses are pursuing careers.

10. Margaret W. Rossiter, "The ~~Matthew~~ Matilda Effect in Science," *Social Studies of Science* 23 (1993): 325–341; Helena M. Pycior, "Reaping the Benefits of Collaboration While Avoiding Its Pitfalls: Marie Curie's Rise to Scientific Prominence," *Social Studies of Science* 23 (1993): 301–323.

11. Phyllis Rose, *Parallel Lives: Five Victorian Marriages* (New York, 1983).

12. Ruth Perry, "Introduction," in *Mothering the Mind: Twelve Studies of Writers and Their Silent Partners,* ed. Ruth Perry and Martine Watson Brownley (New York, 1984), 5 (quote).

13. Mary Catherine Bateson, *Composing a Life* (New York: Grove/Atlantic, Inc., 1989), 78, 9, 13.

14. Whitney Chadwick and Isabelle de Courtivron, eds., *Significant Others: Creativity & Intimate Partnership* (London, 1993), 12, 11. Chadwick and de Courtivron provide a list of additional works informed by the perspective of intimate, creative connectedness (8). In 1994, moreover, there appeared a major study of artistic couples (Renée Riese Hubert, *Magnifying Mirrors: Women, Surrealism, & Partnership* [Lincoln, Neb., 1994]) and a plethora of books on literary couples, including Kate Fullbrook and Edward Fullbrook, *Simone de Beauvoir and Jean-Paul Sartre: The Remaking of a Twentieth-Century Legend* (New York, 1994); Janet Malcolm, *The Silent Woman: Sylvia Plath and Ted Hughes* (New York, 1994); Diana Trilling, *The Marriage of Diana and Lionel Trilling* (New York, 1994); Francine du Plessix Gray, *Rage and Fire: A Life of Louise Colet, Pioneer Feminist, Literary Star, Flaubert's Muse* (New York, 1994); and William L. Shirer, *Love and Hatred: The Troubled Marriage of Leo and Sonya Tolstoy* (New York, 1994). See also Helena Lewis, "Elsa Triolet: The Politics of a Committed Writer," *Women's Studies International Forum* 9 (1986): 385–394, and idem, "Louis Aragon (1897–1982)," in *European Writers: The Twentieth Century,* ed. George Stade, vol. 11 (New York, 1990), 2061–2087. On the history of biographies about women—including dual biographies of spouses and biographies of women in groups—see Linda Wagner-Martin, *Telling Women's Lives: The New Biography* (New Brunswick, N.J., 1994).

15. In a pioneering address to the History of Science Society, Bert Hansen recently offered a "few sketches of lesbian and gay scientists." Stressing that homosexuality can be "a powerful *social* factor in a scientist's life," he called upon scientific biographers to examine the sexual orientation of their subjects. He remarked, moreover, that it was "tempting to argue that [the anthropologist Alice] Fletcher's outsider status and her relationship with Jane Gay shaped her ethnographic research" (Bert Hansen, "One Hundred Years of Homosexuality in the History of Science" [paper presented at the History of Science Society Meeting, New Orleans, October 13, 1994]). See also H. Patricia Hynes, "Toward a Laboratory of One's Own: Lesbians in Science," in *Lesbian Studies: Present and Future,* ed. Margaret Cruikshank (Old Westbury, N.Y., 1982), 174–178.

16. Middle-class and professional backgrounds were typical of scientific couples (especially wives) through the early twentieth century, when higher education was still largely reserved for the upper and middle classes.

17. On the private vs. public spheres, see Linda K. Kerber, "Separate Spheres, Female Worlds, Woman's Place: The Rhetoric of Women's History," *Journal of American History* 75 (1988): 9–39.

18. Hereafter, in offering particular individuals or couples to illustrate general themes, we will dispense with "e.g." and assume that the reader understands that lists are not always comprehensive. Wherever possible, last names alone will be given.

19. On the "control" that a surviving husband can exercise over his wife's life story, see Wagner-Martin, *Telling Women's Lives,* 117–123.

20. A basic definition of "collaborate" is "work jointly (*with*), esp. on a literary or

scientific project" (*The New Shorter Oxford English Dictionary on Historical Principles*, ed. Lesley Brown [Oxford, 1993], 1, 438).

21. In 1980 the sociologist Martha Fowlkes elaborated on the three different "needs" of the contemporary academic man that a wife helped to fulfill: "the need to do the actual work of his profession in a way that will establish him in a favorable competitive position," "the need for a sustained achievement drive," and "the absolutely crucial need for time in which to accomplish the work at hand" (Martha R. Fowlkes, *Behind Every Successful Man: Wives of Medicine and Academe* [New York, 1980], 7–8). We maintain here that in a creative couple these needs are dual, and the most successful of the couples work on reciprocal fulfillment of the needs.

22. Rosalynd Pflaum, *Grand Obsession: Madame Curie and Her World* (New York, 1989), 246.

23. The analysis of each cluster of essays will consider the couples discussed specifically in that cluster as well as couples fitting a similar typology but discussed in the three comparative essays (chapters 15–17) of the collection.

24. As Perry has pointed out, compensation for a partner's deficiencies is sometimes more complicated than first appearances suggest. Elizabeth Barrett and Robert Browning "exaggerated Elizabeth Barrett's weakness because Robert Browning took better care of himself when he was playing caretaker to her. He arranged space for writing, addressed himself to real problems, and set himself to work" (Perry, "Introduction," in *Mothering the Mind*, ed. Perry and Brownley, 17).

25. Compare with Mary Catherine Bateson's remark that "divorce often represents progress rather than failure" in the designing of a life (Bateson, *Composing a Life*, 8).

26. Similarly, André Malraux ignored Clara, once they separated (Isabelle de Courtivron, "Of First Wives and Solitary Heroes: Clara & André Malraux," in *Significant Others*, ed. Chadwick and de Courtivron, 61–62). In contrast, in many of the successful marriages, the surviving spouse (Frieda Cobb Blanchard, Carl Cori, and Marie Curie) carried on the couple's joint work or memorialized the deceased partner through biographies, collected works, and so on.

27. Ludmilla Jordanova asks historians to aim at "subtle, sympathetic understanding" rather than "a big picture." For her distinction as well as a discussion of gender as "an analytical category" in history of science, see Ludmilla Jordanova, "Gender and the Historiography of Science," *British Journal for the History of Science* 26 (1993): 469–483, 483, 474. Major studies pursuing gender in history of science include Donna Haraway, *Primate Visions: Gender, Race, and Nature in the World of Modern Science* (New York, 1989); Ludmilla Jordanova, *Sexual Visions: Images of Gender in Science and Medicine between the Eighteenth and Twentieth Centuries* (New York, 1989); Dorinda Outram, *The Body and the French Revolution: Sex, Class, and Political Culture* (New Haven, 1989); Londa Schiebinger, *The Mind Has No Sex? Women in the Origins of Modern Science* (Cambridge, Mass., 1989); and Londa Schiebinger, *Nature's Body: Gender in the Making of Modern Science* (Boston, 1993). For an early philosophical approach to science and gender, see Evelyn Fox Keller, *Reflections on Gender and Science* (New Haven, Conn., 1985).

28. Marilyn Bailey Ogilvie, Nancy Slack, and Marianne Gosztonyi Ainley helped to pioneer the comparative study of scientific couples. See Ogilvie, "Marital Collaboration"; Nancy G. Slack, "Nineteenth-Century American Women Botanists: Wives, Widows, and Work," in *Uneasy Careers and Intimate Lives*, ed. Abir-Am and Outram, 77–103; and Ainley, "Last in the Field?"

29. See Darlene Clark Hine, "Co-Laborers in the Work of the Lord: Nineteenth-Century Black Women Physicians," in *"Send Us a Lady Physician": Women Doctors in America, 1835–1920*, ed. Ruth J. Abram (New York, 1985), 107–120. Medical couples mentioned by Hine include Verina Morton Jones and W. A. Morton, who shared a medical practice but with different specialties, and Alice Woodby McKane and Cornelius

McKane, whose medical collaboration included establishing hospitals in Liberia and in Savannah, Georgia (117).

30. Shari Benstock has argued that separatism made it easier for immigrant, lesbian women writers to play major roles in the modernist movement in Paris of the 1920s (Shari Benstock, *Women of the Left Bank: Paris, 1900–1940* [Austin, Tex., 1986], 451–452). Many other scholars have suggested that marginality nurtures creative potential (Dean Keith Simonton, *Scientific Genius: A Psychology of Science* [Cambridge, 1988], 126–129).

31. See Bateson's remarks on coming to terms with the models offered by her parents, Margaret Mead and Gregory Bateson (Bateson, *Composing a Life,* 12–13, 62). Bateson wrote that, as Mead's daughter, she was "loving and determined to be different" (62).

32. H. J. Mozans, *Woman in Science* (Cambridge, Mass., 1913), 412.

33. On the opening of the English universities to women, see, e.g., Rita McWilliams-Tullberg, *Women at Cambridge: A Men's University—Though of a Mixed Type* (London, 1975); Annie M. A. H. Rogers, *Degrees by Degrees: The Story of the Admission of Oxford Women Students to Membership of the University* (London, 1938); A. Gardner, *A Short History of Newnham College* (Cambridge, 1921); and Margaret J. Tuke, *A History of Bedford College for Women, 1849–1937* (London, 1939). On Girton, Newnham, and women science students, see Mary R. S. Creese, "British Women of the Nineteenth and Early Twentieth Centuries Who Contributed to Research in the Chemical Sciences," *British Journal for the History of Science* 24 (1991): 277–288. Creese also discusses Mary Christine Tebb, who studied at Bedford from 1882 to 1887 and at Girton from 1887 to 1893, collaborated with the chemist Otto Rosenheim from 1907 to 1910, married him in 1910, and continued her research thereafter (281).

34. On the American women's colleges, see Rossiter, *Women Scientists in America,* 15–25, and Toby A. Appel, "Physiology in American Women's Colleges: The Rise and Decline of a Female Subculture," *Isis* 85 (1994): 26–56.

35. Rossiter, *Women Scientists in America,* xviii.

36. For an appreciation of the revulsion toward the economic dependence associated with marriage, voiced by educated English women of the late nineteenth and early twentieth centuries, see Carol Dyhouse, *Feminism and the Family in England, 1880–1939* (London, 1989), 155–156.

37. As is well known, early women physicians avoided competition with men by confining their practices to such "feminine specialties" as obstetrics, gynecology, pediatrics, and public health (Regina Markell Morantz-Sanchez, *Sympathy and Science: Women Physicians in American Medicine* [New York, 1985], 61–63). On the "strategy" of innovation used by early professional women "who were drawn deliberately or by chance to new fields of interest," see Penina Migdal Glazer and Miriam Slater, *Unequal Colleagues: The Entrance of Women into the Professions, 1890–1940* (New Brunswick, N.J., 1987), 217–219.

38. Well into the twentieth century—perhaps into the 1970s in the United States and Canada—scientific wives found in widowhood professional opportunities that had been denied them so long as their husbands lived (Slack, "Nineteenth-Century American Women Botanists," 81).

39. This contrast suggests the possible fruitfulness of a study of relationships between professional couples (especially wives) and their in-laws. For the opposition of the Mendenhall family to their son's physician-wife (Dorothy Reed Mendenhall), see Glazer and Slater, *Unequal Colleagues,* 95.

40. Many of the chapters (1–3, 7, 13, and 17, e.g.) suggest the children of creative couples, and their relationships to their parents, as fertile topics for future historical analysis.

41. "Intrusive" means "intruding where one is not welcome or invited." With

respect to a couple, an "intrusive third party" can refer to individuals viewed as intrusive by either or both spouses. Although L. L. Nunn enjoyed a welcome relationship with his sister and perhaps even Charles during the early years of their marriage, Charles certainly came to see L. L.'s active involvement in the Whitmans' marriage as unwelcome, uninvited, and divisive.

42. On Curie's relationship with Lippmann, see Pycior, "Reaping the Benefits of Collaboration," 305.

43. The lack of contemporary models of married women scientists remains a problem, which is more serious in some countries than others. A survey of the late 1980s of women "holding (or having held)" lectureships or professorships in biology, chemistry, engineering, mathematics, the medical sciences, and physics at Dutch universities revealed that most of the responding women were either unmarried or divorced with no children. See Esther K. Hicks, "Women at the Top in Science and Technology Fields: Profile of Women Academics at Dutch Universities," in *Women in Science: Token Women or Gender Equality?* ed. Veronica Stolte-Heiskanen et al. (Oxford, 1991), 173–191, 182.

44. For an analytical survey of the secondary literature (through the mid-1980s) on the four different approaches to "women in science" (including "Not So Few: In Search of Lost Women" and "Why So Few? Identifying Structural Barriers"), see Londa Schiebinger, "The History and Philosophy of Women in Science: A Review Essay," *Signs* 12, no. 2 (Winter 1987): 305–332. More recent essays on women and gender in the history of science include Jordanova, "Gender and the Historiography of Science," and Sally Gregory Kohlstedt, "Women's Ambiguous 'Place' in the History of Science" (to appear in *Osiris* 10).

45. See, e.g., Fay Ajzenberg-Selove, *A Matter of Choices: Memoirs of a Female Physicist* (New Brunswick, N.J., 1994); interviews with Cathleen S. Morawetz and Mary Ellen Rudin, as well as an autobiographical sketch of Julia Robinson, in *More Mathematical People: Contemporary Conversations,* ed. Donald J. Albers, Gerald L. Alexanderson, and Constance Reid (Boston, 1990); and interviews with Salome Waelsch, Andrea Dupree, and Sandra Panem in *The Outer Circle,* ed. Zuckerman, Cole, and Bruer.

46. To date, one of the most ambitious studies is a cross-national survey of the state of women scientists in twelve European countries, conducted from 1988 to 1989 by the International Social Science Council, the European Coordination Centre for Research and Documentation in Social Sciences, and UNESCO's Division of Human Rights and Peace. For the twelve case studies and resulting policy recommendations, see *Women in Science,* ed. Stolte-Heiskanen et al. See also Jonathan R. Cole and Harriet Zuckerman, "Marriage, Motherhood, and Research Performance in Science," in *The Outer Circle,* ed. Zuckerman, Cole, and Bruer, 157–170; reprinted from *Scientific American* (1987). Over a hundred scientists, mostly women, shared their experiences with Vivian Gornick for her *Women in Science: Portraits from a World in Transition* (New York, 1983).

47. See, e.g., Jean Langenheim's address as the past president of the Ecological Society of America: Jean H. Langenheim, "The Path and Progress of American Women Ecologists," *Bulletin of the Ecological Society of America* 59, no. 4 (1988): 184–197. For the past three years the journal *Science* has published an annual section on women in science: "Women in Science 1992: Survey Issue," *Science* (March 13, 1992); "Women in Science 1993: Gender & Culture," *Science* (April 16, 1993); and "Women in Science 1994: Comparisons across Cultures," *Science* (March 11, 1994). In another acknowledgment of the centrality of questions of science and family, a symposium at the annual meeting of the American Association for the Advancement of Science (AAAS) in 1994 focused on "dual-career marriages." See Session S104: "Dual-Career Marriages," AAAS Annual Meeting, February 1994, available on audiotape.

48. In addition to the press coverage centering on world-famous dual-career couples, such as Hillary and Bill Clinton, there is specialized literature that analyzes and offers

advice to such couples. See, e.g., Rhona Rapoport and Robert N. Rapoport, *Dual-Career Families Reexamined: New Integrations of Work & Family* (New York, 1977); Fran Pepitone-Rockwell, ed., *Dual-Career Couples* (Beverly Hills, 1980); and Pepper Schwartz, "Thoroughly Modernizing Marriage," *Psychology Today* 27, no. 5 (September/October 1994): 54–59, 86.

49. Modern studies indicate that over 50% of American, Australian, and British married women physicians have physician spouses (Peter Uhlenberg and Teresa M. Cooney, "Male and Female Physicians: Family and Career Comparisons," *Social Science and Medicine* 30 [1990]: 376–377). In 1992, figures from the American Institute of Physics showed that 69% of married women physicists were scientific wives (44% of the married women physicists were married to men physicists). As of 1992, 80% of all American women mathematicians were married to men scientists or engineers. For the latter figures, see Ann Gibbons, "Key Issue: Two-Career Science Marriage," in "Women in Science 1992," 1380. Mary Frank Fox found that, among social scientists in four fields (economics, political science, psychology, and sociology), 40% of married women scientists are married to academic men (Mary Frank Fox, "Gender and Research Productivity," Session S104: "Dual-Career Marriages"). Cole and Zuckerman suggested that as many as four-fifths of married women scientists are married to men scientists (Cole and Zuckerman, "Marriage, Motherhood, and Research Performance," 168–169).

50. Vitalina Koval, "Soviet Women in Science," in *Women in Science,* ed. Stolte-Heiskanen et al., 130–131.

51. See Cole and Zuckerman, "Marriage, Motherhood, and Research Performance," 169–170. Similar conclusions, with somewhat different explanations, were reached by Mary Frank Fox; see Fox, "Gender and Research Productivity."

52. "Interview with Salome Waelsch," in *The Outer Circle,* ed. Zuckerman, Cole, and Bruer, 90. For the impressions of contemporary Russian women scientists and Turkish academic women that endogamous marriages are career enhancing, see Koval, "Soviet Women in Science," 131, and Feride Acar, "Women in Academic Science Careers in Turkey," in *Women in Science,* ed. Stolte-Heiskanen et al., 168. For similar impressions, see Gibbons, "Key Issue," 1381.

53. Fox, "Gender and Research Productivity." As of 1980, moreover, only 6% of American men physicians aged 30 to 49 were married to women physicians (Uhlenberg and Cooney, "Male and Female Physicians," 376).

54. Thus Fox emphasized that her own and related studies of women scientists have centered on those who survived the system (Fox, "Gender and Research Productivity").

55. On the usefulness of longitudinal studies in research on women, along with examples (but no scientific ones), see Diane Tickton Schuster, "Studying Women's Lives through Time," in *Women's Lives through Time: Educated American Women of the Twentieth Century,* ed. Kathleen Day Hulbert and Diane Tickton Schuster (San Francisco, 1993), 9–11. Nancy Slack has been conducting an informal longitudinal study of ten women who received their bachelor's degrees in biology in the 1950s from Cornell University (hereafter referred to as "Undergraduate Women in Biology: Cornell in the 1950s and Afterwards"). Additional information on this study can be obtained directly from Slack.

56. See Helena M. Pycior, "Marie Curie's 'Anti-natural Path': Time Only for Science and Family," in *Uneasy Careers and Intimate Lives,* ed. Abir-Am and Outram, 191–214.

57. See, e.g., "Interview with Salome Waelsch," 85–86, and Koval, "Soviet Women in Science," 132.

58. "Introduction," in *The Outer Circle,* 17 (quote), and Cole and Zuckerman, "Marriage, Motherhood, and Research Performance," 169.

59. Arlie Hochschild, *The Second Shift: Working Parents and the Revolution at Home* (New York, 1989), 104–105, 199.

60. Dorothea Gaudart, "Recommendations," in *Women in Science,* ed. Stolte-Heiskanen et al., 230. For a similar warning, see Cole and Zuckerman, "Marriage, Motherhood, and Research Performance," 169. Women scientists themselves recognize the pitfalls of their rigid schedules (Marcia Barinaga, "Profile of a Field: Neuroscience," in "Women in Science 1992," 1367).

61. Mary Ellen Rudin, an American mathematician who received her Ph.D. in 1949 and a mathematical wife, has separated the women of her generation—for whom a career was "the one thing we didn't demand, in fact it never occurred to us"—from "the young women mathematicians today [who] are thinking in terms of a career from the beginning" ("Mary Ellen Rudin," in *More Mathematical People,* ed. Albers, Alexanderson, and Reid, 302). For a comparison of contemporary Yugoslavian women scientists along generational lines, see Marina Blagojević, "Double-Faced Marginalisation: Women in Science in Yugoslavia," in *Women in Science,* ed. Stolte-Heiskanen et al., 91. According to Blagojević, younger Yugoslavian women "are more prone to view marriage and parenthood as a choice, and their career and own individual development as an obligation (above all towards themselves)" ("Double-Faced Marginalisation," 91).

62. Henry Etzkowitz, Carol Kemelgor, Michael Neuschatz, Brian Uzzi, and Joseph Alonzo, "The Paradox of Critical Mass for Women in Science," *Science* 266 (October 7, 1994): 51–54, 53 (quote).

63. For a survey of the history of contraception and abortion, spanning the lifetimes of the collection's scientific couples through the present, see Suzanne Poirier, "Women's Reproductive Health," in *Women, Health, and Medicine in America: A Historical Handbook,* ed. Rima D. Apple (New York, 1990), 217–245, esp. 221–240.

64. In addition to Frieda Cobb Blanchard, Gerty Cori, and Irène Joliot-Curie, women scientists working during the interwar period who followed this pattern include Dorothy Hodgkin, Cecilia Payne-Gaposchkin, and Dorothy Wrinch. On contemporary women, see, e.g., Cole and Zuckerman, "Marriage, Motherhood, and Research Performance," 168. The pattern of career, then family, should be compared with the histories of the women included in "Undergraduate Women in Biology: Cornell in the 1950s and Afterwards." Whereas three of the ten women went directly to graduate school and earned Ph.D.s early in their marriages, four of the ten women who completed their bachelor's degrees in the 1950s married, had children, and only later all completed Ph.D.s. See also Patricia Farnes's reflections on "the nonfeminist bias" that may lie behind medical warnings against women's postponing childbearing until their careers are established (Patricia Farnes, "Afterword," in *Women of Science,* ed. Kass-Simon and Farnes, 386).

65. See, e.g., Ann Gibbons, "No More Stressed-out Supermom," in "Women in Science 1993," 385.

66. Into the late 1960s (but not as easily since then), some well-off American women scientists were able to hire live-in nannies or nannies with flexible schedules, some of whom cared for their houses as well as children (see, e.g., "Mary Ellen Rudin," 295–296, and "Interview with Salome Waelsch," 86–87). As emphasized in the recent study by Etzkowitz and his colleagues, some younger American women scientists and their husbands want to serve as the primary caregivers of their children. Although these scientists use childcare as a "secondary support system" only, they, too, argue the need for improved childcare (see Etzkowitz et al., "Paradox of Critical Mass," 53).

67. Faye Flam, "Italy: Warm Climate for Women on the Mediterranean," in "Women in Science 1994," 1481. Similarly, leading Russian women social scientists listed "assistance by parents to look after a child" and parents' help with housework as factors promoting their careers (Koval, "Soviet Women in Science," 130–131).

68. On "a new alliance between science policy and gender scholarship . . . coming into being in the 1990s," see Pnina Geraldine Abir-Am, "Science Policy or Social Policy

for Women in Science: From Historical Case-Studies to an Agenda for the 1990s," *Science and Technology Policy* 5, no. 2 (1992): 11–12.

69. Paula M. Rayman and Belle Brett, "Clearing a Path for Women Scientists," *The Scientist* 8, no. 6 (March 21, 1994): 11.

70. For a description of Germany's HSPII program, which includes financial help for women with interrupted careers, see Peter Aldhous, "Germany: The Backbreaking Work of Scientist-Homemakers," in "Women in Science 1994," 1477. Some other governments, as well as professional societies (including the American Association of University Women), offer grants to women with interrupted careers.

71. Gaudart, "Recommendations," 230.

72. Women scientists seem to have made significant inroads into academic ecology, e.g. By the 1980s the percentages of women among students receiving doctorates in ecology from the University of Minnesota and Duke were 44% and 56%, respectively. Moreover, when Langenheim surveyed two hundred women who received their doctorates in ecology after 1975 and subsequently enjoyed compatible employment, she found a high number who reported that they had had a woman ecologist as a role model (42% of the two-thirds who replied to the survey). See Langenheim, "The Path and Progress of American Women Ecologists," 186, 195. Perhaps significantly, however, recent studies suggest that, because of the generational split between women scientists in the United States, there is a shortage of "viable role models" for younger women scientists and graduate students. Thus women scientists can still feel isolated despite the greater female presence in science—hence the "paradox of critical mass" (Etzkowitz et al., "Paradox of Critical Mass," 53).

73. Corinne Manogue, "What Institutions Can Do to Promote and Support Dual-Career Couples," Session S104: "Dual-Career Marriages." Compare with the story of "Annie Morris" (a pseudonym), a chemist who in the 1960s hid her pregnant body in a large laboratory coat (Gornick, *Women in Science,* 102, 150).

74. For data on gender differences in the ranks and salaries of academics, see the annual reports of the American Association of University Professors (e.g., "The Annual Report on the Economic Status of the Profession, 1993–1994," *Academe* 80 [March/April 1994]: 24–25, 30–86).

75. See, e.g., "Interview with Salome Waelsch," 84, and "Interview with Andrea Dupree," in *The Outer Circle,* ed. Zuckerman, Cole, and Bruer, 117–119.

76. Some of these wives secured tenured positions at research universities only in the 1970s or later, when affirmative action laws eased discrimination against women (see, e.g., Ajzenberg-Selove, *A Matter of Choices,* 114–116, 162–166). On the persistence (into the late 1960s) of the idea that women graduate students in the sciences ought to be trained to become research associates and wives of men scientists, see Gornick, *Women in Science,* 78.

77. On shared problems, see Catherine Jay Didion, "Introduction," Session S104: "Dual-Career Marriages." On the persistent gender imbalance in pursuing professional opportunities requiring geographic relocation, see Barinaga, "Neuroscience," 1366. On the resulting, continual underemployment of scientific wives, and its effects on the status of women ecologists, see Langenheim, "The Path and Progress of American Women Ecologists," 189, 192.

78. For this and other options open to dual-career couples, see Donna M. Moore, "Equal Opportunity Laws and Dual-Career Couples," in *Dual-Career Couples,* ed. Pepitone-Rockwell, 236–240. For an example of taking turns in science, see Ann Gibbons, "A Thoroughly Modern Marriage," in "Women in Science 1993," 411.

79. There are still no systematic studies of commuter marriage, with its costs as well as benefits. On commuter marriage, including its implications for parenting, see Naomi Gerstel and Harriet Gross, *Commuter Marriage: A Study of Work and Family* (New York, 1984). For an example of a commuter marriage, with partners (geneticist

Maria Leptin and immunologist Jonathan Howard) in two different countries, see Peter Aldhous, "Fighting for Day Care at the Lab," in "Women in Science 1994," 1475. Commuter marriage may be on the increase because of the failure of other strategies to assure coresidence and compatible employment (Cora B. Marrett, "Commuting and Professional Careers," Session S104: "Dual-Career Marriages").

80. Moore, "Equal Opportunity Laws and Dual-Career Couples," 230–236, 235 (quote). Still, several American universities, including Oregon State and Florida State, have each managed to hire more than a dozen scientific couples. On these universities' efforts and on couples' strategies for compatible, coresident employment, see Manogue, "What Institutions Can Do."

81. See Schwartz, "Thoroughly Modernizing Marriage," 58.

82. For an example of the former, see Gibbons, "A Thoroughly Modern Marriage," 411; for the latter, see Diane Wallander, "A Spotlight on Judy Meyer—Ecologist," *AWIS Magazine* 23 (1994): 10–11.

83. Ajzenberg-Selove, *A Matter of Choices*, 114.

84. Langenheim, "The Path and Progress of American Women Ecologists," 195.

85. Such concerns (see, e.g., "Interview with Salome Waelsch," 88) again underline the need for a study of the children of scientific couples.

86. There is good evidence that the turn-of-the-century physicists Hertha and William Ayrton avoided formal collaboration for this very reason. See Evelyn Sharp, *Hertha Ayrton, 1854–1923: A Memoir* (London, 1926), 117. Recently, Joan Mason has studied the two scientific marriages of William Ayrton (Joan Mason, "Matilda Chaplin Ayrton [1846–1883], William Edward Ayrton [1847–1908], and Hertha Ayrton [1854–1923] [unpublished manuscript, Department of the History and Philosophy of Science, University of Cambridge, CB2 3RH]).

87. Quoted in Gibbons, "Key Issue," 1381. Along similar lines, a study of contemporary creative couples in psychology suggested that the couples who enjoyed the greatest professional and emotional satisfaction were those who shared specialty and institutional affiliation (Matilda Butler and William Paisley, "Coordinated-Career Couples: Convergence and Divergence," in *Dual-Career Couples*, ed. Pepitone-Rockwell, 227–228).

Chapter 1: Pierre Curie and "His Eminent Collaborator Mme Curie"

An earlier version of this chapter was presented at the XIXth International Congress of History of Science, Zaragoza, Spain, in August 1993. The author wishes to thank Lawrence Badash for comments on the chapter as well as the Center for Twentieth Century Studies, the University of Wisconsin-Milwaukee, for a 1990–1991 fellowship to pursue the mythology of Marie Curie.

1. "Prix La Caze," *Comptes rendus des Séances de l'Académie des Sciences* (hereafter abbreviated as *CR*) 133 (1901): 1061. English translations from the *CR* are mine, unless otherwise indicated.

2. *Nobel Lectures Including Presentation Speeches and Laureates' Biographies: Physics, 1901–1921* (Amsterdam, 1967), 50–51.

3. See, e.g., Robert Reid, *Marie Curie* (New York, 1974), 73–74. For a critique of the myth of Marie the chemist and Pierre the physicist, see Helena M. Pycior, "Reaping the Benefits of Collaboration While Avoiding Its Pitfalls: Marie Curie's Rise to Scientific Prominence," *Social Studies of Science* 23 (1993): 301–323.

4. Although the Curies' complementarity has been mentioned in the earlier literature (e.g., Irène Joliot-Curie, "Marie Curie, ma mère," *Europe*, no. 108 [December 1954]: 90, and Rosalynd Pflaum, *Grand Obsession: Madame Curie and Her World* [New York, 1989], 76–77), this thesis has not been elaborated.

5. On the importance of families in science, especially their "central[ity] to the history of women's work in science," see Ann B. Shteir, "Botany in the Breakfast Room: Women and Early Nineteenth-Century British Plant Study," in *Uneasy Careers and Intimate Lives: Women in Science, 1789–1979*, ed. Pnina G. Abir-Am and Dorinda Outram (New Brunswick, N.J., 1987), 31–43, 33 (quote).

6. See, e.g., Marie Curie, *Pierre Curie*, with the *Autobiographical Notes of Marie Curie*, trans. Charlotte and Vernon Kellogg (1923; reprint, New York, 1963), 14, 17, 21–23.

7. Paul Langevin, "Pierre Curie," *La revue du mois* (July 10, 1906): 24; Marie Curie, "Préface," *Oeuvres de Pierre Curie* (Paris, 1908), viii. English translations from these essays are mine.

8. Marie Curie, *Pierre Curie*, 23; Chéneveau quote, 72.

9. Pierre Curie to Marie Skłodowska, August 10, 1894, in Eve Curie, *Madame Curie* (Garden City, N.Y., 1937), 130.

10. Marie Curie, "Préface," viii.

11. Langevin, "Pierre Curie," 15.

12. See, e.g., Eve Curie, *Madame Curie*, 120.

13. Quoted in Marie Curie, *Pierre Curie*, 23.

14. Marie Curie, *Pierre Curie*, 34.

15. Marie Curie, *Autobiographical Notes*, 85.

16. As Marie recorded in her diary, she buried Pierre with a photograph of herself that he had called "the good little student" and "loved" (quoted in Eve Curie, *Madame Curie*, 249).

17. Marie Curie, "Préface," viii.

18. Marie Curie, *Autobiographical Notes*, 85.

19. Reproduced in Eve Curie, *Madame Curie*, (opposite) 293, this portrait bears the caption "Marie Curie's Favorite Portrait of Her Husband."

20. Langevin, "Pierre Curie," 24–25. Langevin reported that only in deciding to marry Marie had Pierre ever acted quickly.

21. Marie Curie, *Pierre Curie*, 13–14.

22. Langevin, "Pierre Curie," 25.

23. Marie Curie, "Préface," x.

24. Pierre and Marie Curie, "Sur les corps radioactifs," *CR* 134 (1902): 87. For the English translation and Marie's attribution of this statement to Pierre, see Marie Curie, *Pierre Curie*, 64–65; see also Langevin, "Pierre Curie," 27.

25. Marie Curie, "Préface," x, xii. Still, as Marie Curie emphasized, Pierre readily framed hypotheses, although he balked at "their premature publication" (Marie Curie, *Pierre Curie*, 65).

26. Langevin, "Pierre Curie," 19–20, 25.

27. Lippmann's prediction, but not its possible effects on Pierre Curie, is mentioned in Marie Curie, *Pierre Curie*, 21.

28. Langevin, "Pierre Curie," 22.

29. Marie Curie, *Pierre Curie*, 70–72; Langevin, "Pierre Curie," 20.

30. Marie Curie, "Préface," ix (quote)–x.

31. Eve Curie, *Madame Curie*, 126.

32. Pierre Curie to Marie Skłodowska, August 14, 1894, in Eve Curie, *Madame Curie*, 131–132.

33. Langevin, "Pierre Curie," 20–22.

34. Pierre Curie to Georges Gouy, July 24, 1905, in Eve Curie, *Madame Curie*, 235.

35. Marie Curie, *Pierre Curie*, 64 (quote); idem, "Préface," xiv. Many of Pierre's traits (e.g., restlessness, distractibility combined with an ability to hyperfocus, and creativity) are suggestive of what today is called attention deficit disorder.

36. Langevin, "Pierre Curie," 27.
37. For this assessment, see Marie Curie, "Préface," xi.
38. On Pierre's early scientific work, see Jean Wyart, "Pierre Curie," in *Dictionary of Scientific Biography*, ed. C. C. Gillispie, vol. 3 (New York, 1970–1980), 504–506, 505 (quote).
39. Marie Curie, "Préface," xxi.
40. Joliot-Curie, "Marie Curie, ma mère," 90. English translations from the essay are mine.
41. Marie Curie, *Pierre Curie*, 21, 17–18. For the photograph, see Eve Curie, *Madame Curie*, opposite 100.
42. In contrasting Pierre and Marie, Langevin stressed Marie's drive toward clarity (Langevin, "Pierre Curie," 15–16 [quote]).
43. See Pycior, "Reaping the Benefits of Collaboration," 302.
44. In her verbal and visual emphasis on "the thinker," Marie Curie seems to have been influenced by Auguste Rodin. *The Thinker*, the male nude that Rodin had first sculpted in 1880 and decided to enlarge in 1901, was discussed widely in Paris from 1889 on (Albert E. Elsen, *Rodin's "Thinker" and the Dilemmas of Modern Public Sculpture* [New Haven, 1985], 61–65, 85–93). In the early 1900s Marie and Pierre Curie numbered Rodin among their few friends outside science (Eve Curie, *Madame Curie*, 232–233). Marie Curie's (in some ways) gender-defiant self-image seems to help explain her identification with "the thinker." Compare with Anne Higonnet's recent comment that "no amount of social power then available to any woman could make her into a *Thinker*" (Anne Higonnet, "Myths of Creation: Camille Claudel and Auguste Rodin," in *Significant Others: Creativity and Intimate Partnership*, ed. Whitney Chadwick and Isabelle de Courtivron [London, 1993], 22).
45. See, e.g., Reid, *Marie Curie*, 51.
46. Marie Skłodowska to Joseph Skłodowski, March 9, 1887, in Eve Curie, *Madame Curie*, 75.
47. Joliot-Curie, "Marie Curie, ma mère," 114.
48. Reid has already briefly contrasted Marie's and Pierre's attitudes toward priority (Reid, *Marie Curie*, 64).
49. Marie Curie, "Préface," x (quotes), xxi.
50. Ibid., xiii–xiv.
51. Marie Curie, *Pierre Curie*, 69.
52. Marie Curie, "Préface," vi.
53. Helena M. Pycior, "Marie Curie's 'Anti-natural Path': Time Only for Science and Family," in *Uneasy Careers and Intimate Lives*, ed. Abir-Am and Outram, 202–214, esp. 202–203.
54. Ibid., esp. 198–199.
55. Marie Curie, *Autobiographical Notes*, 88, 92; idem, *Pierre Curie*, 19, 49.
56. Reid, *Marie Curie*, 73.
57. Eve Curie, *Madame Curie*, 175–177.
58. Langevin, "Pierre Curie," 15.
59. Ibid., 14.
60. Marie Curie, *Pierre Curie*, 30.
61. See Pycior, "Reaping the Benefits of Collaboration," 303–304, 312; "Prix Gegner," *CR* 127 (1898): 1133.
62. Compare with Lawrence Badash, "The Discovery of Thorium's Radioactivity," *Journal of Chemical Education* 43 (1966): 219.
63. Marie Curie, *Radioactive Substances*, trans. Alfred Del Vecchio (Westport, Conn., 1971), 14.
64. For an accessible description of her setup, see Robert L. Wolke, "Marie Curie's

Doctoral Thesis: Prelude to a Nobel Prize," *Journal of Chemical Education* 65 (1988): 564–565.

65. Marie Curie, "Rayons émis par les composés de l'uranium et du thorium," *CR* 126 (1898): 1101–1102.

66. On Schmidt's work, see Badash, "The Discovery of Thorium's Radioactivity."

67. Marie Curie, "Rayons émis," 1103.

68. Ernest Rutherford, "Mme. Curie," *Nature* 134 (July 21, 1934): 90.

69. Ernest Rutherford, "Uranium Radiation and the Electrical Conduction Produced by It," *Philosophical Magazine* 47 (1899): 109–163; reprinted in *The Collected Papers of Lord Rutherford of Nelson* (hereafter *CP*) (New York, 1962), 1: 169–215, at 178.

70. Pierre and Marie Curie, "Sur une substance nouvelle radio-active, contenue dans la pechblende," *CR* 127 (1898): 178. Their new (tracer) method is described in Wolke, "Marie Curie's Doctoral Thesis," 564–565, 568.

71. Pierre and Marie Curie, "Sur une substance nouvelle radio-active," 175–176.

72. Pierre Curie, Marie Curie, and Gustave Bémont, "Sur une nouvelle substance fortement radio-active, contenue dans la pechblende," *CR* 127 (1898): 1216.

73. For a discussion of the Curies' publication policy and its effects, see Pycior, "Reaping the Benefits of Collaboration."

74. On Marie Curie and the "Matthew Effect," see ibid., 315–317. Also see Margaret W. Rossiter, "The ~~Matthew~~ Matilda Effect in Science," *Social Studies of Science* 23 (1993): 325–341.

75. Joliot-Curie, "Marie Curie, ma mère," 92; George Jaffé, "Recollections of Three Great Laboratories," *Journal of Chemical Education* 29 (1952): 237–238.

76. Pycior, "Reaping the Benefits of Collaboration," 309.

77. Pierre Curie, "Action du champ magnétique sur les rayons de Becquerel. Rayons déviés et rayons non déviés," *CR* 130 (1900): 73–76, at 73; Marie Curie, "Sur la pénétration des rayons de Becquerel non déviables par le champ magnétique," *CR* 130 (1900): 76–79.

78. Pierre Curie, "Action du champ magnétique," 73–75. I quote Pierre's major conclusions to give a sense of his scientific style.

79. Rutherford, "Uranium Radiation," 175. Neither of the Curies mentioned this paper.

80. Marie Curie, "Sur la pénétration des rayons," 76.

81. Marie Curie, "Préface," xi.

82. Pierre and Marie Curie, "Les nouvelles substances radioactives et les rayons qu'elles émettent," *Rapports présentés au Congrès international de Physique* 3 (1900); reproduced in *Oeuvres de Pierre Curie,* 392–400, esp. 396.

83. Pierre and Marie Curie, "Sur la charge électrique des rayons déviables du radium," *CR* 130 (1900): 647–650.

84. See, e.g., *Oeuvres de Pierre Curie,* 605–621, and Herbert S. Klickstein, "Pierre Curie—an Appreciation of His Scientific Achievements," *Journal of Chemical Education* 24 (1947): 282 (bibliography).

85. Pierre and Marie Curie, "Sur les corps radioactifs," 87.

86. Marie Curie, *Pierre Curie,* 48.

87. Marie Curie, "Préface," xv.

88. On persistence as a key factor leading to eminence, see Dean Keith Simonton, *Scientific Genius: A Psychology of Science* (Cambridge, 1988), 50–52.

89. On the PCN (*certificat d'études physiques, naturelles et chimiques*), see Harry W. Paul, *From Knowledge to Power: The Rise of the Science Empire in France, 1860–1939* (Cambridge, 1985), 115.

90. For the bare facts of this period, see Marie Curie, *Pierre Curie,* 50–52.

91. Marie Curie, "Sur le poids atomique du radium," *CR* 135 (1902): 161–163.

92. See W. Swietoslawski, "The Legend of Madame Curie," *Bulletin of the Polish Institute of Arts and Sciences in America* 3 (1945): 217–220 (bibliography). During this period Marie Curie also tried to prove that the radioactive element that Willy Marckwald had "discovered" and named radiotellurium was actually polonium. In 1902–1903 she published a note on this topic (see Reid, *Marie Curie*, 139–140, 290), which Swietoslawski does not list.

93. Marie Curie, *Autobiographical Notes*, 94.

94. See Klickstein, "Pierre Curie," 282.

95. See Marjorie Malley, "The Discovery of Atomic Transformation: Scientific Styles and Philosophies in France and Britain," *Isis* 70 (1979): 213–223, esp. 222.

96. Ernest Rutherford, "The Magnetic and Electric Deviation of the Easily Absorbed Rays from Radium," *Philosophical Magazine*, 6th ser., 5 (1903): 177–187; *CP* 1: 549–557. Rutherford thanked Pierre for the radium (550). Here Rutherford showed that alpha rays were slightly deflected by a powerful magnetic field.

97. Pierre Curie and Albert Laborde, "Sur la chaleur dégagée spontanément par les sels de radium," *CR* 136 (1903): 673–675.

98. In a paragraph near the end of her article, Malley began to contrast the scientific style and philosophy of "the frankly speculative" Marie with those of Pierre (Malley, "The Discovery of Atomic Transformation," 222–223).

99. See Marie Curie, *Pierre Curie*, 53 (quote), 57, and idem, "Préface," xx.

100. Marie Curie, "Préface," x.

101. Langevin, "Pierre Curie," 25.

102. Marie Curie, "Préface," x (quote); idem, *Pierre Curie*, 65.

103. Langevin, "Pierre Curie," 26; Marie Curie, "Préface," xiv, xvii.

104. On Pierre's last years, see Reid, *Marie Curie*, 113–123, esp. 121.

105. Marie Curie, "Préface," xiv.

106. Marie Curie, *Pierre Curie*, preface.

107. Langevin, "Pierre Curie," 15.

Chapter 2: Star Scientists in a Nobelist Family

An earlier version of this paper was read on December 2, 1992, in Pnina Abir-Am's course History of Women in Science at The Johns Hopkins University; I wish to thank her for her comments and suggestions on earlier drafts of this paper.

1. Pierre Laszlo, "The Best Part of Beauty," *New Journal of Chemistry* 8, no. 6 (1984): 343–344.

2. From Frédéric and Irène Joliot-Curie, *Oeuvres scientifiques complètes* (Paris, 1961). So strong is the image of a collaborative couple of scientists that in spite of the large amount of separate papers, the French publisher decided to gather the works of both partners in one single volume.

3. On the French milieu of physicists, see Dominique Pestre, *Physique et physiciens en France, 1918–1940* (Paris, 1984).

4. "Le Muséum au premier siècle de son histoire," Colloque organisé dans le cadre du bicentenaire du Muséum national d'histoire naturelle par le Centre Alexandre Koyré, Paris, June 11–14, 1993. Scientific dynasties were also known prior to the nineteenth century; for instance, under the ancien régime the position of director of the Paris Observatory was held by four successive generations of the same family over more than one century: Jean Dominique Cassini (1625–1712), a French astronomer of Italian origin, was succeeded by his son Jacques (Cassini II, 1677–1756), by his grandson César-François Cassini de Thury (Cassini III, 1714–1784), and eventually by his great-grandson, Jean Dominique (Cassini IV, 1748–1845), who was dismissed from his position at the Paris Observatory during the French Revolution.

5. On the female role in the Napoleonic science elite, see Dorinda Outram, "Before Objectivity: Wives, Patronage, and Cultural Reproduction in Early Nineteenth-Century France," in *Uneasy Careers and Intimate Lives: Women in Science, 1789–1979*, ed. P. G. Abir-Am and D. Outram (New Brunswick, N.J., 1987), 19–30. See also the editors' introduction, 1–16.

6. Dorinda Outram, *Georges Cuvier: Vocation, Science and Authority in Post-Revolutionary France* (Dover, N.H., 1984); also Camille Limoges, "The Development of the Museum d'Histoire Naturelle de Paris, c. 1800–1914," in *The Organization of Science and Technology in France, 1808–1914*, ed. R. Fox and G. Weisz (Cambridge, 1980).

7. Marie Curie's and Paul Langevin's love affair became a scandal that had a great impact on Marie Curie; see Robert Reid, *Marie Curie* (New York, 1974), chap. 16, and Susan Quinn, *Marie Curie: A Life* (New York, 1995), chap. 14.

8. This intellectual circle has been described by Camille Marbo, a novelist and the wife of the French mathematician Emile Borel, in *Souvenirs et rencontres (1883–1967)* (Paris, 1967). See also Bernadette Bensaude-Vincent, *Paul Langevin: Science et vigilance* (Paris, 1987).

9. Christiane Olivier, *Les enfants de Jocaste: L'empreinte de la mère* (Paris, 1980). Vehemently protesting against Jacques Lacan, who, denying any individuality to women, used to declare that women existed only collectively as a female genre ("La femme ça ne peut s'écrire qu' à barrer la"), Christiane Olivier emphasizes the mother's dominant figure in a fatherless family.

10. Eve Curie, *Madame Curie* (Paris, 1938); Reid, *Marie Curie*; Helena M. Pycior, "Marie Curie's 'Anti-natural Path': Time Only for Science and Family," in *Uneasy Careers and Intimate Lives,* ed. Abir-Am and Outram, 191–215.

11. Irène's feelings during her early childhood are finely described in the biography by Noelle Loriot, *Irène Joliot-Curie* (Paris, 1991). This biography by a journalist provides a poorly detailed account of Irène Curie's scientific work, but it is extremely interesting for its thoughtful psychological remarks.

12. Eve Curie, "Irène Joliot Curie," *Marianne* (French magazine), 1936; reprinted in *Souvenirs et documents publiés par l'Association Frédéric et Irène Joliot-Curie* (Paris, n.d.), 15–16.

13. In addition to small governmental subsidies, the Radium Institute was able to receive private gifts through the Curie Foundation, created in 1920. The number of collaborators working at the Radium Institute gradually increased to seventeen in 1933. Among them were many women scientists: Mme Cotelle and Mlle Chamié were preparing derivatives of actinium, and later, while investigating the successive disintegrations of actinium, Marguerite Perey discovered a new radioactive element, named francium (atomic number 87).

14. Irène Curie, "Recherches sur les rayons alpha du polonium: Oscillation de parcours, vitesse d'émission, pouvoir ionisant" (Paris, Sorbonne, 1925), *Annales de physique,* 10th series, T.3, May–June 1925, in *Oeuvres scientifiques complètes,* 47–114.

15. Loriot, *Irène Joliot-Curie.*

16. Marie Curie, *Pierre Curie* (Paris, 1921).

17. On the contrast between the physical and the chemical approaches to radioactivity, see Adrienne R. Weill, "Marie Curie," in *Dictionary of Scientific Biography,* ed. C. C. Gillispie (New York, 1972). Bernadette Bensaude-Vincent and Isabelle Stengers, *Histoire de la chimie* (Paris, 1993), chap. 29.

18. Louis de Broglie, *La vie et l'oeuvre de Frédéric Joliot* (Paris, 1959); Pierre Biquart, *Frédéric Joliot-Curie* (Paris, 1961); Spencer Waert, *Scientists in Power* (Cambridge, Mass., 1979). Juan A. Del Regato, "Jean-Frédéric Joliot-Curie," *Journal de biophysique et de médecine nucléaire* 7, no. 2 (1983): 61–74. See also Rosalynd Pflaum, *Grand Obsession: Madame Curie and Her World* (New York, 1989), 283–290.

19. Bertrand Goldschmidt, *Pionniers de l'atome* (Paris, 1989; English trans. by

Georges M. Temmer under the title *Atomic Rivals: A Candid Memoir of Rivalries among the Allies over the Bomb* [New Brunswick, N. J., 1990]).

20. See Ruth Lewin Sime, "Lise Meitner and the Discovery of Fission," *Journal of Chemical Education* 66 (1989): 373–376; "Lise Meitner's Escape from Germany," *American Journal of Physics* 58 (1990): 262–267. See also Xavier Roqué, "Ressemblances et écarts de destins," *Les Cahiers des sciences et vie*, no. 24 (December 1994): 86–90.

21. The initiative of the Front populaire can be dated back to a public demonstration of July 14, 1935, known as the *rassemblement populaire*, organized by Francis Perrin, the son of Jean Perrin. Moreover, Prime Minister Léon Blum was Jean Perrin's close friend. In those days, Jean Perrin was very active in promoting a national science policy, and he was the decisive force behind the creation of the Centre national de la recherche scientifique (CNRS) in 1939. During his campaign, he used to ask Marie Curie to accompany him on his repeated visits to the various ministries to seek funds and grants for pure scientific research. See Waert, *Scientists in Power.*

22. Bensaude-Vincent, *Paul Langevin,* 6.

23. Irène was a close friend of Eugénie Feytis Cotton, a former student of Marie Curie at Sèvres who often came to her home for baby-sitting when Irène was a young child. Eugénie Cotton, herself a physicist, married to a well-known physicist, Aimé Cotton, and a working mother, became a leader of the French movement for women. In 1945 she founded the International Democratic Federation of Women, which she managed until her death in 1967. She was also the founder of the World Peace Council in 1945 (see the volume edited by the Fédération démocratique internationale des femmes, *Eugénie Cotton* [Paris, n.d.]).

24. *Femmes françaises*, no. 12, November 30, 1944; reprinted in *Souvenirs et documents,* 22 (my translation).

25. For a general survey of French intellectuals' attitudes toward the Communist Party, see David Caute, *Le P. C. et les intellectuels* (Paris, 1967); and idem, *Les compagnons de route, 1917–1968* (Paris, 1973).

26. Quoted by Loriot, *Irène Joliot-Curie,* 194.

27. Margaret Rossiter, "L'affaire Curie: Irène Joliot-Curie's Rejection by the American Chemical Society, 1953–1955" (paper read at the XIX International Congress of History of Science, Zaragoza, Spain, August 1993).

Chapter 3: Carl and Gerty Cori

The author wishes to thank Helena M. Pycior for her helpful editing of this chapter and also the Archives of the Medical School library of Washington University for providing the photograph of the Coris. The author is particularly indebted to Thomas Cori for supplying details of his mother's life not available elsewhere.

1. Carl Cori in *Les Prix Nobel en 1947* (Stockholm, 1949), 23.

2. The information concerning Carl Cori's childhood, the early years of his association with Gerty, and their early careers is derived principally from an autobiographical essay, Carl F. Cori, "The Call of Science," *Annual Review of Biochemistry* 38 (1969): 1–20 (on Carl Cori's family, see 2–3).

3. Carl I. Cori's scientific stature is attested to in H. M. Kalckar, "The Isolation of the Cori-ester, 'The Saint Louis Gateway' to a First Approach of a Dynamic Formulation of Macromolecular Synthesis," in *Selected Topics in the History of Biochemistry: Personal Recollections,* vol. 35 of *Comprehensive Biochemistry,* ed. G. Semenza (Amsterdam, 1983), 1–24.

4. O. S. Opfell, *Lady Laureates: Women Who Have Won the Nobel Prize* (Metuchen, N. J., 1986), 213–223.

5. C. F. Cori, "The Call of Science," 5.

6. Carl Cori, *Les Prix Nobel en 1947*, 23.

7. I am indebted to Thomas Cori for this anecdote.

8. C. F. Cori, "The Call of Science," 11–12.

9. F. Peyton Rous, Archives of the American Philosophical Society.

10. Complete lists of faculty are reported in F. E. Hunter, *History of the Pharmacology Department, Washington University Medical School 1900–1980* (St. Louis, 1981).

11. Gerty Cori's research is described in detail by J. Larner, "Gerty Teresa Cori, 1896–1957," *Biographical Memoirs of the National Academy of Sciences* 61 (1992): 111–135. It should be noted that the memoir did not appear until thirty-five years after Gerty's death because, in a letter to the National Academy written shortly after her death, Carl Cori requested that a joint memoir be published after his death since their work was done jointly.

12. Carl Cori's scientific contributions are covered in detail by P. Randle, "Carl Ferdinand Cori," *Biographical Memoirs of Fellows of the Royal Society* 32 (1986): 67–95, and M. Cohn, "Carl Ferdinand Cori," *Biographical Memoirs of the National Academy of Sciences* 61 (1992): 79–109.

13. Gerty Cori, "This I Believe" (interviews with Edward Murrow), Archives of Washington University School of Medicine.

14. C. F. Cori, "The Call of Science," 20.

Chapter 4: John and Elizabeth Gould

I wish to acknowledge helpful suggestions from Maureen Lambourne, great-great-granddaughter of John and Elizabeth Gould, from Dr. Gordon Sauer, the foremost American expert on John Gould, both of whom read an earlier version of this paper and sent photographs to me, and from Brigadier David Edelsten, great-great-grandson of John and Elizabeth Gould, for permission to use the portrait of Elizabeth Gould. Pauline Payne at the University of Adelaide sent reprints unavailable in the United States. Librarians at the University of California at Los Angeles and the Los Angeles City Library were most helpful.

1. Maureen Lambourne, *John Gould: Bird Man* (London, 1987), 27; Hubert Massey Whittell, *The Literature of Australian Birds: A History and a Bibliography of Australian Ornithology* (Perth, 1954), 88.

2. Elizabeth Coxen (later Mrs. John Gould) to her mother, probably late 1827; published in A. H. Chisholm, "Elizabeth Gould—Some 'New' Letters," *Royal Australian Historical Society, Journal and Proceedings* 49, no. 5 (1964): 321–336; letter on 329–330.

3. The Zoological Society was founded on April 29, 1826. Gordon C. Sauer, *John Gould the Bird Man: A Chronology and Bibliography* (Manhattan, Kans., 1982), 92. While there he was commissioned to stuff a giraffe for George IV; Maureen Lambourne, "John Gould and His Illustrators," *Birds* 2 (1969): 168–171 (see 168).

4. John Gould, *A Century of Birds Hitherto Unfigured from the Himalaya Mountains* (London, 1831–1832); one-hundred species are illustrated, hence the title *Century*; Prideaux John Selby mentioned the first four plates in a letter to Sir William Jardine, January 10, 1831, published in Sauer, *John Gould*, 15–17.

5. Nicholas Aylward Vigors, an author on bird classification, was secretary of the Zoological Society from 1826 to 1832; see *Transactions of the Linnaean Society of London* 14 (1825): 395–517, and R. Bowdler Sharpe, "Biographical Memoir," *An Analytical Index to the Works of the Late John Gould, F.R.S.* (London, 1893), xi.

6. Maureen Lambourne, "Birds of a Feather: Edward Lear and Elizabeth Gould," *Country Life*, June 25, 1964, 1656–1657.

7. Lambourne, *John Gould*, 28; K. A. Hindwood, "Mrs. John Gould," *Emu* 38 (1938): 131–138 (see 138).

8. Sauer, *John Gould*, xvii.

9. *A Monograph of the Ramphastidae* [toucans] (London, 1833–1836), 33 colored plates, most by Elizabeth but a few by Lear; *A Monograph of the Trogonidae* [trogons] (London, 1835–1836), 36 plates (Lear worked on the trogons, but all plates are ascribed to "J. & E. Gould"); *A Synopsis of the Birds of Australia* (London, 1837), 73 plates attributed to "E. G."; see Lambourne, *John Gould*, 40; Lambourne, "Birds of a Feather," 1656–1657; Lambourne, "John Gould and His Illustrators," 171; Lynn Barber, *The Heyday of Natural History, 1820–1870* (New York, 1980), 94–96.

10. Gould took Lear to the continent in 1831 or 1832. Elizabeth had a nursing infant in 1831 and was pregnant with a second child in 1832. Sixty-eight of the 448 plates in *Birds of Europe* were drawn by Lear; see Sauer, *John Gould*, 94, 96, 97.

11. Ibid., 19.

12. Lambourne, "Birds of a Feather," 1656.

13. Sauer, *John Gould*, 91–92.

14. Sauer, *John Gould*, 93; Lambourne, *John Gould*, 9.

15. Albert E. Gunther, *A Century of Zoology at the British Museum through the Lives of Two Keepers: 1815–1914* (London, 1975), 267.

16. Neville W. Cayley, "John Gould as an Illustrator," *Emu* 38 (1938): 167–172, quote from 168.

17. Sauer, *John Gould*, xvii–xviii.

18. Quote from Gordon C. Sauer, "Forty Years Association with John Gould the Bird Man," in *From Linnaeus to Darwin: Commentaries on the History of Biology and Geology*, ed. Alwyne Wheeler and James H. Price (London, 1985), 159–166 (see 164); see also Charles Darwin, "Journal," entry for June 26, 1837, *The Correspondence of Charles Darwin*, ed. Frederick Burkhardt and Sydney Smith (Cambridge, 1986), 2:431; F. J. Sulloway, "Darwin and His Finches: The Evolution of a Legend," *Journal of the History of Biology* 15 (1982): 1–53 (see 21–23); and idem, "Darwin's Conversion," *Journal of the History of Biology* 15 (1982): 325–396.

19. Charles Darwin, *The Origin of Species* (London, 1859), 404.

20. C. E. Bryant, "Gould Miscellanea and Some Anecdotes," *Emu* 38 (1938): 226–231 (see 228).

21. A. H. Chisholm, "Mrs. John Gould and Her Relatives," *Emu* 40 (1941): 337–353 (see 337–338).

22. Elizabeth Gould to her mother, ca. 1827; Chisholm, "Elizabeth Gould—Some 'New' Letters," 329–330.

23. Sharpe, "Biographical Memoir," ix–xvi.

24. Lambourne, *John Gould*, 27, 30–37; Christine E. Jackson, "The Changing Relationship between J. J. Audubon and His Friends P. J. Selby, Sir William Jardine and W. H. Lizars," *Archives of Natural History* 18 (1991): 289–307; Barber, *The Heyday of Natural History*, 87–98.

25. Hindwood, "Mrs. John Gould," 132. Thomas Bewick, *History of British Birds*, 2 vols. (Newcastle-upon-Tyne, 1797–1804); William Swainson, *Zoological Illustrations* (London, 1820–1823).

26. Selby to Jardine, January 10, 1831, published in Sauer, *John Gould*, 17.

27. Quote from Jardine's announcement of Gould's proposed but never published "New General History of Birds," in *Magazine of Zoology and Botany* 1 (1837): 108–109; see Sauer, *John Gould*, 37.

28. Bryant, "Gould Miscellanea," quote from 228.

29. Darwin, *Zoology of the* Beagle, vol. 3, *Birds*, 1, quoted in Darwin, *Correspondence*, 2:72, n. 2.

30. Vivian Noakes, *Edward Lear: The Life of a Wanderer* (London, 1968), 39–40, quoted in Barber, *The Heyday of Natural History*, 94.

31. Charles Pickering, unpublished manuscript, quoted in Glenn and Jillian Al-

brecht, "The Goulds in the Hunter Region of N.S.W. 1839–1840," *Naturae* 2 (1992): 18.

32. H. E. Strickland, "Report on the Recent Progress and Present State of Ornithology," *Report of the Fourteenth Meeting of the British Association for the Advancement of Science,* York, September 1844 (London, 1845), 170–221 (see 180, 186).

33. Cayley, "John Gould as an Illustrator", 167–172.

34. Vigors and Horsfield, *Linnaean Society Transactions,* vol. 15.

35. Whittell, *The Literature of Australian Birds,* 87.

36. Chisholm, "Mrs. John Gould," 342; Chisholm, "Elizabeth Gould—Some 'New' Letters," 321.

37. Gregory M. Mathews and Tom Iredale, "Gould as a Systematist," *Emu* 38 (1938): 172–175.

38. John Gould, Prospectus to *The Birds of Australia* (London, 1840–1848), quoted in K. A. Hindwood, "John Gould in Australia," *Emu* 38 (1938): 95–118.

39. Tom Iredale, "John Gould: the Bird Man", *Emu* 38 (1938): 90–95.

40. Sauer, *John Gould,* 99.

41. A. H. Chisholm, "The Story of Eliza Gould," *Victorian Naturalist* 56 (1939–1940): 22–25.

42. Lambourne, *John Gould,* 59.

43. Letter published in Chisholm, "The Story of Eliza Gould," 25.

44. John Gould to William Jardine, April 30, 1838, published in Sauer, *John Gould,* 99.

45. Hindwood, "John Gould in Australia"; Sauer, *John Gould,* 103–104.

46. Lady Franklin to her sister, October 4, 1838, published in Dr. George Mackaness, *Some Private Correspondence of Sir John and Lady Franklin (Tasmania, 1837–1845)* (Sydney, 1947).

47. Entire letter published in Chisholm, "Mrs. John Gould," 347.

48. Hindwood, "Mrs. John Gould," 133.

49. The letter was dated August 27, 1841; Lady Franklin did not know of Elizabeth's death on August 15; quoted in Chisholm, "Mrs. John Gould," 349.

50. Lady Franklin's Diary, published in *Papers and Proceedings of the Royal Society of Tasmania for the Year 1925,* quoted in Sauer, *John Gould,* 105–106.

51. *Hobart Town Courier,* May 24, 1839, quoted in Hindwood, "John Gould in Australia," 97–98.

52. Entire letter published in Chisholm, "Mrs. John Gould," 348.

53. Sauer, *John Gould,* 116–117.

54. Hindwood, "John Gould in Australia," 98–99.

55. Elizabeth Gould's diary is reproduced in its entirety in Hindwood, "Mrs. John Gould," 134–137.

56. Hindwood, "John Gould in Australia," 100; a complete account of this portion of the Goulds' sojourn in Australia is in Albrecht, "The Goulds in the Hunter Region," 7–20.

57. Hindwood, "Mrs. John Gould," 135–136.

58. Chisholm, "Mrs. John Gould," 349; Elizabeth was at her brothers' home when Lady Franklin wrote this letter.

59. Albrecht, "The Goulds in the Hunter Region," 12–13.

60. Hindwood, "Mrs. John Gould," 134–137.

61. Hindwood, "John Gould in Australia," 100.

62. Chisholm, "Elizabeth Gould—Some 'New' Letters," 331–333.

63. Sauer, *John Gould,* 117.

64. Gilbert P. Whitley, "John Gould's Associates," *Emu* 38 (1938): 141–167. Whitley lists Gould's collectors who perished.

65. Sauer, *John Gould,* 119; Hindwood, "Mrs. John Gould," 138.

66. Lambourne, *John Gould*, 15, 26.
67. The male bowerbird builds and decorates a bower for use as a courtship and mating area, with flowers, feathers, shells, etc.
68. Lambourne, *John Gould*, 59.
69. Sauer, *John Gould*, xv, 119.
70. Ibid., 120.
71. Chisholm, "Mrs. John Gould," 350–352.
72. Sharpe, "Biographical Memoir," xii–xiii.
73. Lear's letter is quoted in Lambourne, *John Gould*, 44.
74. Letter quoted in Sauer, *John Gould*, 42.
75. Lambourne, *John Gould*, 15.
76. J. Gould, *Birds of Australia* (1848), 5; quoted in Hindwood, "Mrs. John Gould," 133–134.
77. Chisholm, "Mrs. John Gould," 352.
78. Gregory M. Mathews, "John Gould: An Appreciation" *Emu* 38 (1938): 239–240.

Chapter 5: Dispelling the Myth of the Able Assistant

1. Agnes M. Clerke, *A Popular History of Astronomy during the Nineteenth Century* (Edinburgh, 1887), 421.
2. Margaret Lindsay Huggins to Joseph Larmor, October 17, 1910, Lm. 790, Larmor Papers, Royal Society Library, London. Quoted by permission of the Royal Society Library.
3. See, e.g., William Huggins, "The New Astronomy: A Personal Retrospect," *The Nineteenth Century* 41 (June 1897): 926; Charles E. Mills and C. F. Brooke, *A Sketch of the Life of Sir William Huggins* (London, 1936), 37–42; E. W. Maunder, *Sir William Huggins and Spectroscopic Astronomy* (London, 1913), 64; Marilyn Bailey Ogilvie, "Marital Collaboration: An Approach to Science," in *Uneasy Careers and Intimate Lives: Women in Science, 1789–1979*, ed. Pnina G. Abir-Am and Dorinda Outram (New Brunswick, N.J., 1987), 111–115.
4. Mills and Brooke, *Sketch*, 38–41.
5. Observatory notebooks of William and Margaret Huggins, Whitin Observatory Library, Wellesley College, Wellesley, Mass. There are six known observatory notebooks. For published descriptions of their contents, see Sarah Frances Whiting, "The Tulse Hill Observatory Diaries," *Popular Astronomy* (1916): 158–163, and Julie Morgan, "The Huggins Archives at Wellesley College," *Journal for the History of Astronomy* 11 (1980): 147. All excerpts from the Hugginses' notebooks cited here are used by permission of the Whitin Observatory.
6. Principal repositories for William Huggins's correspondence are the Cambridge University Library Archives; the Royal Astronomical Society Archives and the Royal Society Library, London; the National Museum of American History, Smithsonian Institution; the Mary Lea Shane Archives of the Lick Observatory, University of California, Santa Cruz; the Pusey Library, Harvard University; Dartmouth College Library; and the Library of Congress. Letters and notes from Margaret Huggins can be found throughout her husband's correspondence and in the Huggins Collection at Wellesley College.
7. Frontispiece, Sir William Huggins and Lady Margaret Huggins, *The Scientific Papers of Sir William Huggins* (London, 1909).
8. See Derek de Solla Price, *Little Science, Big Science* (New York, 1963), 86–91; Warren O. Hagstrom, *The Scientific Community* (Carbondale, Ill., 1975), 105–158; Alvin M. Weinberg, *Reflections on Big Science* (Cambridge, Mass., 1967), 47–53; idem, "Scientific Teams and Scientific Laboratories," *Daedalus* 99 (1970): 1056–1075; Lowell

L. Hargens et al., "Research Areas and Stratification Processes in Science," *Social Studies in Science* 10 (1980): 55–74.

9. Ogilvie, "Marital Collaboration," 114.

10. Margaret Huggins's age is given as thirty-two in the 1881 census (1881 Census Return for 90 Upper Tulse Hill Road, Lambeth, Holy Trinity Ecclesiastical District, RG 11/615/f.10). See also Maire Brück and Ian Elliott, "The Family Background of Lady Huggins (Margaret Lindsay Murray)," *Irish Astronomical Journal* 20 (1992): 210–211.

11. Personal communication, Ian Elliott to the author, October 3, 1991.

12. Sarah Frances Whiting, "Lady Huggins," *Science* 51 (1915): 854. See also Whiting, "Margaret Lindsay Huggins," *Astrophysical Journal* 42 (1915): 1–3; Louise Manning Hodgkins, "Lady Huggins: Astronomer," *The Christian Advocate* (October 21, 1915): 1417–1418; H. F. Newall, "Dame Margaret Lindsay Huggins," *Monthly Notices of the Royal Astronomical Society* 76 (1916): 278–282; Ogilvie, "Marital Collaboration," 110.

13. "God's Glory in the Heavens," *Good Words* 1 (1860): 23, 116, 161, 225, 289, 465, 513, 577, 625, 729.

14. Charles Pritchard, "A True Story of the Atmosphere of a World on Fire," *Good Words* 8 (1867): 250–251.

15. Marriage Record, September 8, 1875, Dublin County.

16. Hodgkins, "Lady Huggins," 1417.

17. 1851 Census, 97 Gracechurch St., London, St. Peter-upon-Cornhill Parish, HO 107/1531/f.1.

18. Mills and Brooke, *Sketch*, 11.

19. *Memoirs of the Royal Astronomical Society* 24 (1856): 253.

20. William Huggins, "Description of an Observatory Erected at Upper Tulse Hill," *Monthly Notices of the Royal Astronomical Society* 16 (1856): 175–176.

21. Huggins's involvement in lunar photography is alluded to in an address by Warren De la Rue on celestial photography. See Warren De la Rue, "Report on the Present State of Celestial Photography in England," *Report of the British Association* (Aberdeen, 1859): 132.

22. The wet collodion process was not the only photographic process available in 1863, although it is the one with which Miller was most familiar. See W. A. Miller, *Elements of Chemistry: Theoretical and Practical, Part II. Inorganic Chemistry,* 2d ed. (London, 1860), 825–846.

23. Agnes M. Clerke, "William Allen Miller," *Dictionary of National Biography* 37 (1894): 429–430.

24. William Huggins and W. A. Miller, "On the Spectra of Some of the Fixed Stars," *Proceedings of the Royal Society* 13 (1864): 244. Idem, "On the Spectra of Some of the Fixed Stars," *Philosophical Transactions* 154 (1864): 428.

25. William Huggins, "Note on the Photographic Spectra of Stars," *Proceedings of the Royal Society* 25 (1876): 445–446.

26. See, e.g., "Lady Margaret Huggins," *Who Was Who: 1897–1916* (London, 1935); Alice E. Donkin, "Margaret Lindsay Huggins," *The Englishwoman* (May 1915): 152.

27. Bernard V. and Pauline F. Heathcote, "The Feminine Influence: Aspects of the Role of Women in the Evolution of Photography in the British Isles," *History of Photography* 12 (1988): 260.

28. Margaret Huggins, March 31, 1876, notebook 2.

29. Margaret Huggins, April 3, 1876, notebook 2.

30. Margaret Huggins, May 7, 1876, notebook 2.

31. Margaret rarely took dictation from William, but when she did, she made this point clear to the reader.

32. Margaret Huggins, July 28, 1876, notebook 2.

33. Margaret Huggins, May 9, 1876, notebook 2.

34. Margaret Huggins, "June" 1876, notebook 2.

35. Note added by Margaret Huggins, William Huggins to David Gill, October 7, 1879, Gill Papers, South African Astronomical Observatory Archives. Quoted by permission of the South African Astronomical Observatory Archives.

36. Margaret Huggins, March 21, 1887, notebook 2.

37. Margaret Huggins, November 12, 1893, notebook 5.

38. William Huggins, "On the Spectra of Some of the Nebulae," *Philosophical Transactions* 154 (1864): 437–444.

39. See, e.g., Ogilvie, "Marital Collaboration," 111–114.

40. Margaret's early involvement in the work of the observatory raises provocative, but unanswerable, questions about her role in the writing and editing of papers that William published during 1876–1889. Whether or not he was the sole author of these papers, every effort would have been made to insure that it appeared that way. If Margaret participated in this process, she did so knowing that she was effectively writing herself out of the accounts. On the dynamics of this important aspect of the construction of William Huggins's public image, see Barbara J. Becker, "Electicism, Opportunism, and the Evolution of a New Research Agenda: William and Margaret Huggins and the Origins of Astrophysics" (Ph.D. diss., The Johns Hopkins University, 1993).

41. William Huggins and Mrs. Huggins, "On the Spectrum, Visible and Photographic, of the Great Nebula in Orion," *Proceedings of the Royal Society* 46 (1889): 40.

42. William Huggins to George Stokes, April 27, 1889, Stokes Papers, Add MS 7656.H1243, Manuscripts Room, Cambridge University Library. Quoted by permission of the Syndics of the University Library, Cambridge.

43. Margaret Huggins, October 12, 1888, notebook 2.

44. J. Norman Lockyer, "Researches on the Spectra of Meteorites: A Report to the Solar Physics Committee," *Proceedings of the Royal Society* 43 (1887): 133–139; idem, "Suggestions on the Classification of the Various Species of Heavenly Bodies," Bakerian Lecture, *Proceedings of the Royal Society* 44 (1888): 2–4. On the broader issues of the controversy over the identity of this spectral line, see A. J. Meadows, *Science and Controversy: A Biography of Sir Norman Lockyer* (Cambridge, Mass., 1972): chap. 7, esp. 183–187.

45. William Huggins, "Spectra of Nebulae," 444.

46. This suspected new element was popularly called "nebulium," apparently at Agnes Clerke's suggestion. However, William Huggins called it nebulum, while Margaret preferred nephelium or nephium. See Margaret Huggins, ". . . Teach me how to name the . . . light," *Astrophysical Journal* 8 (1898): 54; Richard F. Hirsh, "The Riddle of the Gaseous Nebulae," *Isis* 70 (1970): 203.

47. Margaret Huggins, October 12, 1888, notebook 2.

48. William Huggins, February 18, 1889, notebook 3.

49. William Huggins, March 6, 1889, notebook 3.

50. Margaret Huggins, March 9, 1889, notebook 2.

51. William Huggins, March 9, 1889, notebook 3.

52. Margaret Huggins, March 11, 1889, notebook 1.

53. William Huggins, March 11, 1889, notebook 3. Unfortunately, Margaret stopped making her separate notebook entries at this point and did not resume them until September 1889.

54. This would become clearer to the Hugginses when, following the appearance of their paper, Lockyer questioned William's claims to measurement accuracy. See J. Norman Lockyer, "On the Chief Line in the Spectrum of the Nebulae," *Proceedings of the Royal Society* 48 (1890): 167–198.

55. Margaret Huggins, March 6, 1889, notebook 2.

56. At the very outset of this research, Margaret expressed hope that they would

obtain photographs of the nebula's spectrum because, "It would be very important not to have to depend on *eye* observations in anything so difficult and important as that identity of Mg lines question" (Margaret Huggins, October 24, 1888, notebook 2).

57. Margaret Huggins, March 9, 1889, notebook 2.

58. Huggins and Huggins, "On Spectrum of Nebula in Orion," 48–49.

59. That this may not necessarily be the case can be seen in William Huggins's negative reaction to the news of Hertha Ayrton's receipt of the prestigious Hughes Medal. See Joan Mason, "Hertha Ayrton (1854–1923) and the Admission of Women to the Royal Society of London," *Notes and Records of the Royal Society of London* 45 (1991): 214–216.

60. J. Norman Lockyer, "Appendix to Bakerian Lecture: Suggestions on the Classification of the Various Species of Heavenly Bodies," *Proceedings of the Royal Society* 45 (1889): 200–201.

61. William Huggins, "Preliminary Note on the Photographic Spectrum of Comet b, 1881," *Proceedings of the Royal Society* 33 (1881): 1–3.

62. Margaret Huggins, n.d., added under June 30, 1881, notebook 2.

63. William Huggins to George Stokes, January 22, 1889, Stokes Papers, Add MS 7656.H1235, Manuscript Room, Cambridge University Library. Quoted by permission of the Syndics of the University Library, Cambridge.

64. John Ruskin, "Of Queens' Gardens," *Sesame and Lilies* (London, 1904), 107.

Chapter 6: The Comstocks of Cornell

1. Anna Botsford Comstock, *The Comstocks of Cornell: John Henry Comstock and Anna Botsford Comstock* (Ithaca, N.Y., 1953), 53–60. Phoebe Irish Botsford scrapbook, box 7, John Henry and Anna Botsford Comstock Papers, 21/23/25, Department of Manuscripts and University Archives, Cornell University Library (hereafter DMUA, CUL).

2. Comstock, *Comstocks*, 69–74. Anna Botsford diaries, newsclippings, and teaching certificate, box 7, Comstock Papers.

3. Morris Bishop, *A History of Cornell* (Ithaca, N.Y., 1962), 143–152; Comstock, *Comstocks*, 74–85; Charlotte Williams Conable, *Women at Cornell: The Myth of Equal Education* (Ithaca, N.Y., 1977), 54–61, 66–68, 74–76, 85–86; Roberta Frankfort, *Collegiate Women: Domesticity and Career in Turn-of-the-Century America* (New York, 1977), 29–31; Marilyn B. Ogilvie, *Women in Science: Antiquity through the Nineteenth Century* (Cambridge, Mass. 1986), 61–62, 182. Anna Botsford correspondence, 1871–1876, box 1, folders 5–12, Comstock Papers.

4. Comstock, *Comstocks*, 75–99.

5. Ibid., 1–35, 93–94. John Henry Comstock to Susan Allen Comstock, 1859–1868, box 1, folders 1–4, Comstock Papers.

6. Comstock, *Comstocks*, 36–46. See also John Henry Comstock's correspondence, 1859–1868, box 1, folders 1–4, Comstock Papers.

7. Comstock, *Comstocks*, 94–106, 139–140. Oral history interview of Albert Hazen Wright by Gould P. Colman, 12/19/62, 47/2/OH22, DMUA, CUL.

8. Comstock, *Comstocks*, 103, 105, 151.

9. Ibid., 105–136; John Henry Comstock, "Plan of Formation of the Biological Collection of Insects in the Department of Agriculture in Washington," *Canadian Entomologist* 11 (1879): 202–203, idem, *Report on Cotton Insects* (Washington, D.C., 1879), and idem, *Report on Scale Insects* (Washington, D.C., 1880).

10. Comstock, *Comstocks*, 137–138, 143–144, 147, 149–150.

11. Ibid., 106–108, 143–144, 147, 152–154, 158, 201–208; Thomas Woody, *A History of Women's Education in the United States* (New York, [1929], 1980), 247–249.

12. Comstock, *Comstocks*, 189; John Henry Comstock, *An Introduction to Entomology*

(Ithaca, N.Y., 1888); idem, "Evolution and Taxonomy: An Essay on the Application of the Theory of Natural Selection in the Classification of Animals and Plants," in *The Wilder Quarter-Century Book* (Ithaca, N.Y., 1893); idem, *Report of the Entomologist of the United States Department of Agriculture for the Year* 1880 (Washington, D.C., 1881); idem, *Report on Cotton Insects*; and Simon H. Gage, "A History of the Comstock Publishing Company," 1944, unpublished manuscript in the Simon H. Gage Papers, 14/26/30, DMUA, CUL.

13. Margaret W. Rossiter, *Women Scientists in America: Struggles and Strategies to 1940* (Baltimore, 1982), 290–292, 393.

14. Comstock, *Comstocks*, 150, 165, 179, 182, 184–185.

15. John Henry and Anna Botsford Comstock, *A Manual for the Study of Insects* (Ithaca, N.Y., 1895).

16. Ibid.

17. Ibid., 666–667.

18. Ibid., 672–673.

19. Ibid.; A. J. Meadows, "The Evolution of Graphics in Scientific Articles," *Publishing Research Quarterly* 7 (1991): 24–26; Alpheus Spring Packard, *Guide to the Study of Insects, and a Treatise on Those Injurious and Beneficial to Crops* (Salem, Mass., 1869).

20. Comstock, *Comstocks*, 188; Pamela M. Henson, "Evolution and Taxonomy: John Henry Comstock's Research School in Evolutionary Entomology at Cornell University, 1874–1930" (Ph.D. diss., University of Maryland, 1990), 185–188, 230–237, 257–259.

21. Comstock, *Comstocks*, 179–187, 198, 200–208, 216, 241–243.

22. Ibid., 82, 86, 139–140, 249; Henson, "Evolution and Taxonomy," 410–411; Michael M. Sokal, "Companions in Zealous Research, 1886–1986," *American Scientist* 74 (1986): 486–489.

23. James McKeen Cattell, ed., *American Men of Science* (New York, 1906); Comstock, *Comstocks*, 249; Conable, *Women at Cornell*, 89; Henson, "Evolution and Taxonomy," 346–347; Sally G. Kohlstedt, "In from the Periphery: American Women in Science, 1830–1880," *Signs* 4 (1978): 86–91; Ogilvie, *Women in Science*, 182; and Rossiter, *Women Scientists*, 59, 285.

24. Comstock, *Comstocks*, 138–139, 151–158, 177, 192, 196, 217–218; Henson, "Evolution and Taxonomy," 323–352; James G. Needham, "The Lengthened Shadow of a Man and his Wife," *Scientific Monthly* 62 (1946): 219–220, 227; Ogilvie, *Women in Science*, 61–62; Edward H. Smith, "The Comstocks and Cornell: In the People's Service," *Annual Review of Entomology* 21 (1976): 17–25.

25. Marian Lee [Anna Comstock], *Confessions to a Heathen Idol* (New York, 1906).

26. Comstock, *Comstocks*, 217–218, and Lee, *Confessions*, passim.

27. Comstock, *Comstocks*, 108–109, 139, 146, 168–171; J. H. Comstock correspondence with Benedict W. Law, Comstock Papers.

28. Comstock, *Comstocks*, 101–102; Needham, "Lengthened Shadow," 226.

29. Comstock, *Comstocks*, 172–177, 183–187. See correspondence between the Comstocks and Jordan, Comstock Papers, and the David Starr Jordan Papers in the Stanford University Archives.

30. Comstock, *Comstocks*, 189–190; E. Laurence Palmer, "The Cornell Nature Study Philosophy," *Cornell Rural School Leaflet* 38 (1944): 39–40; Ruby Green Smith, *The People's Colleges: A History of the New York State Extension Service in Cornell University and the State, 1876–1948* (Ithaca, N.Y., 1949), 37.

31. Comstock, *Comstocks*, 190–191; Palmer, "Nature Study," 39–40; R. Smith, *People's Colleges*, 31–33, 37.

32. Comstock, *Comstocks*, 191–192; Palmer, "Nature Study," 40; R. Smith, *People's Colleges*, 43–50.

33. Palmer, "Nature Study," 7–8, 17–22.

34. Liberty Hyde Bailey, *The Nature-Study Idea: An Interpretation of the New School-movement to Put the Young into Relation and Sympathy with Nature* (New York, 1903).

35. Bailey, *Nature-Study Idea*; Gould P. Colman, *Education and Agriculture: A History of the New York State College of Agriculture at Cornell University* (Ithaca, N.Y., 1963), 129–130; Comstock, *Comstocks*, 191–192; Palmer, "Nature Study," 40–43.

36. Colman, *Education and Agriculture*, 155–158; Comstock, *Comstocks*, 196, 209–215, 228; Palmer, "Nature Study," 26–27, 30–32, 43–49, 72–75; Rossiter, *Women Scientists*, 64–65; R. Smith, *People's Colleges*, 63–64, 75, 87, 222.

37. Comstock, *Comstocks*, 197; Palmer, "Nature Study," 48, 68–69.

38. John Henry Comstock, *Insect Life: An Introduction to Nature Study and a Guide for Teachers, Students, and Others Interested in Out-of-Door Life* (New York, 1897).

39. Comstock, *Comstocks*, 210–212, idem, *How to Keep Bees* (New York, 1905), and idem, *The Ways of the Six-Footed* (Boston, 1903); John Henry and Anna Botsford Comstock, *How to Know the Butterflies: A Manual of the Butterflies of the Eastern United States* (New York, 1904).

40. Comstock, *Comstocks*, 192, 210–212, 274.

41. Ibid., 229–234, and Anna Botsford Comstock, *Handbook of Nature Study* (Ithaca, N.Y., 1911); Palmer, "Nature Study," 62–63.

42. Comstock, *Handbook of Nature Study*.

43. Comstock, *Comstocks*, 192–244, 254, 259; Conable, *Women at Cornell*, 89–91, 127, 130; Rossiter, *Women Scientists*, 62–65; E. Smith, "Comstocks and Cornell," 17–20; R. Smith, *People's Colleges*, 84.

44. Comstock, *Comstocks*, 255.

45. Abraham Flexner, *Henry S. Pritchett, A Biography* (New York, 1943), 87–92; Howard J. Savage, *Fruits of an Impulse: Forty-Five Years of the Carnegie Foundation, 1905–1950* (New York, 1953), 11, 13, 81, 168–170. Simon H. Gage to Albert Mann, 6/17/21, box 10, and James G. Needham to Simon H. Gage, 10/18/27, box 9, Gage Papers.

46. Comstock, *Comstocks*, 251–256, idem, *The Pet Book* (Ithaca, N.Y., 1914), and idem, *Trees at Leisure* (Ithaca, N.Y., 1916); Kathleen Jacklin, "Anna Botsford Comstock," in *Notable American Women*, ed. Edward T. James, Janet W. James, and Paul S. Boyer (Cambridge, Mass., 1971), 367–369; Rossiter, *Women Scientists*, 282.

47. Comstock, *Comstocks*, 210, 213, 262–264.

48. Ibid., 245–259, 267; R. Smith, *People's Colleges*, 41.

Chapter 7: *Grace Chisholm Young and William Henry Young*

The author would like to thank David Wiegand and the editors for their assistance in editing this manuscript. She is grateful to Ivor Grattan-Guinness, who initially organized the letters and information about the Youngs. In addition she appreciates the assistance of the personnel at the University of Liverpool Archives.

1. In addition to this chapter Sylvia Wiegand has written two earlier articles on Grace Chisholm Young. The first appeared in *The Association for Women in Mathematics Newsletter* 7, no. 3 (May-June 1977): 5–10; and the second in *Women of Mathematics: A Biobibliographic Sourcebook*, ed. L. S. Grinstein and P. J. Campbell (Westport, Conn., 1987), 247–254. For further information the reader could consult M. L. Cartwright, "Grace Chisholm Young," *Journal of the London Mathematical Society* 19 (1944): 185–192.

2. Grace's autobiographical notes (D140/12/1). Hereafter all references D—refer to the numbering in the archives at the University of Liverpool.

3. Cecily Tanner's notes mention Will's coaching: "For he became, as Fellow of Peterhouse, the most exacting & successful coach of the day, with the reputation of getting his pupils into the class above the one they legitimately belonged to. He obtained

a virtual monopoly of the math. coaching at Girton College, where comparatively large numbers of girls successfully took up mathematics, quite contrary to Girton policy today" (D140/2/2.1).

4. Grace's autobiographical notes include this description of one of her first conversations at college: " 'Who are you going to coach with?' they asked. 'I haven't an idea. What is coaching?' she asked. 'Oh, don't you know? You have two hours of Mr. Young or three hours of any other lecturer.' 'You had much better have Mr. Young,' said another. 'He has such a lovely mouth. Madge calls it a lemniscate mouth,' said a third. Grace turned away in disgust & went up the stairs to her room. How could they talk like that about a lecturer! Well! If she had a choice she would *not* choose Mr. Young!" (D140/12/23).

5. (D140/2/2.2).

6. An article from the *Daily Graphic* newspaper of May 1895 mentions Grace's achievements (performance on the Oxford examination and study in Germany) and includes a picture of her (D140/12/33).

7. Grace to Hugh Chisholm, March 1895, quoted in I. Grattan-Guinness, "William Henry and Grace Chisholm Young: A Mathematical Union,"*Annals of Science* 29 (1972): 105–186, 126–129. Excerpts from G. C. Young's autobiographical notes appear in this article.

8. Cecily Tanner's notes refer to articles in German and English newspapers about Grace setting a precedent among women receiving a Ph.D. degree (D140/2/2.2).

9. Grace compared her situation to Kovalevskaia's as follows: "It seems she was never in Göttingen she simply sent her thesis, had no examination & was awarded the degree; this is not possible now as the regulations are much stricter & a great deal of stress is laid on the Exam. which is the only test of one's 'Minors' & as Prof. Klein says they want their Doctors to know something outside their 'Major' not to be one idea people" (Grace to Hugh Chisholm, February 20, 1894 [D140/2/–]). Klein refers to Grace as "the first woman to pass the normal examination in Prussia for the doctor's degree" in F. Klein, *Elementary Mathematics from an Advanced Standpoint: Arithmetic, Algebra, Analysis*, 3d ed., trans. E. R. Hedrick and C. A. Noble (New York, 1939), 179–180.

10. The following letters illustrate the development of this correspondence and include the quotations given: Will to Grace, July 21, 1895 (D140/8/76a), Will to Grace, October 11, 1895 (D140/8/1–95), Grace to Will, October 13, 1895 (D140/8/1–95), and Grace to Will, May 7, 1896 (D140/8/236a).

11. Grace to Frances de Evans, a friend from Girton College, January 1897 (D140/6/158 and D140/6/159a).

12. Grace's autobiographical notes, quoted in Grattan-Guinness, "Union," 131.

13. Quoted in G. H. Hardy, "William Henry Young," *Journal of the London Mathematical Society* 17 (1942): 309.

14. So Grace wrote in a letter to Frances de Evans, February 7, 1897 (D140/4/2). Grace's paper was received on March 2, 1897, according to I. Grattan-Guinness, "A Joint Bibliography of W. H. and G. C. Young," *Historia Mathematica* 2 (1975): 47.

15. Grace to Frances de Evans (D140/6/223).

16. W. H. Young, "On Integration with Respect to a Function of Bounded Variation," *Proceedings of the London Mathematical Society* 2, no. 13 (1914): 110 (footnote).

17. Grace (Göttingen) to Will, March 9, 1907 (D140/6/1076).

18. The surviving family members also have an authoritarian manner of speaking, even though they welcome other viewpoints. The Youngs' daughter Janet Michael explained to Sylvia Wiegand (June 1992) that the children, including the girls, were taught at an early age to speak forcefully and with conviction.

19. Daughter-in-law Elizabeth Young has told Sylvia Wiegand (probably about 1975) that Will was unfriendly and abrupt, while Grace was kind, patient, and well

loved by all who knew her; she never became angry with anyone unless they crossed Will.

20. Grace to Will, May 7, 1896 (D140/8/236a).

21. Will (Sieci) to Grace (Göttingen), February 28, 1907 (D140/6/1072).

22. L. C. Young, "The Life and Work of W. H. and G. C. Young," in preparation for a series of volumes on the complete works of W. H. and G. C. Young, 30.

23. Parts of the report have been published in I. Grattan-Guinness, "University Mathematics at the Turn of the Century, Unpublished Recollections of W. H. Young," *Annals of Science* 29 (1972): 369–384.

24. G. Sutton, "The Centenary of the Birth of W. H. Young (20th October 1863)," *Mathematics Gazette* 49 (1965): 20.

25. Helen (Marian) Canu, who had studied graduate-level mathematics, died early—at age 38 in 1941. Second child Rosalind (Cecily) Tanner, a mathematician and a historian, died at age 92 in 1992. (Cecily was awarded Girton College's Gamble Prize about 1937.) Daughter Janet Dorothea Michael turned 92 in December 1993; she fulfilled her mother's dream of becoming a medical doctor and still practices medicine in Croydon, Surrey. Laurence Chisholm Young, a mathematician at the University of Wisconsin in Madison for many years, retired in 1975 at age 70, but in 1994 he is still there, actively writing, doing mathematics, and editing his parents' works. Formerly a chemist and international civil servant, Patrick Chisholm Young (born in 1908) lives near his sister Janet and is busy with many volunteer organizations.

26. Grattan-Guinness, "Bibliography," 43.

27. Hardy, "William Henry Young," 224.

28. H. Lebesgue, *Leçons sur l'integration et la recherche des functions primitives, professées au Collège de France,* 3d ed. (New York, 1973), 263.

29. Hardy, "William Henry Young," 228.

30. In a letter to her sister Helen in 1905 (D140/6/893), Grace wrote, "I have got 150 figures & photos to prepare carefully for the engraver so that our little 'Introduction to Geometry' may go off . . . this week." In other letters she seems to be working very hard by herself on the book. She seems more interested in the teaching of children than Will was. Although Grace preferred not to take credit, she was persuaded that the German edition should be in her name by the suggestion that it would help sales: "[The publisher] is writing a preface & wants to put me rather in the foreground as his idea is to get it into the *girl's* schools" (Grace [Göttingen] to Will, February 27, 1907 [D140/6/1067]). She might also have been persuaded by the fact that she had essentially written the book herself.

31. G. C. Young and W. H. Young, *The Theory of Sets of Points* (Cambridge, 1906), x.

32. Georg Cantor to Grace, January 23, 1907 (D140/6/1028).

33. E. F. Collingwood and A. J. Lohwater, *The Theory of Cluster Sets* (Cambridge, 1966). The bibliography lists two joint articles and two articles by Will.

34. F. Riesz and B. Nagy, *Functional Analysis* (New York, 1955), 17. (Many women mathematicians who were graduate students during those years have told Sylvia Wiegand of their joy at seeing a reference to a *Mrs.* Young in their graduate textbook in real analysis—because women mathematicians were so seldom mentioned in books.) Another discussion of this theorem with pictures of Dini, Grace Chisholm Young, Denjoy, and Saks appeared more recently; cf. P. S. Bullen, "Bad Functions Can't Be Too Bad," *Menemui Mathematica* 8 (1986): 36–50.

35. Klein, *Elementary Mathematics,* 179–180, and E. W. Hobson, *Theory of Functions of a Real Variable and the Theory of Fourier's Series* (Cambridge, 1927). The latter book contains many references to the work of Grace and Will.

36. G. C. Young, "On the Curve $y = \left(\frac{1}{x^2 + \sin^2 \phi}\right)^{3/2}$ and Its Connection with an

Astronomical Problem," *Monthly Notices* of the *Royal Astronomical Society* 57 (1897): 379–387; "On the Solution of a Pair of Simultaneous Diophantine Equations Connected with the Nuptial Number of Plato," *Proceedings of the London Mathematical Society* 2, no. 23 (1925): 27–44; "Pythagore, comment a-t-il trouve son théorème?" *Enseignement mathematica* 25 (1926): 248–255.

37. The Youngs left over six thousand letters, correspondence between themselves and with other prominent people of their times. The letters, which are stored in the archives of the University of Liverpool, are not yet accessible to the general public, to protect the privacy of living family members.

38. Will to Grace, September 6, 1905, postcard (D140/6/892).

39. Will (Hanover) to Grace (Göttingen), September 21, 1905, postcard (D140/6/895).

40. Will to Grace, September 5, 1906, postcard (D140/6/989).

41. Grace (Göttingen) to Will (Cheshire), October 26, 1906, postcard (D140/6/980[2]), and Grace (Göttingen) to Will (Cheshire), November 2, 1906, (D140/6/982[2]).

42. Will (Cromer) to Grace (Göttingen), July 2, 1908 (D140/6/1169).

43. Will to Grace, about 1906, quoted in Grattan-Guinness, "Union," 141–142.

44. Cecily Tanner to Lida Barret, letter of about 1968.

45. Laurence Young to Sylvia Wiegand, letter of about 1991.

46. Grattan-Guinness, "Union," 141, 151.

47. Laurence Young and Cecily Tanner have both mentioned this.

48. Apparently neither of the other two women students studying at Göttingen at the same time as Grace, Americans Isabel Maltby and May Winston, continued to do research. Winston gave up mathematics when she married a professor at the University of Kansas, according to a letter from her daughter, Caroline Beschers, to Sylvia Wiegand (January 30, 1979). See also Betsey S. Whitman, "Mary Frances Winston Newson (1869–1959)," in *Women of Mathematics: A Biobibliographic Sourcebook,* ed. Grinstein and Campbell, 161–164.

49. C. Reid, *Hilbert,* (Berlin, 1970), 143.

50. Grace (Göttingen) to Helen, April 2, 1905, postcard (D140/6/893).

51. Will (on Ayrshire hotel stationery) to Grace (Geneva), September 15, 1909 (D140/6/1198).

Chapter 8: Marriage and Scientific Work in Twentieth-Century Canada

Research was funded by the Social Science and Humanities Research Council of Canada. I thank Dr. Mary Arai, Dr. A.W.H. Needler, and the late Dr. Helen Sawyer Hogg for personal information, David Ainley and Margaret Silk for sharing my visits to the Pacific Biological Station and the David Dunlap Observatory respectively, and Susan Drysdale, Barbara Meadowcroft, Helena Pycior, Nancy Slack, and Judy Whitehead for their input on the project.

1. In the early literature on the professionalization of science, there is little mention of the implications of this process for women, but see Margaret W. Rossiter, *Women Scientists in America: Struggles and Strategies to 1940* (Baltimore, 1982), Pnina Abir-Am and Dorinda Outram, eds., *Uneasy Careers and Intimate Lives: Women in Science, 1789–1979* (New Brunswick, N. J., 1987), and G. Kass-Simon and Patricia Farnes, eds., *Women of Science: Righting the Record* (Bloomington, 1990). On the Canadian situation see Marianne G. Ainley, ed., *Despite the Odds: Essays on Canadian Women and Science* (Montreal, 1990).

2. Information on this topic may be found in the correspondence of Canadian university presidents, e.g., the H. M. Tory Papers, University of Alberta, and the Sir Arthur Currie Papers, McGill University, in the correspondence of numerous scientists, and in various Canadian university histories.

3. The implications of the Matthew Effect for women scientists need to be explored. W. L Bilby, "Sex Differences in Careers," in Harriet Zuckerman, Jonathan R. Cole, and John T. Bruer, eds., *The Outer Circle: Women in the Scientific Community* (New York, 1991); on Canadian women see Joan Pinner Scott, "The Disadvantagement of Women by the Ordinary Processes of Science: The Case of Informal Collaborations," in *Despite the Odds,* ed. Ainley, 316–328; I have analyzed the career paths of more than a dozen Canadian scientific couples between 1920 and 1970, and of nearly two hundred women scientists between 1890 and 1970, and found that Merton's "Matthew Effect" is a useful concept in understanding women's scientific careers. See M. G. Ainley, "Women and the Matthew Effect: A Century of Cumulative Disadvantages for Canadian Women in Science" (paper presented at the Canadian Sociology and Anthropology Association Conference, Kingston, Ontario, 1991). See also Margaret W. Rossiter, "The MatthewMatilda Effect in Science," *Social Studies of Science* 23 (1993): 325–341.

4. Rossiter, *Women Scientists*; M. F. Rayner-Canham and G. W. Rayner-Canham, *Harriet Brooks: Pioneer Nuclear Scientist* (Montreal, 1992); Ainley, *Despite the Odds.*

5. Rossiter, *Women Scientists*; Judith Fingard, "Gender and Inequality at Dalhousie: Faculty Women before 1950," *Dalhousie Review* (winter 1984–1985): 687–703.

6. Lorraine C. Smith, "Canadian Women Natural Scientists—Why Not?" *The Canadian Field-Naturalist* 90 (January–March 1976): 2–3.

7. Biographical information on the Berkeleys was provided by their son-in-law, Dr. A.W.H. Needler, and their granddaughter, Dr. Mary Needler Arai; I thank Dr. A.W.H. Needler for lending me a copy of Cyril Berkeley, *My Autobiography* (n.p., n.d., privately published). In 1892 Edith obtained a £15 sterling/year scholarship, for two years; in 1894 she was awarded a three-year Surrey Council Scholarship worth £60 sterling/year. Undated list of Edith's educational background, in the possession of Mary Needler Arai; Medical Student Registry Certificate No. 24309, October 2, 1895. University College, London, admitted and granted degrees to women long before Oxford and Cambridge. See H. H. Bellot, *University College London* (London, 1929).

8. Students took junior, intermediate, and advanced courses. B.S. candidates had to pass the matriculation, the intermediate (taken at the earliest one academic year after the matriculation), and the degree examinations. At the degree level, Edith obtained first-class honors in zoology, and a pass in chemistry, but failed geology. There is no evidence that she ever received her B.S. The *General Register* of the University of London "makes no reference to a degree." B. C. Weedon to M. G. Ainley, April 15, 1992.

9. Berkeley, *Autobiography.* Only the most promising students could conduct research in Ramsay's laboratory.

10. Sir William A. Tilden, *Sir William Ramsay, K.C.B., F.R.S.: Memorials of His Life and Work* (London, 1918); Morris W. Travers, *A Life of Sir William Ramsay* (London, 1956); Cyril Berkeley, "Recollections of Sir William Ramsay and His Work," *Chemistry in Canada* 14 (February 1962): 31, 34, and (August 1962): 16, 18, 19.

11. Berkeley, *Autobiography.*

12. There is a growing literature on the contradictory position of British wives in the colonies. See, for instance, Nupur Chaudhuri and Margaret Strobel, eds., *Western Women and Imperialism: Complicity and Resistance* (Bloomington, 1992).

13. He collected specimens at every thousand feet on the Hooker Trail. Later, one species was named *Megascolides bergtheili* in his honor.

14. Berkeley, *Autobiography.*

15. J. P. McMurrich, "Presidential Address—Fifty Years of Canadian Zoology," *Transactions of the Royal Society of Canada*, section 6, 2 (1917): 1–14.

16. Edith Berkeley Memorial Lectures flier, Department of Zoology, University of British Columbia; the Board of Governors Minutes, give no job description; Chris Hives, personal communication.

17. Lee Stewart, *"It's up to You": Women at UBC in the Early Years* (Vancouver, 1990).

18. See A. W. H. Needler, "Biological Station, Nanaimo, B.C., 1908–1958," *Journal of the Fisheries Research Board of Canada* 15 (1958): 759–796.

19. The *Autobiography* did not mention their financial arrangements. "There was some family money and they saved some in India." Dr. Mary Needler Arai to M. G. Ainley, April 1, 1992.

20. Edith Berkeley Memorial Lectures flier.

21. Marian H. Pettibone, "Type-Specimens of Polychaetes Described by Edith and Cyril Berkeley (1923–1964)," *Proceedings of the United States National Museum* 119 (1967): 1–23.

22. Berkeley, *Autobiography.*

23. Ibid.

24. J. Cameron Stevenson, "Edith and Cyril Berkeley—An Appreciation," *Journal of the Fisheries Research Board of Canada* 28, no. 10 (October 1971): 1363, 1364; Doug Peck, "The Berkeleys: Octogenarian Couple Recall Active Life," *Nanaimo Daily Free Press,* January 17, 1959. Interviews with Mary Needler Arai, August 9, 1985; A.W.H. Needler, July 22, 1986; Joan Marsden, November 28, 1988.

25. Stevenson, "Edith and Cyril Berkeley"; Pettibone, "Type-Specimens."

26. Interview with Marianne G. Ainley, June 25, 1991; by 1951 she had been elected to fellowship in the Royal Society of Canada (1946). See also R. A. Jarrell, *The Cold Light of Dawn: A History of Canadian Astronomy* (Toronto, 1988).

27. Nicole Morgan, *The Equality Game: Women in the Federal Public Service (1908–1987)* (Ottawa, 1988). Apparently, antinepotism rules were more strictly enforced in Canadian government institutions than at universities; larger universities, such as McGill and Toronto, were more likely to use such rules than were the smaller provincial ones, perhaps because the latter could not afford to hire men (M. G. Ainley, research notes). See Rossiter, *Women Scientists,* for antinepotism rules in the United States.

28. Jarrell, *Cold Light.* Helen Hogg disagreed with Jarrell's statement that amateur astronomy was "poorly developed" in Canada; personal communication.

29. Helen Hogg, conversations with the author.

30. Helen Sawyer Hogg, "Shapley's Era," in *The Harlow Shapley Symposium on Globular Cluster Systems in Galaxies,* ed. J. E. Grindlay and A. G. Davis Philip (Cambridge, Mass., 1988).

31. Ibid.

32. H. S. Hogg to M. G. Ainley, April 28, 1992.

33. University of Toronto Press release, January 4, 1967.

34. See Margaret W. Rossiter, "Women's Work in Science, 1880–1910," *Isis* 71 (1980): 391–398; also John Lankford and Rickey L. Slavings, "Gender and Science: Women in American Astronomy, 1859–1940," *Physics Today* (March 1990): 58–61.

35. H. S. Hogg to M. G. Ainley, April 28, 1992. On Payne-Gaposchkin, see Peggy A. Kidwell, "Cecilia Payne-Gaposchkin: Astronomy in the Family" in *Uneasy Careers and Intimate Lives,* ed. Abir-Am and Outram, 216–238, and Cecilia Payne-Gaposchkin, *An Autobiography and Other Recollections* ed. K. Haramundanis (Cambridge, 1984); Peggy A. Kidwell, personal communication.

36. Helen Hogg, conversations with the author.

37. Helen Hogg, "Memories of the Plaskett Era of the Dominion Astrophysical

Observatory, 1931–1934," *Journal of the Royal Astronomical Society of Canada* 82, no. 6 (1988): 328–335; Jarrell, *Cold Light.*

38. Dr. Alice V. Douglas of McGill University obtained access to it in 1932.
39. Hogg, "Memories."
40. H. S. Hogg to M. G. Ainley, April 28, 1992.
41. Ibid.
42. Ibid.
43. Ibid.
44. Hogg, "Memories."
45. Interview with M. G. Ainley, June 25, 1991.
46. H. S. Hogg to M. G. Ainley, April 28, 1992.
47. H. S. Hogg to M. G. Ainley, July 19, 1991.
48. Ruth Northcott (M.A. 1935, University of Toronto) was hired by the Department of Astronomy as a lecturer in 1944. Although Northcott never married, her career advanced slower than Hogg's; she became assistant professor in 1954 and associate professor in 1962; University of Toronto Archives. Dr. Alice V. Douglas, an internationally known astrophysicist (another single woman), remained a lecturer at McGill 1926–1939; in 1939 she went to Queen's, where her position was dean of women. She was paid an extra $400 for the one astronomy course she taught from 1941 to 1946, when her stipend became $500. Douglas became full-time professor of astronomy only in 1959; Queen's University Archives. On women at the University of Toronto, see Alison Prentice, "Bluestockings, Feminists, or Women Workers? A Preliminary Look at Women's Early Employment in the University of Toronto," *Journal of the Canadian Historical Association,* n.s., 2 (1991): 231–261.
49. H. S. Hogg, conversations with the author.
50. Ibid.
51. Ibid.
52. Ibid.
53. Invitation to "Dedication Ceremony," June 19, 1992.
54. Judith A. Laroque to Helen S. Hogg, November 30, 1992; copy in the author's possession.
55. Sally Hogg MacDonald to M. G. Ainley, March 28, 1993.

Chapter 9: Unusually Close Companions

I wish to thank Dorothy Blanchard and Grace Blanchard Iverson for graciously sharing with me information about their family and for permitting me to use the family papers, the staff of the Bentley Historical Library at the University of Michigan for their courteous help, and Stephen F. Austin State University for a faculty development leave which provided time for research.

The Blanchard Family Papers, housed in the Michigan Historical Collections, Bentley Library (BL), are referred to in the following notes as BFP-BL. Other papers and letters which the family has not yet deposited in the library are referred to as BFP. Unpublished letters and papers which Frieda wrote are cited as FC or FCB; those Frank wrote are cited as FNB.

1. Margaret Rossiter, *Women Scientists in America: Struggles and Strategies to 1940* (Baltimore, 1982), xv–xviii.
2. Nematodes are small, unsegmented roundworms, also called threadworms ("nematode" means threadlike), eelworms, or nemas. Many are free-living, although the most commonly known are parasites of humans, other animals, and plants. They are in a class by themselves and have no close relatives.

3. There are a few essays about Cobb, including J. R. Christie, "Obituary, Nathan Augustus Cobb," *Transactions of the American Microscopical Society* 51 (1932): 276–278; Edna M. Buhrer, "Nathan Augustus Cobb (1859–1932), a Tribute," *Journal of Nematology* 1 (January 1969): 2–3; Sylvia McGrath, "Nathan Augustus Cobb," *Encyclopedia USA* 13 (1991): 81–83; and R. N. Huettel and A. M. Golden, "Nathan Augustus Cobb: The Father of Nematology in the United States," *Annual Review of Phytopathology* 29 (1991): 15–26. However, the best source of information about his early life and impact on his children is Frieda Cobb Blanchard, "Nathan A. Cobb, Botanist and Zoologist: A Pioneer Scientist in Australia," *The Asa Gray Bulletin*, n.s., 3 (spring 1959): 205–272.

4. Blanchard, "Nathan A. Cobb," 205–232, 251–255; McGrath, "Cobb," 81; Alice Cobb to Lizzie (Harriet Elizabeth Proctor Greenwood), June 7, 1890, BFP.

5. Blanchard, "Nathan A. Cobb," 232–238.

6. Ibid., 238.

7. Ibid., 238, 251–255. Blanchard's devotion to her parents and admiration for them is clear in the tone of numerous letters in the Blanchard Papers.

8. Ibid., 235–236; FCB, undated notes on irregular early schooling, BFP.

9. See Huettel and Golden, "Nathan Augustus Cobb," for the most recent discussion of Cobb's continuing influence; see also *American Men of Science,* 1st ed. (1906) through 4th ed. (1927).

10. Record of Frieda Cobb, Radcliffe College Archives; "Class of 1913, Yearbook," Radcliffe College Archives.

11. FCB to A. L. Taylor, April 15, 1968, BFP.

12. FC to Nathan Cobb, August 9, 1916; FC to Alice Cobb, September 10, 1916, BFP-BL.

13. Kenneth L. Jones, *The Harley Harris Bartlett Diaries* (Ann Arbor, 1975), 3–4, 295. Margaret Cobb became a psychologist and Dorothy Cobb Adams a medical doctor.

14. No biography of Frank Blanchard exists, but there are several good memorials prepared shortly after his death. The most comprehensive are Howard K. Gloyd, "Frank Nelson Blanchard, Scholar and Teacher," *Herpetologica* 1 (1940): 197–211; M. Graham Netting, "Frank Nelson Blanchard, as He Appeared to His Students," *Bios* 10 (1939): 131–135; and a memorial by A. G. Ruthven, G. R. LaRue, and H. H. Bartlett read at the October 4, 1937, meeting of the Faculty of the College of Literature, Science, and Arts at the University of Michigan, copy in BFP-BL. A more recent appraisal, based on the earlier memorials, is in Kraig Adler, *Contributions to Herpetology*, no. 5 (1989), 95–96.

15. Frank N. Blanchard, "Two New Species of Stigonema," *Tufts College Studies* 3 (1914): 117–124.

16. FNB, diaries for 1915 and 1916, quotations from February 9, 1915, and April 8, 1916, BFP-BL; C. E. Gordon to George R. LaRue, December 14, 1937, BFP.

17. FNB, diary, 1916, BFP-BL.

18. See FNB, diaries for 1916, 1917, and 1918; FC letters to her parents for those same years, esp. FC to family, March 15, 1917, November 7, 1917, and February 8, 1918, BFP-BL; and "Frieda Cobb Blanchard," in *Twenty-Fifth Anniversary Report, Radcliffe Class of 1913* (Cambridge, 1938), 20–21.

19. FNB, diaries for 1917, 1918, and 1919. See esp. May 16, 19, 28, July 26, and August 25, 1917; February 18, 19, and November 18, 1918; April 19 and October 2, 7, 1919, BFP-BL.

20. Hugo de Vries and H. H. Bartlett, "The Evening Primroses of Dixie Landing, Alabama," *Science* 35 (1912): 599–601. Evening primroses are New World plants carried to Europe during the European colonization of America.

21. See Ralph E. Cleland, *Oenothera: Cytogenetics and Evolution* (London and New

York, 1972) esp. v–ix, 216–220; Jones, *Bartlett Diaries*, 1–3, 31; author's interviews with Kenneth Jones and Erich Steiner, July 11, 1990. See also H. H. Bartlett, "The Experimental Study of Genetic Relationships," *American Journal of Botany* 11 (1915): 132–155, as an example of Bartlett's early work on botanical genetics.

22. L. R. Flook and H. A. Gleason, "The University's Botanical Garden," reprinted from *The Michigan Alumnus* (May 1917), in William S. Benninghoff, ed., "The University of Michigan's Botanical Gardens, the First 75 Years," *Bartlettia* 2, no. 3 (February 1982): 5–8; H. H. Bartlett, "The Botanical Garden at the University of Michigan," reprinted from *The Michigan Quarterly Review* (February 1943), in Benninghoff, "Botanical Gardens," 9–20; FC to family, October 9 and 29, 1916, and March 15, 1917, BFP-BL.

23. Bartlett, "The Botanical Garden," 17–19; Recommendations to Executive Committee of the Botanical Gardens, May 21, 1918; Minutes of a Meeting of the Administrative Committee of the Botanical Garden of the University of Michigan, April 3, 1919; H. H. Bartlett to the Administrative Committee of the Botanical Gardens, March 13, 1920, all in Michigan University Botanical Gardens file, BL; FC to Father, August 6, 1919, BFP-BL. Women had held the rank of instructor at the University of Michigan beginning in the late 1800s, but the proportion of women on the university faculty decreased in the first decades of the twentieth century; during Frieda's active career there, women had almost no representation at policy-making levels. See Dorothy Gies McGuigan, *A Dangerous Experiment: 100 Years of Women at the University of Michigan* (Ann Arbor, 1970), 74–75, 86, 111–112.

24. Jones, *Bartlett Diaries*. Several people whom the author interviewed between 1982 and 1992 indicated that Blanchard was the active administrator at the gardens; some thought she was the director.

25. See FC letters to her family, 1916–1918, esp. October 8, 1916, November 15, 1916, February 10, 1917, March 25, 1917, October 27, 1917, and December 16, 1917; also "Botanical Society of America, Program for 12th Annual Meeting, Pittsburgh, December 28, 1917–January 1, 1918," all in BFP-BL.

26. FC to family, October 4, 1917, December 18, 1917, and September 12, 1918; Nathan Cobb to FC, October 1, 1917, December 5, 1917, and February 17, 1918; FC to Nathan Cobb, December 12, 1917, BFP-BL.

27. Frieda Cobb, "A Case of Mendelian Inheritance Complicated by Heterogametism and Mutation in *Oenothera pratincola*," *Genetics* 6 (January 1921): 1–42. Interview with Erich Steiner, July 11, 1990.

28. FC to family, September 12, 1918, November 9, 1919, May 3, 1920, and June 19, 1920, BFP-BL.

29. FC to FNB, August 10, 1920, and December 6, 1920, BFP.

30. FC to family, March 15, 1917, April 26, 1918, February 1, 1919, September 28, 1919, October 28, 1920, March 5, 1921, May 1, 1921, and May 15, 1921, BFP-BL.

31. FNB to the Cobbs, June 13, 1922; FCB to Mother, June 22, 1922, June 23, 1922, September 2, 1923, September 20, 1924, January 12, 1925, and February 16, 1925; FCB to family, July 17, 1922, October 28, 1922, and June 5, 1923, BFP-BL.

32. Netting, "Frank Nelson Blanchard," 131–135; FCB to Mother, June 1, 1925, BFP-BL. See also Gloyd, "Frank Nelson Blanchard," and Adler, *Contributions*, 95–96. Frank Blanchard was listed in the 3d (1921), 4th (1927), and 5th (1933) editions of *American Men of Science*. A letter from Cattell telling him he had received a "star" for the forthcoming 6th edition was one of the last things he understood before his death.

33. See Botanical Gardens file in the Bentley Library as well as the BFP.

34. FCB to family, February 11, 1923; FCB to Mother, February 23, 1923; FCB to Mother Blanchard, July 11, 1924, BFP-BL. See also FNB, diary for April 26, April 27, and May 26, 1926, BFP-BL, for typical examples.

35. FC to family, May 14, 1922; FCB to family, October 11, 1922, October 28, 1922, and November 12, 1922; FCB to Mother, September 20, 1924, BFP-BL; Frank N. Blanchard and Frieda Cobb Blanchard, "Factors Determining Time of Birth in the Garter Snake *Thamnophis sirtalis sirtalis* (Linnaeus)," *Papers of the Michigan Academy of Science, Arts, and Letters* 26 (1940): 161–176; Frank N. Blanchard and Frieda Cobb Blanchard, "The Inheritance of Melanism in the Garter Snake *Thamnophis sirtalis sirtalis* (Linnaeus), and Some Evidence of Effective Autumn Mating," *Papers of the Michigan Academy of Science, Arts, and Letters* 26 (1940): 177–193.

36. FCB to Mother, October 23, 1925; FNB, diary, November 6, 1925, BFP-BL; University of Michigan Regents Minutes and Working Papers, October 8, 1925, BL.

37. FNB, diary for November and December 1925; FCB to Mother, November 13, 1925; FCB to Victor and Maude Cobb, January 3, 1926; FCB to Father, July 5, 1927, BFP-BL.

38. FCB to Mother, May 12, 1926, August 13, 1929, November 20, 1929, and December 20, 1929; FCB to FNB, July 12, 1929, and July 11, 1930; FCB to Margaret Cobb, January 15, 1932, and May 12, 1932; directions about house and babies, undated, BFP-BL.

39. Bartlett to Dean John R. Effinger, August 4, 1926; FNB to Effinger, March 2, 1927; FCB to Effinger, March 10, 1927; Effinger to Honorable Board of Regents, for March 1927 meeting; Regents Minutes, March 25, 1927, all in University of Michigan Regents Minutes and Working Papers, BL. Bartlett had written his recommendation before he left for a year's leave in Asia, leaving Frieda as acting director of the Gardens. He pointed out that Frank, recently promoted to assistant professor, had not been in professorial rank long enough for a leave but urged that the leave be granted anyway because the work the Blanchards planned to do would be very valuable to the university.

40. See FCB, letters to family for 1927–1928; FCB to Mrs. Charles Barrett, October 29, 1929, BFP-BL; and Frieda Cobb Blanchard, "Tuatara," *The National Geographic Magazine* 67 (1935): 649–662. Bartlett's diary (manuscript) for 1928 contains several entries mentioning accessioning seeds that Frieda had sent.

41. FCB to Mrs. Taylor, December 17, 1935; FCB to Joe Fry, February 18, 1936; FNB, diary for 1937, BFP-BL; Bartlett, diary for 1937, BL.

42. Blanchard and Blanchard, "Factors Determining Time of Birth"; idem, "The Inheritance of Melanism"; idem, "Mating of the Garter Snake *Thamnophis sirtalis sirtalis* (Linnaeus)," *Papers of the Michigan Academy of Science, Arts, and Letters* 27 (1941): 215–234; Frieda Cobb Blanchard, "A Test of Fecundity of the Garter Snake *Thamnophis sirtalis* (Linnaeus) in the Year Following the Year of Insemination," *Papers of the Michigan Academy of Science, Arts, and Letters,* 28 (1942): 313–316; FCB to Gilbert Archey, January 12, 1939, BFP-BL.

43. "Frieda Cobb Blanchard," in *Twenty-Fifth Anniversary Report, Radcliffe Class of 1913.*

Chapter 10: Kathleen and Thomas Lonsdale

The author had the privilege of being a postdoctoral fellow for Kathleen Lonsdale. The following people, through conversations over the past several years, have contributed to this chapter: Thomas Lonsdale, Jane Lonsdale Goodwin, Nan Lonsdale Dawson, Dorothy Hodgkin, Isabella Karle, Judith Milledge, June Sutor, Doris Evans, Caroline H. MacGillavry, P. P. Ewald, Anne and David Sayre, Marjorie Senechal, Jenny Glusker, Enid and James V. Silverton, Gabrielle and Jose Donnay, Kathleen and Francis Julian, and Carl Julian. Portions of this chapter are adapted from the author's works listed in note 1.

1. Maureen M. Julian, "X-ray Crystallography and the Work of Dame Kathleen

Lonsdale," *The Physics Teacher* 19 (1981): 159–165; idem, "Kathleen Lonsdale, 1903–1971," *Journal of Chemical Education* 59 (1982): 965–966; idem, "Women in Crystallography," in *Women of Science: Righting the Record*, ed. G. Kass-Simon and P. Farnes (Bloomington, 1990), 335–383; idem, "Kathleen Yardley Lonsdale (1903–1971)," in *Women in Chemistry and Physics: A Biobibliographic Sourcebook*, ed. L. S. Grinstein, R. K. Rose, and M. H. Rafailovich (Westport, Conn., 1993), 329–336; idem, "Four Women Crystallographers," in *Women's Contribution to Chemistry, An Historical Perspective and Women at the Forefront*, ed. V. I. Birss, P. W. Codding, and G. Rayner-Canham (Quebec, 1993), 1–29.

2. Kathleen Lonsdale, *The Christian Life—Lived Experimentally, An Anthology of the Writings of Kathleen Lonsdale* (London, 1976), 8.

3. Kathleen Lonsdale, *Is Peace Possible?* (London, 1957), 13.

4. Kathleen Lonsdale, *I Believe . . ., The Eighteenth Arthur Stanley Eddington Memorial Lecture* (Cambridge, 1964), 8.

5. Kathleen Lonsdale, "Women in Science: Reminiscences and Reflections," *Impact of Science on Society* 20 (1970): 45–59.

6. J. J. Lonsdale, "The Ionization Produced by the Splashing of Mercury," *Philosophical Magazine*, 6th ser., 20 (1910): 464–474.

7. The latter is based on the author's conversations with Thomas Lonsdale.

8. Kathleen Yardley, "The Crystalline Structure of Succinic Acid, Succinic Anhydride and Succinimide," *Proceedings of the Royal Society, London A* 105 (1924): 451–467.

9. Kathleen Yardley and W. T. Astbury, "Tabulated Data for the Examination of the 230 Space Groups by Homogeneous X-rays," *Philosophical Transactions of the Royal Society, London A* 224 (1924): 221–257.

10. Kathleen Lonsdale, *International Tables for X-ray Crystallography*, vol. 1, with N.F.M. Henry (Birmingham, 1952); vol. 2, with J. Kasper (1959); vol. 3, with C. H. MacGillavry and G. D. Rieck (1962).

11. Dorothy Hodgkin, "Kathleen Lonsdale," *Biographical Memoirs of Fellows of the Royal Society* 21 (1976): 456.

12. See Maureen M. Julian, "Crystallography in the Laboratory of William Henry Bragg," *Chemistry in Britain* 72, no. 8 (1986): 729–732.

13. W. S. Denham and Thomas Lonsdale, "Properties of the Silk Fibre," *Transactions of the Faraday Society* 20 (1924): 259–268.

14. Thomas Lonsdale, "Recording Extensometer," *Journal of the Textile Institute* 17 (1926): 248–253; W. S. Denham and Thomas Lonsdale, "Testing Instruments for Yarns and Fibres," *Journal of Scientific Instruments* 5 (1928): 348–354.

15. Kathleen Lonsdale, *Is Peace Possible?* 13.

16. The latter is based on the author's conversations with Thomas Lonsdale.

17. Hodgkin, "Kathleen Lonsdale," 452.

18. Kathleen Yardley (Mrs. Lonsdale), "An X-ray Study of Some Simple Derivatives of Ethane, Parts I and II," *Proceeding of the Royal Society, London A* 118 (1928): 449–497.

19. Kathleen Lonsdale, "Evidence of the Anisotropy of the Carbon Atom," *Philosophical Magazine* 6 (1928): 433.

20. Kathleen Lonsdale, "Crystallography at the Royal Institution," in *Fifty Years of X-ray Crystallography*, ed. P. P. Ewald (Utrecht, 1962), 414. In 1929 she returned the balance of £46 3s to the Royal Society and received a receipt, which she carefully saved (idem, "Crystallography," 414).

21. Kathleen Lonsdale, "The Structure of the Benzene Ring in $C_6 (CH_3)_6$," *Proceedings of the Royal Society, London A* 123 (1929): 494–515.

22. Thomas Lonsdale, "Changes in the Dimensions of Metallic Wires Produced by Torsion; II Silver, Gold, Aluminum and Nickel; III Lead," *Philosophical Magazine* 11 (1931): 1169–1187.

23. Quoted in Kathleen Lonsdale, "Reminiscences," in *Fifty Years of X-ray Crystallography*, 599.

24. Kathleen Lonsdale, "The Structure of the Benzene Ring," *Nature* 122 (1928): 810; idem, "The Structure of the Benzene Nucleus," *Transactions of the Faraday Society* 25 (1929): 352–366. See Maureen M. Julian, "Kathleen Lonsdale and the Planarity of the Benzene Ring," *Journal of Chemical Education* 58 (1981): 365–366.

25. Hodgkin, "Kathleen Lonsdale," 451; Kathleen Lonsdale, "Reminiscences," 599.

26. W. S. Denham and T. Lonsdale, "The Tensile Properties of Silk Filaments," *Transactions of the Faraday Society* 29 (1933): 305–316.

27. Kathleen Lonsdale, "An X-ray Analysis of the Structure of Hexachlorobenzene Using the Fourier Method," *Proceedings of the Royal Society, London* A 133 (1931): 536.

28. Quoted in Kathleen Lonsdale, "Reminiscences," 600. Note the attitude of the organic chemists toward the crystallographers in the use of the word "professional" with the implication that all others are amateurs.

29. Hodgkin, "Kathleen Lonsdale," 458.

30. Kathleen Lonsdale, "Magnetic Anisotropy and Electronic Structure of Aromatic Molecules," *Proceedings of the Royal Society, London* A 159 (1937): 149–161.

31. Hodgkin, "Kathleen Lonsdale," 458.

32. Kathleen Lonsdale and I. Ellie Knaggs, "Structure of Benzil," *Nature* 143 (1939): 1023.

33. Kathleen Lonsdale and H. A. Jahn, "Diffuse X-ray Reflections from Diamond," *Nature* 147 (1941): 88.

34. Kathleen Lonsdale and H. Smith, "Diffuse X-ray Diffraction from the Two Types of Diamond," *Nature* 148 (1941): 112.

35. Kathleen Lonsdale, "Formation of Lonsdaleite from Single-Crystal Graphite," *American Mineralogist* 56 (1971): 333–336.

36. Private correspondence of Kathleen Lonsdale to Clifford Frondel (given to the author by C. F.). Quoted in Julian, "Women in Crystallography," 356.

37. Kathleen Lonsdale, *I Believe*, 54.

38. Hodgkin, "Kathleen Lonsdale," 453.

39. See Maureen M. Julian, "Edmond de Valera and the Founding of the Dublin Institute," *Journal of Chemical Education* 60 (1983): 199–200.

40. Joan Mason, "The Admission of the First Women to the Royal Society of London," *Notes and Records of the Royal Society, London* 46 (1992): 279–300.

41. Quoted in Hodgkin, "Kathleen Lonsdale," 454.

42. Kathleen Lonsdale, "The Training of Modern Crystallographers," *Acta Crystallographia* 6 (1953): 874.

43. Kathleen Lonsdale, *Crystals and X-rays* (London, 1948).

44. Kathleen Lonsdale, H. Judith Milledge, and E. Nave, "X-ray Studies of Synthetic Diamonds," *Mineralogy Magazine* 32 (1959): 185–201.

45. Kathleen Lonsdale and H. Judith Grenville-Wells, "Study of Nickel Inclusions in Laboratory-Made Diamond," *Bulletin of the National Institute of Science, India* 14 (1958): 130.

46. Kathleen Lonsdale, E. Nave, and J. F. Stephens, "X-ray Studies of a Single Crystal Chemical Reaction: Photo-Oxide of Anthracene to (Anthraquinone, Anthrone)," *Philosophical Transactions of the Royal Society, London* A 261 (1966): 1–31; Kathleen Lonsdale, H. Judith Milledge, and K. El Sayed, "The Crystal Structure (at Five Temperatures) and Anisotropic Thermal Expansion of Anthraquinone," *Acta Crystallographia* 20 (1966): 1.

47. Kathleen Lonsdale, "Human Stones," *Scientific American* 219 (1968): 104–111.

48. Quoted in Hodgkin, "Kathleen Lonsdale," 470.

49. Kathleen Lonsdale, *Is Peace Possible?* viii.

50. Thomas Lonsdale, "The Road Research Laboratory, Open House," *Nature* (1958): 23–24.

51. Hodgkin, "Kathleen Lonsdale," 474–475.

52. Kathleen Lonsdale, "Women Scientists—Why So Few?" *Laboratory Equipment Digest* (1971): 85.

Chapter 11: Clanging Eagles

1. Mary Putnam Jacobi, "Shall Women Practice Medicine?" *North American Review* (1882), in *Mary Putnam Jacobi, A Pathfinder in Medicine,* ed. Women's Medical Association of New York (New York, 1925), 367–390, quotation 388.

2. Ibid., 384.

3. Ibid., 389. She has slightly misquoted these lines from Alfred Lord Tennyson's "The Princess," which read "The crane," I said "may chatter of the crane / The dove murmur of the dove, but I / An eagle clang an eagle to the sphere."

4. Joy Harvey, "Dr. Mary Putnam Jacobi and the Paris Commune," in *Dialectical Anthropology,* special issue "Women in Revolution," ed. Josephine Diamond (fall–winter 1990): 107–117.

5. Mary Putnam Jacobi, "Women in Medicine," in *Women's Work in America,* ed. Annie Nathan Meyer (New York, 1891), 139–205, citation 159. Her reference to Michelet is probably to Jules Michelet, *La Femme* (Paris, 1860).

6. Putnam Jacobi, "Shall Women Practice Medicine?" 385.

7. Ibid., 389.

8. Among the rich biographical sources are Roy Lubove, "Mary Corinna Jacobi," in *Notable American Women* (Cambridge, Mass., 1971), 2:263–265; Rhoda Truax, *The Doctors Jacobi* (Boston, 1952), and the companion volumes, *Life and Letters of Mary Putnam Jacobi,* ed. Ruth Putnam (New York, 1925), and *Mary Putnam Jacobi,* ed. Women's Medical Association of New York (New York, 1925). The latter includes her major papers and a list of most of her publications. The largest group of manuscript materials is in the Mary Putnam Jacobi Collection in Schlesinger Library, Radcliffe College.

9. For a study of her experiences in the French clinics, see Joy Harvey, "La Visite: Mary Putnam Jacobi and the Paris Medical Clinics," in *French Medical Culture in the Nineteenth Century,* ed. Ann La Berge and Mordechai Feingold, Clio Medica 25/The Wellcome Institute Series in the History of Medicine (Amsterdam/Atlanta, 1994).

10. Harvey, "Dr. Mary Putnam Jacobi and the Paris Commune."

11. For the complexities of the relationship between Elizabeth Blackwell and Mary Putnam Jacobi, see Regina Markell Morantz Sanchez, *Sympathy and Science: Women Physicians in American Science* (New York, 1985).

12. Victor Robinson, "The Life of Abraham Jacobi," *Medical Life* 35 (1928): 213–306.

13. Putnam Jacobi, "Women in Medicine."

14. Ferdinand F. Meyer, a professor of chemistry at the College of Pharmacy, was Jewish, like Jacobi. He disappeared, probably a suicide, in 1870. The name of the French doctor has never been determined. Truax, *The Doctors Jacobi,* 261.

15. "Dr. Jacobi's first wife was a traditional woman who died in the third year of their marriage and the second a sickly patient of his own whom he was unable to rescue from her eight years of wretched invalidism when she also died." Mary Putnam Jacobi to George McAneny, October 27, 1898, Mary Putnam Jacobi Collection, Addenda 84–M86.

16. Mary Putnam Jacobi to her mother, Victorine Haven Putnam, August 15, 1875, Mary Putnam Jacobi Collection, A-26.

17. Mary Putnam Jacobi to Victorine Haven Putnam, Zurich, July 23, 1876, ibid.

18. Abraham Jacobi, *A Treatise on Diphtheria* (New York, 1880). Putnam Jacobi had written a review on the topic a few years before: Mary Putnam Jacobi, "Croup and Diphtheria," *Medical Record* 12 (1877): 125.

19. John Allen Wyeth cited in "Member of Medical Societies," in *Mary Putnam Jacobi*, ed. Women's Medical Association of New York, xxix.

20. This was published the following year as *The Question of Rest for Women During Menstruation* (New York, 1876).

21. Putnam Jacobi, "Women in Medicine," 190–191. She mentioned that Dr. Annie Angell, one of her students, participated in this clinic with her.

22. Mary Putnam Jacobi, "Opening Lecture on Disease of Children at the Post-Graduate Medical School," *Boston Medical & Surgical Journal* (1882); reprinted in *Mary Putnam Jacobi*, ed. Women's Medical Association of New York, 403–418.

23. Mary Putnam Jacobi wrote a small book expressing her positivist philosophy: *The Value of Life: A Reply to Mr. Mallock's Essay "Is Life Worth Living?"* (New York, 1879).

24. Eugene P. Link, "Abraham and Mary P. Jacobi, Humanitarian Physicians," *Journal of the History of Medicine* 4 (1949): 382–392.

25. Hans L. Trefousse, *Carl Schurz, A Biography* (Knoxville, Tenn., 1982). A recent bilingual German-English edition contains many portraits and caricatures published in Munich: *Carl Schurz, Revolutionary and Statesman*, ed. Ruediger Wesich.

26. Ibid. Trefousse mentions Jacobi thirteen times throughout this biography. Putnam Jacobi is mentioned only once in connection with women's suffrage.

27. Abraham Jacobi to Mary Putnam Jacobi [spring 1883], Mary Putnam Jacobi Collection.

28. Abraham Jacobi, *Treatise on Diphtheria*, and idem, "Rudolf Virchow," *Medical Record* 20 (1881): 449–457.

29. Evelyn Hammonds, "The Search for Perfect Control: A Social History of Diphtheria, 1880–1930" (Ph.D. diss., Harvard University, 1993).

30. Truax describes Jacobi, even in his old age, rowing around Lake George on his son's birthday in honor of the moment and the place where he first heard the newborn child cry. Truax, *The Doctors Jacobi*.

31. Ibid., 215.

32. Mary Putnam Jacobi to Mrs. Edward Curtis, Sept. [1883], Cliff House, Rye, Mary Putnam Jacobi Collection, Addenda 84–M86.

33. Elizabeth Robinton, "Annie Wessel Williams," in *Notable American Women: The Modern Period*, ed. Barbara Sicherman and Carol Hurd Green (Cambridge, Mass., 1980), 737–739.

34. Mary Putnam Jacobi to George McAneny, Dec. 19, 1898, Mary Putnam Jacobi Collection, M84–M86.

35. Sydney Halpern, *American Pediatrics* (Los Angeles, 1989).

36. Morantz Sanchez, *Sympathy and Science*.

37. Mary Putnam Jacobi, "An Address Delivered at the Commencement of the Woman's Medical College of the N.Y. Infirmary, May 30, 1883" in *Mary Putnam Jacobi*, ed. Women's Medical Association of New York, 391–402.

38. Mary Putnam Jacobi, *Common Sense Applied to Woman's Suffrage* (New York, 1894).

39. Ibid., 86.

40. Carl Schurz to Mary Putnam Jacobi, July 26, 1894, Mary Putnam Jacobi Collection.

41. Mary Putnam Jacobi to Carl Schurz, August 2, 1894, ibid., A-26. This is the same letter quoted in *Life and Letters of Mary Putnam Jacobi*, ed. Ruth Putnam, 327, but this postscript has been edited from the published version.

42. Mary Putnam Jacobi to Agatha Schurz, August 10, 1888 (typescript excerpt), Mary Putnam Jacobi Collection.

43. Mary Putnam Jacobi to George McAneny, May 16, 1899, ibid.

44. Ibid.

45. Ibid.

46. Cited by Truax, *The Doctors Jacobi*, 240.

47. Abraham Jacobi, "Women Physicians in America," translated from a German article that appeared in *Deutsche Medicinische Wochenschrift*, no. 25 (1896); reprinted in Abraham Jacobi, *Collecteana Jacobi*, vol. 6 (New York, 1909). Bonner has read this article more favorably in the context of the German medicine of its day. Thomas Bonner, *To the Ends of the Earth: Women's Search for Education in Medicine* (Cambridge Mass., 1992), 110.

48. He collected his papers and addresses in a series of twelve volumes, published in 1909 as *Collecteana Jacobi*.

49. Mary Putnam Jacobi, "Address before the Women's Medical Association about 1900," in *Mary Putnam Jacobi*, ed., Women's Medical Association of New York, 494–500.

50. Ibid., 494. In her commencement address of 1883 she had also urged women physicians to think of themselves as a class and work with each other on research topics (ibid., 391–402).

51. For example, see the account of the complex sexual, social, and political relationships of the exiled Russian romantic revolutionary Alexander Herzen, his family, and friends in Edward Hallett Carr, *The Romantic Exiles: A Nineteenth-Century Portrait Gallery* (1933; reprint, Boston, 1961). Dorinda Outram reminded me of the differences in expectations of the two generations. On the Réclus family, see Hélène Sarrazin, *Élisée Réclus ou la passion du monde* (Paris, 1985).

Chapter 12: *"My Life Is a Thing of the Past"*

1. Philip J. Pauly, "The Appearance of Academic Biology in Late Nineteenth-Century America," *Journal of the History of Biology* 17 (1984): 383; Jane Maienschein, ed., *Defining Biology: Lectures from the 1890s* (Cambridge, Mass., and London, 1986), 16–19.

2. Margaret W. Rossiter, *Women Scientists in America: Struggles and Strategies to 1940* (Baltimore and London, 1982), notes on 323, 334, and 345; Julia Morgan, *Women at The Johns Hopkins University: A History* (Baltimore, 1986), 2–3.

3. Nunn family background is based on Stephen A. Bailey, *L. L. Nunn, A Memoir* (Ithaca, N.Y., 1933), and the Lucien L. Nunn Papers 37/4/1770, Rare and Manuscript Collections, Cornell University Library (henceforth LLNP).

4. Heckman to Sweeting, September 22, 1969, LLNP.

5. C. R. Nunn, handwritten diary and accounts, 1848–1871, LLNP.

6. *Newnham College Register* (1963), 57.

7. H. F. Durant to D. C. Gilman, November 12, 1877, Johns Hopkins University Collection, Ms. 137, Special Collections, Milton S. Eisenhower Library, The Johns Hopkins University (henceforth JHUC).

8. Ibid.

9. Personal communication, Staatsarchiv des Kantons Zürich to Christiane Groeben, May 2, 1986.

10. Personal communication, Stadtarchiv Zürich to Christiane Groeben, May 23, 1986.

11. Gilman to Durant, November 6, 1877, Daniel Coit Gilman Papers, Ms. 1, Special Collections, Milton S. Eisenhower Library, The Johns Hopkins University (henceforth DCGP).

12. Frances Zirngiebel, "Teachers' School of Science," *Popular Science Monthly* 55 (1899): 451–465 and 640–652.

13. "Report of the Custodian," *Proceedings of the Boston Society of Natural History* (henceforth *PBSNH*) 20 (1878–1880): 225.

14. *PBSNH* 21 (1880–1882): 299.

15. Patricia Ann Palmieri, *In Adamless Eden: A Social Portrait of the Academic Community at Wellesley College 1875–1920* (Ann Arbor, 1981), 26–28, and esp. 386 and 451 on Nunn.

16. E. A. Nunn to Gilman, October 5, 1877, DCGP.

17. Memorandum, Gilman to trustees, November 6, 1877, JHUC.

18. Gilman to E. A. Nunn, November 6, 1877, JHUC; Gilman to Durant, November 6, 1877, DCGP.

19. *Third Annual Report of The Johns Hopkins University* (1878): 34–35; Martin to Gilman, March 18, 1878, and March 20, 1878, DCGP.

20. T. H. Huxley to A. Dohrn, January 18, 1882, Thomas Henry Huxley Papers (henceforth THHP), The Archives of Imperial College of Science, Technology and Medicine, London (henceforth AICSTM).

21. Personal communication, Tanya Holton to Linda Tucker, August 26, 1991.

22. Helen Magill to Eva Channing, July 7, 1878, Helen Magill White Papers 4107, Rare and Manuscript Collections, Cornell University Library.

23. E. A. Nunn, "The Structural Changes in the Epidermis of the Frog, Brought About by Poisoning with Arsenic and with Antimony," *Journal of Physiology* 1 (1878–1879): 247–256 and plate 8.

24. *Wellesley College Calendar,* 1878–1879 through 1880–1881.

25. W. K. Brooks, "Roll of the Chesapeake Laboratory, 1878–83," *Johns Hopkins University Circular* 3 (1883–1884): 93.

26. *Johns Hopkins University Circular* 1 (1879–1882): 16.

27. K. Mitsukuri to Dohrn, April 2, 1883, Archives of the Stazione Zoologica "Anton Dohrn" (henceforth ASZN).

28. S. F. Clarke to Gilman, August 19, 1879, DCGP.

29. *Johns Hopkins University Circular* 1 (1879–1882): 16.

30. Durant to Gilman, November 12, 1877, JHUC.

31. Mary A. Willcox to Marian Hubbard, December 2, 1927, Mary A. Willcox Papers, Wellesley College Archives.

32. Huxley to Dohrn, January 18, 1882, THHP; Royal School of Mines Biological Laboratory, Record of Laboratory Work for sessions 1876–77 to 1883–84; 1881–82: 249, AICSTM.

33. E. A. Nunn, "On the Development of the Enamel of the Teeth of Vertebrates" (received June 14, 1882), *Proceedings of the Royal Society* 34 (1883): 156–165 and plates 2–4.

34. M. Foster to Dohrn, October 24, 1882, ASZN.

35. Dohrn to Huxley, January 25, 1883, ASZN; trans. from German.

36. E. A. Nunn to Dohrn, two letters, May 3, 1883, ASZN.

37. Dohrn to Cunningham, May 25, 1883, ASZN.

38. Sources for the narrative of Whitman's life, except where noted, are F. R. Lillie, "Charles Otis Whitman," *Journal of Morphology* 22 (1911): xv–lxxvii; and E. S. Morse, "Charles Otis Whitman," *National Academy of Sciences Biographical Memoirs* 7 (1912): 269–288.

39. *Quarterly Journal of Microscopical Science* 18 (1878): 215–315.

40. W. K. Brooks, "Report of the Chesapeake Zoological Laboratory, Summer of 1879," in *Fourth Annual Report of The Johns Hopkins University* (1879): 64.

41. H. K. Frost to F. R. Lillie, October 4, 1911, box 1, folder 3, Charles Otis Whitman Papers, University of Chicago Library (henceforth COWP/UC).

42. E. A. Nunn, "The Naples Zoological Station," *Science* 1 (1883): 479–481 and 507–510; C. O. Whitman, "The Advantages of Study at the Naples Zoological Station," *Science* 2 (1883): 93–97.

43. E. L. Mark to F. R. Lillie, October 28, 1911, box 1, folder 3, COWP/UC.

44. Massachusetts State Archives, Vital Records, vol. 343, p. 151, September 4, and vol. 344, p. 212, September 15.

45. C. O. Whitman to Paul Mayer, October 21, 1883, ASZN.

46. *Annual Report . . . Museum of Comparative Zoology at Harvard College* (1883–1884): 4.

47. *PBSNH* 23 (1884–1888): 1.

48. E. A. Nunn, "The Zoological Station at Naples," *Century Illustrated Magazine* 32 (1886): 791–799.

49. Lillie, "Charles Otis Whitman," includes Whitman's bibliography.

50. Notebook 1 (March 1884), Papers of M. Carey Thomas, Bryn Mawr College Archives (henceforth MCTP), microfilm reel 166.

51. Thomas, "Conversations about College Organization in 1884," MCTP, microfilm reel 166.

52. Thomas, "Candidates for Positions at Bryn Mawr" [1866], p. 11, MCTP, microfilm reel 166; E. N. Whitman to L. L. Nunn, May 10, 1894, LLNP.

53. Ernest J. Dornfeld, "The Allis Lake Laboratory 1886–1893," *Marquette Medical Review* 21 (1956): 120–124.

54. F. R. Lillie, *The Woods Hole Marine Biological Laboratory* (Chicago, 1944), 44 and 26–31.

55. C. O. Whitman to W. E. Frost, May 27, 1892, quoted in H. K. Frost to F. R. Lillie, October 4, 1911, box 1, folder 3, COWP/UC.

56. Mary P. Winsor, *Reading the Shape of Nature: Comparative Zoology at the Agassiz Museum* (Chicago and London, 1991), 210–211.

57. C. B. Davenport, "The Personality, Heredity and Work of Charles Otis Whitman, 1843–1910," *American Naturalist* 51 (1917): 17.

58. E. N. Whitman to G. F. Hoar, June 15, 1890, quoted in personal communication from Koelsch to Christiane Groeben, March 28, 1986.

59. E. N. Whitman to L. L. Nunn, April 9, 1894, LLNP.

60. A. Dohrn to Marie Dohrn, August 3, 1897, ASZN.

61. E. N. Whitman to A. Dohrn, May 14, 1898; July 17, 1900; and December 10, 1901, ASZN.

62. Horatio Hackett Newman, "The History of the Zoology Department in the University of Chicago," *Bios* 19 (1948): 220–224.

63. Maienschein, *Defining Biology,* 19 and 45–47.

64. C. O. Whitman, "The Natural History Work at the Marine Biological Laboratory, Woods Hole," *Science,* n.s., 13 (1904): 539; idem, "Some of the Functions and Features of a Biological Station," *Science,* n.s., 7 (1898): 40–42.

65. Lillie, *Marine Biological Laboratory,* 60–61.

66. Cited in Maienschein, *Defining Biology,* 20 (n. 46).

67. E. N. Whitman to R. S. Woodward, n.d., and November 4, 1912, Charles Otis Whitman file, Carnegie Institute of Washington Archives; Richard W. Burkhardt, "Charles Otis Whitman, Wallace Craig, and the Biological Study of Behavior in the United States, 1898–1925," in Ronald Rainger, Jane Maienschein, and Keith Benson, eds., *The American Development of Biology* (Philadelphia, 1988), 194.

68. F. R. Lillie, "Charles Otis Whitman," *University of Chicago Magazine* (1910–1911): 144; G. W. Bartelmez, *National Academy of Sciences Biographical Memoirs* 43 (1973): 4; Whitman to E. G. Conklin, November 4, 1899, Marine Biological Laboratory Archives; E. P. Allis to F. R. Lillie, August 22, 1911, box 1 folder 1, COWP/UC.

69. Bailey, *L. L. Nunn,* 46–82.

70. L. L. Nunn to Francis Nunn Whitman, June 16, 1909, LLNP.

71. L. L. Nunn to E. N. Whitman, October 6, 1906, and L. L. Nunn to C. O. Whitman, October 25, 1905, LLNP.

72. Lillie, *Marine Biological Laboratory*, 43–46.

73. Ibid., 48–60.

74. L. L. Nunn to E. N. Whitman, October 6, 1906, LLNP.

75. L. L. Nunn to F. N. Whitman, June 16, 1909, LLNP.

76. E. N. Whitman to L. L. Nunn, May 10, 1894, LLNP.

77. E. N. Whitman to L. L. Nunn, March 10 [1901?], LLNP.

78. Deed, L. L. Nunn to Florence Tinkham, January 27, 1904, and L. L. Nunn to Carroll Nunn Whitman, September 28, 1912, LLNP.

79. C. O. Whitman to L. L. Nunn, October 14, 1905, LLNP.

80. Ibid.

81. L. L. Nunn to E. N. Whitman, October 6, 1906. LLNP.

82. Heckman to Sweeting, n.d., LLNP.

83. C. O. Whitman to C. N. Whitman, June 8, 1909, LLNP.

84. E. N. Whitman to L. L. Nunn, July 31 [1909], LLNP.

85. Burkhardt, "Whitman and Craig," 194–196.

Chapter 13: Albert Einstein and Mileva Marić

I am grateful to the editors of *The Collected Papers of Albert Einstein* for making available drafts of volumes 3, 4, and 5; and to A. J. Kox, Jürgen Renn, and Robert Schulmann for discussions of a number of these documents. I am also grateful to Gerald Holton, Françoise Balibar, and the editors of this volume for comments on earlier drafts of this chapter.

1. She sometimes used Marity, the Hungarian form of her last name; she followed Swiss custom after her marriage, using Einstein-Marić or Einstein-Marity.

2. *Albert Einstein and Mileva Marić, The Love Letters*, trans. Shawn Smith, ed. Jürgen Renn and Robert Schulmann (Princeton, 1992), 72–73, cited hereafter as *The Love Letters*. Einstein's correspondence, including letters to and from Marić, will also be cited from *The Collected Papers of Albert Einstein*, vol. 1, *The Early Years, 1879–1902*, ed. John Stachel et al. (Princeton, 1987), and vol. 5, *The Swiss Years: Correspondence, 1902–1914*, ed. Martin Klein et al. (Princeton, 1993); cited hereafter as *Collected Papers*, vols. 1 and 5.

3. They met in 1896, married in 1903, separated in 1914, and divorced in 1919.

4. For his publications during this period, see *The Collected Papers of Albert Einstein*, vol. 2, *The Swiss Years: Writings, 1900–1909*, ed. John Stachel et al. (Princeton, 1989); vol. 3, *The Swiss Years: Writings, 1909–1911*, ed. Martin Klein et al. (Princeton, 1993); and vol. 4, *The Swiss Years: Writings, 1912–1914*, ed. Martin Klein et al. (Princeton, 1995); cited hereafter as *Collected Papers*, vols. 2, 3, and 4.

5. See Desanka Trbuhović-Gjurić, *Im Schatten Albert Einsteins/Das tragische Leben der Mileva Einstein-Marić* (Bern/Stuttgart, 1983), cited hereafter as *Im Schatten Albert Einsteins*; Senta Troemel-Ploetz, "Mileva Einstein-Marić: The Woman Who Did Einstein's Mathematics," *Women's Studies International Forum* 13 (1990): 415–432; Evan Harris Walker, "Did Einstein Espouse His Spouse's Ideas?" *Physics Today* 42, no. 2 (February 1989), 9–11 (for my comments, see ibid., 11–13); idem, "Ms. Einstein" (paper presented at the AAAS meeting, New Orleans, February 1990); and idem, "Mileva Marić's Relativistic Role" (presented at the AAAS Meeting, Washington, D.C., February 1991).

6. "Einstein and Marić: The Early Years," in *Einstein's Early Years: 1879–1905*, ed. Don Howard and John Stachel (Boston/Basel/Berlin, forthcoming), cited hereafter

as "Einstein and Marić." See also Roger Highfield and Paul Carter, *The Private Lives of Albert Einstein* (London/Boston, 1993), cited hereafter as *Private Lives,* and Abraham Pais, *Einstein Lived Here* (Oxford/New York, 1994).

7. Sources for information on her life include *Im Schatten Albert Einsteins*; Dorde [George] Krstić, "Mileva Einstein-Marić," Appendix A in Elizabeth Roboz Einstein, *Hans Albert Einstein: Reminiscences of His Life and Our Life Together* (Iowa City, 1991); her correspondence with Einstein in *Collected Papers,* vols. 1 and 5; and her letters to her friend and confidante, Helene Savić, née Kaufler. Some excerpts from the Savić letters are cited from *Collected Papers,* vol. 1, and unpublished excerpts are cited (in my translations) from photocopies of originals presented by Savić's grandson, Professor Milan Popović (Belgrade), to the editors of *The Collected Papers.* These copies will be cited as in the Einstein Papers Project Archives, Boston University. A useful synthesis of this material is found in *Private Lives.*

8. Einstein is discussed here only insofar as is relevant to their intellectual relationship. For a fuller discussion of their relationship up to 1905, see "Einstein and Marić." For a differing account of their relationship, more skeptical of Einstein's early devotion to Marić, see *Private Lives.*

9. See Phyllis Stock, *Better Than Rubies: A History of Women's Education* (New York, 1978), 166; cited hereafter as *Better Than Rubies.* There also may have been medical reasons for Marić's move, since she had been very ill with a lung disorder.

10. See Schweizer Verband der Akademikerinnen, *Die Frauenstudium an der Schweizer Hochschulen* (Zurich, 1928), cited hereafter as *Die Frauenstudium.*

11. For a discussion of the first generation of Russian women to study in Zurich, see Christine Johanson, *Women's Struggle for Higher Education in Russia, 1850–1900* (Kingston/Montreal, 1987), 51–58. According to Johanson, while many male students were hostile, "most professors allowed no sexual discrimination in the classroom" (53).

12. Indeed, pressure from Russian women prompted Zurich to open its doors (see *Better Than Rubies,* 145). In the first decades after the Swiss universities admitted women, the large majority were non-Swiss, mainly Slavs (see *Die Frauenstudium*).

13. For his *Matrikel* (official record), see *Collected Papers,* vol. 1, doc. 28, pp. 45–50. Her *Matrikel* is in file no. 85, *Rektoratsarchiv,* Eidgenössische Technische Hochschule (ETH).

14. Trbuhović-Gjurić suggests, without any evidence, that Marić left the Poly in flight from her intense romantic relationship with Einstein (see *Im Schatten Albert Einsteins*). Their letters suggest that the relationship was not yet very intense (see *Collected Papers,* vol. 1, esp. docs. 36 and 39). The brevity of Marić's stay in Heidelberg may be explained by Kaplan's observation that "the first women students at Heidelberg . . . suffered from extraordinary gender discrimination" (Marion Kaplan, *The Making of the Jewish Middle Class: Women, Families, and Identity in Imperial Germany* [New York, 1991], 149).

15. For this information, see *Collected Papers,* vol. 1, esp. docs. 50, 52, and 53.

16. His parents' opposition was based on Marić's age (she was four years older than Einstein), her intellectuality, and probably her Slavic origins. His mother made the first two objections explicit: "By the time you're 30 she'll be an old witch." "Like you, she is a book—but you ought to have a wife" (*The Love Letters,* 20). Anti-Slav prejudices are still common in Germany, and Einstein's parents had not objected to his earlier romance with a young teacher of Swiss-German background who was also slightly older than he (see *Collected Papers,* vol. 1, docs. 15, 18, and 32).

17. Einstein's letters to Marić mention treatises by Boltzmann, Drude, Helmholtz, Kirchhoff, and Mach (see *Collected Papers,* vol. 1).

18. See *Collected Papers,* vol. 1, doc. 67, p. 247. The three mathematics students in VIA took different exams. Trbuhović-Gjurić (*Im Schatten Albert Einsteins*) does not mention her failure to graduate; Troemel-Ploetz ("The Woman Who Did Einstein's

Mathematics") ascribes it to discrimination against women at the Poly without mentioning her grades; while Walker ("Ms. Einstein") states, without citing evidence, that "Marks below 5.00 were probably customarily below the passing grade." Einstein, with a total of 54 points out of a possible 66, was one point short of that average, while Marić, with a total of 44 points, was 11 points short.

19. In mid-1900, she mentions "a large work . . . that I have chosen for myself as a Diploma Thesis and probably also a Doctoral Thesis" (*Collected Papers*, vol. 1, 260, n. 5). In May 1901, Einstein asks about her doctoral thesis, advising her to use some of Weber's work in it, "even if you only seem to" (ibid., 305).

20. In May 1901, Marić wrote Savić, "I have already quarreled a couple of times with Weber, but we're already used to that" (*Collected Papers*, vol. 1, doc. 109, p. 303, my translation).

21. See *Collected Papers*, vol. 1, doc. 87, p. 275.

22. See Protocol of Section VIA, July 26, 1901, ETH Library (Zurich). Her average was again 4.

23. Einstein first mentions Kleiner in October 1900 (*Collected Papers*, vol. 1, 267); a year later, he discussed the complete dissertation (ibid., 321). He withdrew it in February 1902 (see ibid., doc. 132, p. 331), probably because of objections by Kleiner, but they stayed in contact. Einstein's successful 1905 doctoral dissertation was approved by Kleiner, who helped him obtain his first full-time academic post in 1909 (see below).

24. Presumably, Lieserl was born at Marić's home. However, recent efforts to find civil or church records of the birth in her hometown or nearby failed.

25. The delay was connected with the opposition of his family (see *Collected Papers*, vol. 1, doc. 138, p. 336). On his deathbed, Einstein's father gave his consent in October 1902, according to Abraham Pais, *"Subtle is the Lord . . .": The Science and the Life of Albert Einstein* (Oxford, 1982), 47.

26. See *Private Lives*, 90.

27. Late in 1901, after he was assured of a Patent Office job, he wrote Marić: "The only problem that still needs to be resolved is how to keep our Lieserl with us; I wouldn't want to have to give her up. Ask your Papa, he's an experienced man and knows the world better than your overworked, impractical Johnny" (*Collected Papers*, vol. 1, doc. 127, p. 324, translation from *The Love Letters*, 68).

28. Peter Michelmore, *Einstein: Profile of the Man* (New York, 1962), states: "Hans Albert Einstein . . . had never discussed his father before with any writer, at least not in depth. But he answered all my questions, and waited while I wrote down all the answers" (vii). Hans Albert inherited his mother's papers, and his first wife, Frieda Einstein-Knecht, transcribed excerpts from Einstein's letters discussing Lieserl. So, if not told earlier by either parent, Hans Albert knew about his sister by the time he spoke to Michelmore.

29. Michelmore, *Einstein*, 42.

30. Leo Tolstoy, *Anna Karenina*, trans. Louise and Aylmer Maude (London, 1965), 1.

31. *Collected Papers*, vol. 5, doc. 5, letter of January 22, 1903, p. 10 (my translation).

32. Marić to Savić, March 20, 1903, copy in Einstein Papers Project Archives, Boston University.

33. *Collected Papers*, vol. 5, doc. 13, p. 22, translation modified from *The Love Letters*, 53.

34. For further speculation, see *Private Lives*, 88–91.

35. *Collected Papers*, vol. 5, doc. 13, p. 22, translation from *The Love Letters*, 53.

36. Marić to Savić, September 3, 1909, copy in Einstein Papers Project Archives, Boston University.

37. The flirtatious nature of their earlier relationship is apparent from a poem Albert wrote for her (*Collected Papers*, vol. 1, doc. 49, p. 220).

38. See *Collected Papers*, vol. 5, 181, 198–199; Einstein-Marić to Georg Meyer, May 23, 1909, copy in the Archive of the Einstein-Gesellschaft, Swiss National Library (Bern). For a fuller account, see *Private Lives*, 124–126. Einstein's anger flared up again over forty years later, when he blamed Marić's pathological jealousy on "uncommon ugliness" (Einstein to Erika Schaerer-Meyer [Meyer-Schmid's daughter], cited in *Collected Papers*, vol. 5, 199, n. 4).

39. Marić to Savić, September 3, 1909, copy in Einstein Papers Project Archives, Boston University.

40. Marić to Savić, n.d. [c. October 1909], copy in Einstein Papers Project Archives, Boston University.

41. By this point, the Poly had been renamed the Eidgenössische Technische Hochschule, or ETH for short.

42. Marić to Savić, n.d. [c. January 1911], copy in Einstein Papers Project Archives, Boston University.

43. Michelmore, *Einstein*, 57.

44. Marić to Savić, n.d. [c. October 1909], copy in Einstein Papers Project Archives, Boston University.

45. As children, they were well acquainted, and her father (nicknamed "Rudolf the rich" by Einstein) was the chief creditor of his father's debts (see *Collected Papers*, vol. 1, doc. 93, p. 281). for their relationship, see his letters to her in *Collected Papers*, vol. 5; for her poetry reading, see Pais, *Einstein Lived Here*, 145.

46. *Collected Papers*, vol. 5, 585, 587.

47. Ibid., 558.

48. After their divorce he regularly stayed at Marić's house when visiting Zurich.

49. See *Collected Papers*, vol. 1. For a more detailed discussion of their relationship up to 1905, see "Einstein and Marić."

50. For her most extensive comment on physics, see *Collected Papers*, vol. 1, doc. 36, last paragraph, p. 59 For an example of her descriptive powers, see ibid., doc. 109, pp. 301–302.

51. *The Love Letters*, 9.

52. Ibid., 12–13.

53. See *Collected Papers*, vol. 1, doc. 37, p. 139.

54. Ibid., vol. 1, xxxix–xi.

55. "On the Electrodynamics of Moving Bodies" is the title of his famous 1905 paper on special relativity (*Collected Papers*, vol. 2, doc. 28). See the next section for further discussion of this topic.

56. Philipp Frank, *Einstein: His Life and Times* (New York, 1953), 21.

57. Albert Einstein, *Lettres à Maurice Solovine*, ed. Maurice Solovine (Paris, 1956), introduction, xii.

58. This has sometimes been confused with a doctoral thesis. Marić hoped to use her diploma thesis work as the basis for a doctorate, but she was never a candidate for that degree.

59. *Collected Papers*, vol. 1, doc. 63, pp. 243–244; translation from the supplementary *English Translation*, trans. Anna Beck (Princeton, 1987), 138.

60. *The Love Letters*, 30.

61. See *Collected Papers*, vol. 1, doc. 67.

62. See ibid., note 33, 244.

63. I.e., the *Annalen der Physik*; it became his first publication (see *Collected Papers*, vol. 2, doc. 1).

64. *Collected Papers*, vol. 1, doc. 85, p. 273, my translation.

65. Ibid., doc. 79, p. 267, my translation..

66. See *Collected Papers*, vol. 1, doc. 132, p. 331.

67. Ibid., doc. 125, p. 320, my translation.

68. It has been suggested that she attributed *her* work to him. But it is hard to see why she would do so in private letters to a close personal friend. If the expressions of admiration in these letters were meant to characterize her own work, they would give a most unpleasant impression of her character. If we accept her word that she picked her final diploma thesis topic, I see no reason to doubt it when she says he wrote the articles in question.

69. See the articles by Walker and Troemel-Ploetz cited in note 5.

70. *The Love Letters,* 54.

71. Ibid., 39.

72. Ibid., 69.

73. Michelmore, *Einstein,* 45–46. Such comments, and similar (but less reliable) anecdotal accounts by Marić's relatives in the Vojvodina (see *Im Schatten Albert Einsteins*), led to Senta Troemel-Ploetz's appellation: "Mileva Marić: The Woman Who Did Einstein's Mathematics."

74. See *Collected Papers,* vol. 2, doc. 23, pp. 276–306.

75. Ibid., 306. Besso's role is explained more precisely in later reminiscences by Einstein, notably his 1922 Kyoto lecture (see ibid., 264), and Michelmore also mentions it (*Einstein,* 45).

76. *Collected Papers,* vol. 3, doc. 1, p. 125, descriptive note.

77. Ibid., doc. 3, pp. 177–178.

78. Ibid., doc. 11, p. 321.

79. Mileva Marić to Albert Einstein, October 4, 1911, in Einstein, *Collected Papers,* vol. 5, doc. 290, p. 331.

80. Einstein and Marić met Marie Curie only after Pierre's death. For her life, see Eve Curie, *Madame Curie,* trans. Vincent Sheean (New York, 1937); Rosalind Pflaum, *Grand Obsession: Madame Curie and Her World* (New York, 1989); and Helena M. Pycior, "Marie Curie's 'Anti-natural Path': Time Only for Science and Family," in *Uneasy Careers and Intimate Lives: Women in Science, 1798–1979,* ed. Pnina G. Abir-Am and Dorinda Outram (New Brunswick, N.J., 1989), 191–214.

81. Both Einstein and Marić knew Ehrenfest and Afanasieva. For his life and their relationship, see Martin Klein, *Paul Ehrenfest,* vol. 1, *The Making of a Theoretical Physicist* (Amsterdam, 1970). Klein cites an obituary in Dutch, but there is no biography of Afanasieva.

82. Speaking of the German milieu, Kaplan notes "the popular stereotype of the Russian female student, who was portrayed as a radical, both politically and personally" (*The Making of the Jewish Middle Class,* 147); and she writes that "bourgeois parents displayed extraordinary ambivalence regarding their daughters' aspirations. . . . the fear lingered that educated daughters would educate themselves right out of the marriage market" (142).

83. Pierre had a well-established career in physics when he met Marie.

84. A few years later he referred to his first two papers as "worthless beginner's works" (see *Collected Papers,* vol. 5, doc. 66, p. 79).

85. "[O]ut of about one thousand [male] students there is hardly a single one who has the abilities for independent scientific accomplishment in the higher sense, so the demands on women at the least should not be set any higher" (Ella Wild, Einleitung to *Die Frauenstudium,* 15–16).

86. It seems plausible that he used Marić to help him break free of his family, especially his mother.

87. See, e.g., Lewis Pyenson, "Einstein's Early Scientific Collaboration," *Historical Studies in the Physical Sciences* 7 (1976): 84–123.

88. I am indebted to Pnina Abir-Am for this insight.

89. See, e.g., the account by his son Hans Albert, cited in *Private Lives,* 129.

90. For the Curies, See Helena M. Pycior, "Reaping the Benefits of Collaboration

While Avoiding Its Pitfalls: Marie Curie's Rise to Scientific Prominence," *Social Studies of Science* 3 (1993): 301–323. There is no study of the collaboration between the Ehrenfests, but I can cite a few indications of his efforts. Of the two articles they wrote jointly in 1906, the first is signed Tatiana and Paul Ehrenfest, the second is signed Paul and Tatiana Ehrenfest (see Paul Ehrenfest, *Collected Scientific Papers*, ed. Martin Klein [Amsterdam/New York, 1959], 107, 127). Their joint article on the foundations of statistical mechanics in the prestigious *Encyklopaedie der Mathematischen Wissenschaften* states: "The critical review and systematization of the results of all fundamental investigations was carried out by the authors in common work. P. Ehrenfest bears the ultimate responsibility for the final editing" (213).

Chapter 14: Sociologists in the Vineyard

The author wishes to thank the Special Collections Department, Regenstein Library, University of Chicago, for permission to quote from the Everett C. Hughes Papers. The research was funded by Concordia University and the Social Science and Humanities Research Council of Canada. Several individuals have contributed valuable insights and criticisms: Elizabeth Hughes Schneewind, Robin Ostow, Barrie Thorne, Kurt Jonassohn, Pnina Abir-Am, Helena Pycior, Nancy Slack, Marianne Ainley, Rick Helmes-Hayes, Bill Buxton, John Drysdale, and Howard Becker. The analysis presented here, however, is the author's and may not, in a comprehensive way, reflect the views of these individuals. This article is dedicated to David Drysdale in gratitude for his assistance, love, and sense of humor.

1. Everett Hughes, "Professional and Career Problems in Sociology," *Transactions of the Second World Congress of Sociology* 1 (1954): 182.
2. Helen Hughes, "Wasp/Woman/Sociologist," *Society*, July-August 1977, 80.
3. See especially Mary Jo Deegan, *Jane Addams and the Men of the Chicago School, 1892–1918* (New Brunswick, N.J., 1988), 55–69, 194–195; Rosalind Rosenberg, *Beyond Separate Spheres: Intellectual Roots of Modern Feminism* (New Haven, Conn., 1982), 28–53, 62–68. See also Marlene Shore, *The Science of Social Redemption: McGill, the Chicago School, and the Origins of Social Research in Canada* (Toronto, 1987), and Dorothy Ross, *The Origins of American Social Science* (Cambridge, 1991). For details on Small, see Harry Elmer Barnes, "Albion Woodbury Small: Promoter of American Sociology and Expositor of Social Interests," in *An Introduction to the History of Sociology*, ed. H. E. Barnes (Chicago, 1948), 432.
4. See Deegan, *Jane Addams*, 152–155, 213–214, for evidence of Park's hostility to women including those who were his colleagues; he opposed women's suffrage as well.
5. "Recollections," Everett C. Hughes Papers, Joseph Regenstein Library, University of Chicago, box 1.
6. David Riesman, "The Legacy of Everett Hughes," *Contemporary Sociology* 12 (1983): 477.
7. E. C. Hughes Papers, autobiographical writings, box 23, "Notes on Student Days," January 1976, 1.
8. He studied with Jacob Viner, then a young economist; Albion Small, a "thoughtful man who taught a scholarly theory and history of sociology course"; Ellsworth Faris; and Robert Park, who used his own new *Introduction to the Science of Sociology*, written with Ernest Burgess, which defined sociology in terms of the second-generational perspective described earlier. Ibid., 1–4.
9. Ibid., 7.
10. E. C. Hughes Papers, box 23, graduate school memoirs, 3.
11. Ibid., 4.

12. Ibid.

13. Helen Hughes, "Wasp/Woman/Sociologist," 69–80.

14. Ibid., 72.

15. Ibid., 73; interview with Elizabeth Hughes Schneewind, March 27, 1993.

16. Helen's younger sister, Elsie Gregory MacGill, entered the field of engineering and became the first woman to graduate in electrical engineering at the University of Toronto (1927), and in aeronautical engineering from the University of Michigan (1929). She designed a special-duty trainer plane, the Maple Leaf, for the Mexican air force. She married late and had stepchildren.

17. Winifred Raushenbush contributed a great deal to Park's *Immigrant Press and Its Control* (1946). She later wrote an informative biography of Park, *Robert E. Park: Biography of a Sociologist* (Durham, N.C., 1979).

18. E. C. Hughes Papers, box 23, autobiographical writings, "Years in Graduate School," 6. Neither Helen nor Everett, it seems, has left any detailed description of their meeting, courtship, or marital relationship.

19. "Park enjoyed companionship with his brighter students. One of his prolonged and close relations was with Helen MacGill Hughes and her husband, Everett Hughes. He saw them through their theses. When he went on exploratory sociological trips, they sometimes served as his drivers. If he woke up in the morning wanting to begin a new book, they cooperated—they were his favorite playmates. 'They know how to enjoy life; they have a genius for it,' he said." Raushenbush, *Robert E. Park*, 103.

20. Park's wife, Clara Cahill, was an intelligent, energetic, artistic, and cultured woman who supplemented the modest family income at intervals by writing and painting, while taking care of the home and rearing four children.

21. Helen Hughes, "Wasp/Woman/Sociologist," 76.

22. Helen Hughes, "On Becoming a Sociologist," *International Journal of the History of Sociology* 3 (fall–winter 1980–1981): 36.

23. Helen MacGill Hughes, *News and the Human Interest Story* (University of Chicago, 1940). She published articles on her doctoral research in the *Journalism Quarterly*, the *Public Opinion Quarterly*, and the *American Journal of Sociology*. See "The Lindbergh Case: A Study of Human Interest and Politics," *American Journal of Sociology* 42 (1936): 32–45.

24. Robin Ostow, "Everett Hughes: The McGill Years," *Society/Sociète* 8 (1984): 12. During the previous decade he had published a dozen articles in the leading sociology journals on the division of labor, institutions, industry, and religion in Germany, personality types, and French-English relations in Quebec. His dissertation on the Chicago Real Estate Board had been published by the Society of Social Research of the University of Chicago in 1931 as a monograph in a series.

25. Ibid., 13.

26. In keeping with the increasing trend at that time of antinepotism policies in universities, McGill principal Sir Arthur Currie refused to employ couples.

27. "Hughes had originally thought of *French Canada in Transition* as but one phase of a larger project whose goal was, not only to expose some of the major social changes occurring as a result of the industrialization of French Canada, but also, and equally important, to interest Canadian students—both French and English—in studying French Canada's problems" (Ostow, "Everett Hughes: The McGill Years," 14).

28. Everett C. Hughes, *French Canada in Transition* (Chicago, 1943), xi.

29. Ibid., 69, 70, 72.

30. Helen Hughes, "Wasp/Woman/Sociologist," 76.

31. Everett did in fact coauthor publications with Margaret McDonald and other students during that period. The Hughes and McDonald article, "French and English in the Economic Structure of Montreal," *Canadian Journal of Economics and Political*

Science 7 (1941): 493–505, is however reprinted in Hughes's *The Sociological Eye: Selected Papers* (Chicago, 1971) as single-authored by Hughes.

32. See, for example, Ostow, "Everett Hughes: The McGill Years" and Alan Anderson et al., "Sociology in Canada: A Developmental Overview," in *Handbook of Contemporary Developments in World Sociology,* ed. Raj Mohan and Don Martindale (Westport, Conn., 1975), 160.

33. This investigation of social and economic change in a small industrial city in Quebec explored the theme which fascinated Hughes throughout his professional life: "the way that industrialism and urbanization mix together people of different religions, ethnic groups, races and cultures" (Hughes, *The Sociological Eye,* 242). "Virtually all histories of French Canadian sociology trace the beginning of modern sociological scholarship in French Canada to Hughes's *French Canada in Transition* project" (Robin Ostow, "Everett Hughes: From Chicago to Boston," *Society/Sociète* 9 [1985]: 9).

34. Riesman, "The Legacy of Everett Hughes," 477.

35. Hughes moved "with a 20% increase over his salary at McGill (which had been raised only once since he had been hired in 1927). McGill refused to match Chicago's offer." Ostow, "Everett Hughes: The McGill Years," 15.

36. Helen Hughes, "Wasp/Woman/Sociologist," 76. This is not entirely true, since she published her dissertation in 1940, and probably worked on the preparation of *French Canada in Transition,* which was published in 1943.

37. Ibid., 76–77.

38. Helen Hughes, letter to Messrs. Shugg, Hauser, and Blau, July 31, 1961, E. C. Hughes Papers, box 30. See also Helen Hughes, "Maid of All Work or Departmental Sister-in-Law? The Faculty Wife Employed on Campus," *American Journal of Sociology* 78 (1975): 767–772.

39. Helen Hughes, "Maid of All Work?" 770. Helen was always reluctant to seek a raise in salary from the university press, especially while Everett was editor of the *Journal*; so she did not.

40. Ibid., 772.

41. Margaret W. Rossiter, *Women Scientists in America: Struggles and Strategies to 1940* (Baltimore, 1982), 260. Chicago was second only to Yale from 1877 to 1900 in awarding Ph.D.s to women (Rossiter, *Women Scientists,* 35).

42. Helen Hughes, "Maid of All Work?" 770.

43. Ibid., 771.

44. Helen Hughes, "Wasp/Woman/Sociologist," 76.

45. Letter to David Riesman, November 28, 1960, E. C. Hughes Papers, box 9.

46. As the Rapoports point out, friendships as a couple are important in the dual-career relationship particularly since they help to reduce role conflict and perhaps also role deprivation, which women so often experience. Rhona Rapoport and Robert N. Rapoport, "The Dual-Career Family: A Variant Pattern and Social Change," in *Toward a Sociology of Women,* ed. Constantina Safilios-Rothschild (Lexington, Mass., 1972), 216–244. See esp. 230–233.

47. "Dear Everett and Helen: I am glad to hear that Helen has fooled the medical profession. [Helen had had a miscarriage before the birth of her first child]. I appreciate the reviews that Helen must have written under great difficulties. Such sacrifice should be rewarded" (letter from Louis Wirth to E. C. and Helen Hughes, February 9, 1938, E. C. Hughes Papers, box 30).

48. Interview with Elizabeth Hughes Schneewind, June 21, 1989.

49. The research shows that role conflict and stress in the family are greater for women than for men, and that men have always considered the resolution of these problems to be women's responsibility. There is no clear pattern by social class. In fact, middle-class and professional marriages are shared or equal partnerships often to a lesser extent than working-class marriages. See P. Willmott, "The Influence of Some

Social Trends of Regional Planning," mimeographed paper (London, 1968), cited by the Rapoports.

50. "La Mere de Famille," memorandum to Margaret Westley, E. C. Hughes Papers, 2d ser., box 30.

51. Helen Hughes, *The Fantastic Lodge: The Autobiography of a Girl Drug Addict* (Chicago, 1961), was the result of an interview by Howard Becker in a study of drug addicts commissioned in Chicago. Helen's hard work in editing and rewriting the material resulted in a book which was published in England and America, in hardcover and paperback, and translated into French for publication in 1972.

52. Interview with Elizabeth Hughes Schneewind, June 21, 1989.

53. Letter of E. C. Hughes to Morin and Becker, July 9, 1969, Hughes Papers, 2d ser., box 23. The chapter was not used. The Rapoports note that among their dual-career couples certain degrees of strain and defensiveness resulted in "often unrecognized [and unconscious?] undercutting behavior by the husband toward the wife," even when "subordinated to the more dominant aspect of their relationship, which was that the husband did in fact support, sponsor, encourage, and otherwise facilitate his wife's career." Rapoport and Rapoport, "The Dual-Career Family," 229–230.

54. Letter from Lillian Levi to ECH, May 21, 1964, Hughes Papers, box 1.

55. Helen also was a contributor to *Time* magazine on social science, education, science, and medicine. She was engaged by the National Opinion Research Center to write grant proposals and introductions to reports and was local arrangements person for the Public Opinion Research Societies. See Helen Hughes, "Wasp/Woman/Sociologist," 77.

56. "Elsie was Helen's alter ego." Interview with Robin Ostow, October 1992. Elsie had an unusually good career for a woman in that period (see n. 16).

57. For more details on Helen's assessment of her career, see S. Hoecker-Drysdale, "Women Sociologists in Canada: The Careers of Helen MacGill Hughes, Aileen Dansken Ross, and Jean Robertson Burnet," in *Despite the Odds: Essays on Canadian Women and Science,* ed. Marianne G. Ainley (Montreal, 1990), 152–176.

58. Helen Hughes, "Woman/Wasp/Sociologist," 80. Significantly, in the 1970s Helen worked with Rose Coser, another sociologist married to a sociologist, to produce the first report on the status of women in sociology for the American Sociological Association. Helen did considerable research and writing on the topic.

59. She was a woman of "private humanity, a peacemaker and a healer, generous in her recognition of others." Elizabeth Hughes Schneewind, reported by Barrie Thorne and Arlene Kaplan Daniels, "Remembering Helen MacGill Hughes, 1903–1992," *Sociologists for Women in Society Network* (newsletter), 1992. Helen was elected vice president of SWS in 1973 and edited the SWS newsletter from 1973 to 1978 (Thorne and Daniels, "Remembering," 3).

60. Conversation with Kurt Jonassohn.

61. In the late 1960s and early 1970s, when a series of guest lectures honoring Everett was held at Boston College, on each occasion Helen cooked dinner for the guests. She complained but followed through with what had been her lifelong hostess role. Interview with Elizabeth Hughes Schneewind, October, 1992.

62. The author thanks Rick Helmes-Hayes for this helpful suggestion.

63. An observation by Bill Buxton regarding his research on Everett Hughes.

64. Everett adored and esteemed Helen in several respects, but he did not question the order of things. Nevertheless, he had observed his own father's domination of his mother, whom he felt had been imposed upon. Interview with Elizabeth Hughes Schneewind, June 21, 1989. In fact, Everett seems to have been "quietly supportive" of the women's movement of the 1960s and of Helen's activity on behalf of women in sociology. Everett encouraged many women graduate students toward their professional goals. "Everett's approach was deeply collaborative in nature, engaging others

in a way of seeing, and helping them find that vision in themselves." Letter of Barrie Thorne, former student of Everett Hughes, to the author, March 26, 1994.

Chapter 15: Botanical and Ecological Couples

1. Geraldine Kaye, "Violetta S. White: A Mycologist Who Got Away," *Boston Mycological Club Bulletin* 39, no. 2 (1984): 10–11.

2. Emmanuel D. Rudolph, "Women in Nineteenth-Century Botany: A Generally Unrecognized Constituency," *American Journal of Botany* 69 (1982): 1346–1355.

3. Emma L. Bolzau, *Almira Hart Lincoln Phelps: Her Life and Work* (Philadelphia, 1941), 44.

4. Nancy G. Slack, "Nineteenth-Century American Women Botanists: Wives, Widows, and Work," in *Uneasy Careers and Intimate Lives,* ed. P. G. Abir-Am and D. Outram (New Brunswick, N.J., 1987), 77–103, 298–310.

5. Anne B. Shteir, "Botany in the Breakfast Room: Women and Early Nineteenth-Century British Plant Study," in *Uneasy Careers and Intimate Lives,* ed. Abir-Am and Outram, 31–43; 288–292.

6. Marilyn B. Ogilvie, "Marital Collaboration: An Approach to Science," in *Uneasy Careers and Intimate Lives,* ed. Abir-Am and Outram, 104–125, 311–314.

7. William Sullivant to Asa Gray, 1850 and October 12, 1851, quoted in Andrew D. Rodgers III, *"Noble Fellow" William Starling Sullivant* (New York, 1968), 208–209.

8. Elmer D. Merrill, "Memoir of Nathaniel Lord Britton: 1859–1934," *National Academy of Sciences, Biographical Memoirs* 19 (1938): 147–201. Nancy G. Slack in "Nineteenth-Century American Women Botanists" provided basic biographical information as well as further references on Elizabeth Knight Britton.

9. Margaret R. Rossiter, *Women Scientists in America: Struggles and Strategies to 1940* (Baltimore, 1982), 85.

10. Letter of appointment as "Honorary Curator of Mosses," May 29, 1912, addressed to Mrs. Elizabeth Britton, New York Botanical Gardens; signed "N. L. Britton, Secretary" (in Elizabeth Britton Archives, New York Botanical Garden).

11. Letter from Elizabeth Britton to Mr. Pollard, December 8, 1897, Smithsonian Institution Archives. Ogden Tanner and Adele Auchincloss, *The New York Botanical Garden: An Illustrated Chronicle of Plants and People* (New York, 1991).

12. Nancy G. Slack, "Charles Horton Peck, Bryologist, and the Legitimation of Botany in New York State", *Memoirs of the New York Botanical Garden* 45 (1987): 28–45.

13. Nancy G. Slack, "The Botanical Exploration of California from Menzies to Muir (1786–1900) with Special Emphasis on the Sierra Nevada," in *John Muir: Life and Works,* ed. Sally Miller (Albuquerque, N.M., 1993), 194–242.

14. Arlie Hochschild with Anne Machung, *The Second Shift* (New York, 1989). Edith S. Clements, *Adventures in Ecology: Half a Million Miles . . . from Mud to Macadam* (New York, 1960).

15. Joseph A. Ewan, "Clements, Frederic Edward," *Dictionary of Scientific Biography* (New York, 1971), 3:317–318. Quotes from an unpublished six-page biographical account of Frederic E. Clements (FEC) by Edith S. Clements (ESC) in the Edith and Frederic Clements Archives, University of Wyoming. Also in these archives are long, very literary love letters Frederic wrote "Miss Schwartz" in 1898 and 1899 before their marriage in May 1899.

16. Eugene Cittadino in "Ecology and the Professionalization of Botany in America, 1890–1905," *Studies in the History of Biology* 4 (1980): 171–198, and Joel B. Hagen in *An Entangled Bank: The Origins of Ecosystem Ecology* (New Brunswick, N.J., 1992), 33–49, both discuss Frederic Clements's importance in the ecology of his day. Hagen mentions Edith only as the author of *Adventures in Ecology,* in which "her late husband consistently comes off as a cold fish."

17. E. S. Clements, *Adventures in Ecology*, 222.

18. Biographical account of FEC; see note 15.

19. ESC journal, November 14–December 10, 1939, ESC and FEC archives, University of Wyoming. Interviews by Nancy Slack with F. E. Clements's former field assistants Dr. Paul Lemon and Dr. Joseph Ewan, and with Ewan's wife, Nesta, confirm Edith's helpmate role. Lemon, then a University of Nebraska student, said that Edith did a great deal of typing for Frederic and painted Colorado wildflowers but did not do anything "original" in ecology.

20. Frances L. Long graduated from the University of Nebraska, received her doctorate at the University of Minnesota with F. E. Clements, and became a permanent staff member of the Carnegie Institution in 1918. She is often mentioned in Edith Clements's journal from 1919 to 1943. An April 10, 1921, letter from FEC to FLL at the Tucson laboratory discusses pollination and photosynthesis experiments. Clements Archives, University of Wyoming.

21. Frederic E. and Edith S. Clements, *Flower Families and Ancestors* (White Plains, N.Y., 1928). They also coauthored *Rocky Mountain Flowers* (White Plains, N.Y., 1928), illustrated with paintings by ESC; E. S. Clements, *Adventures in Ecology*.

22. Joseph Ewan, "San Francisco as a Mecca for Nineteenth-Century Naturalists," in *A Century of Progress in the Natural Sciences, 1853–1953.* (San Francisco, 1955): 1–56.

23. Slack, "Botanical Exploration of California."

24. From "Reminiscences by Mrs. Brandegee," prepared at the request of P. B. Kennedy for the California Botanical Society, April 1916, University Herbarium, University of California, Berkeley.

25. T. S. Brandegee's two unpublished autobiographical accounts written in 1916 and 1921, both in the University Herbarium, Berkeley.

26. William A. Setchel, "Townshend and Mary Brandegee," *University of California Publications in Botany* 13 (1926): 156–158.

27. Slack, "Botanical Exploration of California"; A. Hunter Dupree and Marian L. Gade, "Brandegee, Katharine Curran," in *Notable American Women*, (Cambridge, Mass., 1971), 1:228–229. Letters from Katharine Brandegee to Asa Gray and Sereno Watson in the Gray Herbarium, Harvard University.

28. Marcus E. Jones, "Katherine Brandegee," *Desert* 4 (1932): 65–70, and Albert W.C.T. Herre, "Katherine Brandegee: A Reply to a Fantasy by J. Ewan," privately published by the author ca. 1960, quoted in Jones. Secondary sources often spell her name as above, but she signed her letters "Katharine." Herre, an ichthyologist and lichenologist who knew both the Brandegees and Edward Lee Greene, came to Kate's defense in a dispute about the latter 40 years after her death when Herre himself was over ninety. Kate had critiqued Greene's botanical methods and results in print, but not him personally, although, according to Herre, she knew he was gay and the reasons for which he was defrocked. Jones also discussed Kate Brandegee and Greene in "Katherine Brandegee" and in "The Brandegees," *Contributions to Western Botany* 15 (1929): 15–18. In the latter, Jones attacked Greene (in print!) as "a moral retrobate, . . . Episcopalian minister, kicked out of the pulpit because of sexual vices." Greene, nevertheless, subsequently became a botany professor at Berkeley and the president of the 1893 International Botanical Congress.

29. Jones, "Katherine Brandegee."

30. Letter from Robert B. Brandegee to T. S. Brandegee, October 6, 1913, University Herbarium, University of California, Berkeley.

31. Marcus E. Jones, "Mrs. T. S. Brandegee," *Contributions to Western Botany* 18 (1933): 12–18.

32. "Off on an Odd Expedition: Two Men and a Woman Chasing after Snakes and Bugs," *San Francisco Examiner*, 1893, California Academy of Sciences Archives.

33. Letter from T. S. Brandegee to Katharine Brandegee, July 18, 1897, University Herbarium, University of California, Berkeley

34. Letters from Katharine Brandegee to Townshend S. Brandegee, July 25 and July 29 or 30 (her notation), no year but before 1906, and July 24, 1915, and from T. S. to Katharine Brandegee, March 1, 1893, and October 22, 1905, all in the University Herbarium, Berkeley.

35. Letter from K. to T. S. Brandegee, October 3, 1917, University Herbarium, Berkeley.

36. Letter from K. to T. S. Brandegee from the Alta Bates Sanatorium (no date) in the University Herbarium, Berkeley.

37. Herre, quoted in Jones, "Katherine Brandegee" (see n. 28.)

38. Janice E. Bowers, *A Sense of Place: The Life and Work of Forrest Shreve* (Tucson, 1988), 30.

39. "Shreve enjoyed tequila in moderation and smoked a pipe; Clements neither smoked nor drank." The British ecologist Arthur Tansley reported that "it gave him real pain to see other people do so." Bowers, *A Sense of Place,* 60 and n. 3.

40. Bowers, *A Sense of Place,* 66–67.

41. Robert L. Burgess, "The Ecological Society of America: Historical Data and Some Preliminary Analyses," in *History of American Ecology,* ed. F. N. Egerton and R. P. McIntosh (New York, 1977).

42. Jean H. Langenheim, "The Path and Progress of American Women Ecologists," *Bulletin of the Ecological Society of America* 69 (1988), 184–197 (address of the past president of ESA at Davis, California, August, 1988).

43. Ibid. Hazel Delcourt is currently a tenure-track associate professor at the University of Tennessee. According to her husband, this change was fostered by Langenheim's address and article (Paul Delcourt, personal communication). Some institutions still have antinepotism rules, which have historically kept wives in secondary positions.

44. Grace Pickford was completely left out of *The Kindly Fruits of the Earth,* Hutchinson's autobiography of his early years, including all the time he was married to Grace. This was done in deference to his second wife, Margaret Seal Hutchinson. Remarkably, Grace Pickford's marriage to Hutchinson is left out of her obituary by Ball, although their collaborative work is included there. J. N. Ball, "In Memoriam, Grace E. Pickford (1902–1986)," *General and Comparative Endocrinology* 65 (1987): 162–165.

45. G. E. Hutchinson, G. E. Pickford, and J.F.M. Schuurman," The Inland Waters of South Africa," *Nature* 123 (1929): 832, and idem, "A Contribution to the Hydrobiology of Pans and Other Inland Waters of South Africa," *Archiv für Hydrobiologie* 24 (1932): 1–154.

46. Letters from G. Evelyn Hutchinson to Grace Pickford from Ladakh in the Yale University Archives, made available only in 1991 after the death of both. The nature of their marriage and divorce is discussed further in a forthcoming book by Nancy G. Slack on G. Evelyn Hutchinson and the development of modern ecology.

47. Ball, "In Memoriam." See also "Grace Pickford," *Newnham College Register* 1 (1987): 1877–1923 (obituary).

48. Ball, "In Memoriam," and personal communications to Nancy G. Slack from Anna Bidder, Penelope Jenkin, and Dorothea Hutchinson. Bidder and Jenkin, who have made major contributions to biological science, were her classmates at Newnham College, Cambridge; D. Hutchinson was her sister-in-law, G. Evelyn Hutchinson's sister.

49. Letter from J. S. Nicholas to Charles Seymour, both of Yale University, November 22, 1945. In the Nicholas Archives, Yale University Archives.

50. Slack, "Nineteenth-Century American Women Botanists."

Chapter 16: Patterns of Collaboration
in Turn-of-the-Century Astronomy

1. Donald E. Osterbrock, John R. Gustafson, and W. J. Shiloh Unruh, *Eye on the Sky: Lick Observatory's First Century* (Berkeley, 1988), 152–153, 156–160.

2. Throughout her diaries, Elizabeth writes that her husband is the scientist, and she, the organizer and planner. Elizabeth Campbell, "In the Shadow of the Moon," Mary Lea Shane Archives of the Lick Observatory, The University Library, Santa Cruz, California (hereafter Shane Archives).

3. W. H. Wright, *Biographical Memoir of William Wallace Campbell, 1862–1938,* National Academy of Sciences of the United States of America, Biographical Memoirs 25, 3d Memoir (Washington, D.C., 1947), 35; Elizabeth Campbell, "Notes about W.W.C.," Shane Archives.

4. Wright, *William Wallace Campbell,* 35–37; Osterbrock et al., *Eye on the Sky,* 130–131.

5. Osterbrock et al., *Eye on the Sky,* 131.

6. E. Campbell, "Notes about W.W.C."

7. "Programme, Class of '90," Carnegie Branch Library for Local History, Boulder, Colo.; Osterbrock et al., *Eye on the Sky,* 133.

8. E. Campbell, "Spain: In the Shadow of the Moon," Shane Archives, 84, 87, 70, 110.

9. W. Campbell, "A General Account of the Lick Observatory–Crocker Eclipse Expedition to India," *Publication of the Astronomical Society of the Pacific* 10 (August 1, 1898): 138.

10. E. Campbell, "India: In the Shadow of the Moon," Shane Archives, 1.

11. Osterbrock et al., *Eye on the Sky,* 155.

12. E. Campbell, "India: In the Shadow of the Moon."

13. Ibid., 2.

14. Ibid., 91.

15. Ibid.

16. E. Campbell, "Spain: In the Shadow of the Moon," 52.

17. E. Campbell, "Nineteen to the Dozen," Russian eclipse, August 21, 1914, Shane Archives, photo file E-11, folder 2; E. Campbell (under the name of Elizabeth Ballard Thompson), "With Eclipse Expedition in Russia," *San Francisco Chronicle,* 1915, Russian eclipse, Shane Archives, photo file E-ll, folder 4.

18. W. Campbell, "General Account of the Expedition to India," 127–140, 131.

19. Wright, *William Wallace Campbell,* 43.

20. W. Campbell, "The Lick Observatory–Crocker Eclipse Expeditions of 1905," 2, 3, Shane Archives, box 2, Spanish eclipse, photo file E-9.

21. E. Campbell, "Flint Island Eclipse," Shane Archives, photo file E-10, folder 9.

22. W. Campbell, R. G. Aitken, C. D. Perrine, and E. P. Lewis, "The Crocker Expedition of 1908," *Lick Observatory Bulletin,* no. 131; 1–14; Osterbrock et al., *Eye on the Sky,* 162.

23. E. Campbell, "Nineteen to the Dozen"; "With Eclipse Expedition in Russia," *San Francisco Chronicle,* 1915.

24. "Instruments of Dr. Campbell Delayed; Time Wins the Race," *San Francisco Bulletin,* June 3, 1918, Shane Archives, photo file E-12, folder 5.

25. W. Campbell, "Observations on the Deflection of Light in Passing through the Sun's Gravitational Field. Made during the Total Solar Eclipse of Sept. 21, 1992," *Lick Observatory,* University of California Publications, Astronomy, no. 346, 41–54, Shane Archives, photo file E-13, folder 8; E. Campbell, Australian eclipse, parts of diary, Shane Archives, photo file E-13, folder 3.

26. Wright, *William Wallace Campbell,* 50.

27. Donald E. Osterbrock, "To Climb the Highest Mountain: W.W. Campbell's 1909 Mars Expedition to Mount Whitney," *Journal for the History of Astronomy* 20 (1989): 77–96.

28. Ibid., 53–55.

29. *The Call-Bulletin* (San Francisco) 163, no. 123, June 14, 1938. Elizabeth explained that Wallace, who was so talented mathematically, could no longer do even the simplest arithmetic.

30. Margaret W. Rossiter, *Women Scientists in America: Struggles and Strategies to 1940* (Baltimore, 1982), 73–99.

31. Family tree, constructed by Alan Maunder, great-grandson of Walter.

32. Marriage certificate of Edward Walter Maunder and Edith Hannah Bustin, General Register Office, London, application number G69958 (obtained by Alan Maunder).

33. Alan Maunder's aunt, Violet Doreen (1924–), recalled talking to her father, Edward Arthur (1886–1966) about her grandfather, Walter. Edward Arthur often felt ignored and neglected by his father. Interview with Alan Maunder, June 1991. Walter, however, took his grown daughters on an eclipse expedition and also described his concern about his youngest son's health.

34. Certified copy of an entry of death, pursuant to the Births and Deaths Registration Act, 1953, registration district, Greenwich, death in the sub-district of Greenwich East in the county of Kent, HC158054 (contributed by Alan Maunder).

35. A. J. Meadows, *Greenwich Observatory: One of Three Volumes by Different Authors Telling the Story of Britain's Oldest Scientific Institution, the Royal Observatory at Greenwich and Herstmonceux, 1675–1975*, vol. 2, *Recent History, 1836–1975* (London, 1975), 10; Edward Walter Maunder, *The Royal Observatory, Greenwich: A Glance at Its History and Work* (London, 1900), 117.

36. Meadows, *Greenwich Observatory*, 14. Most of the boys had not even attended the university.

37. *College Register, Girton College, 1869–1946* (Cambridge, 1948), 1886.

38. Recommendation, W. H. Young, Esq., M.A., Fellow of Peterhouse, Cambridge. Royal Greenwich Observatory Archives 7/14 (hereafter RGO).

39. Although women could take the examinations, they were not eligible for Cambridge titular degrees until 1923 and were not full members of the university until 1948. Russell's position was described as senior optime (between brackets 41 and 43). *The Girton Review*, July 1889, examination lists, 8, Girton College Archives.

40. Barbara Stephen, *Girton College, 1869–1932* (Cambridge, 1933), 88, 188. Shortly after her marriage to Walter, Annie Maunder received the Pfeiffer Research Student Fellowship. As the first recipient of this fellowship (1896), she used the money to make a photographic study of the Milky Way.

41. Annie Scott Dill Russell (Maunder) to H. H. Turner, [1891], RGO 7.

42. A.S.D. Russell (Maunder) to Turner, January 13, 1890, RGO 7; recommendation from Miss Welsh, Mistress, Girton College, Cambridge, Aug. 15, 1889, RGO 7; recommendation from W. H., Young, Esq., M.A., Fellow of Peterhouse, Cambridge, RGO 7.

43. Ball continued by explaining that he did not "know the young lady personally," but "her two brothers whom I do know are both able men, the elder is Professor of Astronomy and mathematics at Pekin, where he has acquired himself with much distinction" (Robert Ball [F.R.S.] to Christie, July 29, 1891, RGO 7).

44. Turner to A.S.D. Russell (Maunder), July 28, 1891, RGO 7; A.S.D. Russell (Maunder) to Turner, July 30, 1891, RGO 7.

45. Maunder did not receive the promotions he thought he deserved. See, for example, E. W. Maunder to W. H. Christie, Aug 15, 1883, and Christie to E. W. Maunder, Aug. 16, 1883, RGO 7/8. When the position of Astronomer Royal for Scotland opened, Maunder did everything possible to get it. See series of letters, 1905, RGO 7. At

one point, Maunder was tempted to leave England to accept a position at the Lick Observatory.

46. Deborah Jean Warner, "Maunder, Edward Walter," *Dictionary of Scientific Biography,* ed. Charles Coulston Gillispie, 9:184.

47. E. W. Maunder, obituary, *The Journal of the British Astronomical Association* 38 (1928): 165–169.

48. E. W. Maunder, obituary; Mrs. Walter Maunder, obituary, *The Journal of the British Astronomical Association* (1947): 57.

49. A.S.D. Maunder to Molesworth, Sept. 14, 1900, Archives, Royal Astronomical Society (hereafter RAS), MSS Molesworth 1, folio 2, letter 4; A.S.D. Maunder to Molesworth, Dec. 28, 1901, RAS, MSS Molesworth 1, folio 2, letter 7.

50. Michael J. Crowe, *The Extraterrestrial Life Debate, 1750–1900: The Idea of a Plurality of Worlds from Kant to Lowell* (Cambridge, 1986), 491, 630.

51. E. W. Maunder, *Guide to the Constellations, and Introduction to the Study of the Heavens with the Unassisted Sight* (London, 1904), iii.

52. E. W. Maunder, *The Astronomy of the Bible: An Elementary Commentary on the Astronomical Reference of Holy Scripture,* 4th ed. (London [1922]). An Example: "Notes on an Early Astronomical Observation in the Book of Joshua," *Observatory* (January 1904).

53. Examples of A.S.D. Maunder's interest in eastern astronomy are found in A.S.D. Maunder, "Iranian Migrations before History," *Scientia* 19 (February 1916): 115–124; idem, "Astronomical Allusions in Sacred Books of the East, being a paper read before the Victoria Institute at the 567th Meeting, held April 12th, 1915" (in the Girton College Archives); idem, "When the Snow-White Bull with Gilded Horns Ushers in the Year," *The Journal of the British Astronomical Association* 41 (October 1930–September 1931): 127–133; and idem, "The Four Star Champions of Iran," *The Journal of the British Astronomical Association* 23 (June 1913): 425–429.

54. E. W. Maunder, *The Science of the Stars* (London, n.d.); E. W. Maunder and A.S.D. Maunder, "The Oldest Astronomy," *The Journal of the British Astronomical Association* 14 (1903–1904): 241; A.S.D. Maunder, "The Sothic Cycle of the Nakshatras," *The Journal of the British Astronomical Association* 43 (January 1933): 121–125.

Chapter 17: *Collaborative Couples Who Wanted to Change the World*

Comments by Elizabeth Crawford, Aant Elzinga, Susan Hoecker-Drysdale, Karen Reeds, Margaret Rossiter, Brigitte Schroder Gudehus, and participants in colloquiums at The Johns Hopkins University, UCLA, Caltech, the Nineteenth International Congress for the History of Science in Zaragoza, and the Annual Meeting of the History of Science Society in Santa Fe are gratefully acknowledged. Discussions with students in my course History of Women in Science (fall 1991, fall 1992), especially Ursula McVeigh, class of '94, Susan Pae, class of '95, and Kadisha Rapp, class of '94, proved very helpful in clarifying the collaborative dimension of these couples. Russell archivist Kenneth Blackwell of McMaster University was most helpful with tracing archival sources on the Russells and checking their section. The support of the NSF-VPW program in the form of a visiting professorship award and of the Johns Hopkins Department of History of Science is gratefully acknowledged for the academic years 1991–1993.

1. See the chapters on couples in the natural sciences in this collection; see also the essays by Marilyn Ogilvie, Nancy Slack, Helena Pycior, Peggy Kidwell, and Pnina G. Abir-Am in *Uneasy Careers and Intimate Lives: Women in Science, 1789–1979,* ed. Pnina

G. Abir-Am and Dorinda Outram (New Brunswick, N.J., 1987). The prevalence of collaborative couples in science has already been pointed out by Margaret W. Rossiter's pioneering *Women Scientists in America: Struggles and Strategies to 1940* (Baltimore, 1982). For recent compelling examinations of collaborative couples in the arts and humanities, see Renee Riese Hubert, *Magnifying Mirrors: Women, Surrealism and Partnerships* (Lincoln, Neb., 1994); Whitney Chadwick and Isabelle de Courtivron, eds., *Significant Others: Creativity and Intimate Partnership* (London, 1993); Kate Fullbrook and Edward Full-brook, *Simone de Beauvoir and Jean-Paul Sartre: The Remaking of a Twentieth-Century Legend* (Hemel Hempstead, Herts, 1994); and Diana Trilling, *The Marriage of Diana and Lionel Trilling* (New York, 1994).

2. On the public-private dichotomy and the role of gender in the constitution of the social order, see Linda Kerber, "Separate Spheres, Female World, Woman's Place," *Journal of American History* 75 (June 1988): 9–39; see also Rosalind Rosenberg, *Beyond Separate Spheres* (New York, 1982).

3. For a similar approach of drawing lessons from historical case studies for the purpose of framing a policy agenda for redressing the underrepresentation of women in science, see Pnina G. Abir-Am, "Science Policy or Social Policy for Women in Science: From Historical Case-Studies to an Agenda for the 1990s," report on a Work-shop held at Johns Hopkins University, December 6–8, 1991, with support from NSF-VPW program, *Science and Technology Policy*, April 1992, 11–12; idem, *EASST Newsletter*, February 1990, 14–17. See also Elizabeth Cody, "The Buck Stops with Him. . . . and Her," *Atlanta Constitution*, February 21, 1993, F-1, F-3. The article was subtitled "Bill and Hillary Clinton are breaking down the old taboos against couples working in tandem, high-powered jobs. The country doesn't know what to make of it."

4. On their versions of their encounter and life together see their respective autobi-ographies: Dora Russell, *The Tamarisk Tree: My Quest for Liberty and Love* (London, 1975); Bertrand Russell, *The Autobiography of Bertrand Russell*, vol. 2, *1914–1944* (Lon-don, 1951). On Dorothy Wrinch, the common friend that first introduced them, see Pnina G. Abir-Am, "Synergy or Clash: Disciplinary and Marital Strategies in the Career of Mathematical Biologist Dorothy Maud Wrinch, 1894–1976," in *Uneasy Careers and Intimate Lives*, ed. Abir-Am and Outram, 338–394. The Russells' correspondence with each other was not yet cataloged as of April 1993 (Kenneth Blackwell, personal commu-nication to the author). See also the insightful biography by the Russells' daughter, Katherine Tait, *My Father Bertrand Russell* (New York, 1976), and those by Ronald W. Clark, *The Life of Bertrand Russell* (New York, 1976); S. P. Rosenbaum, "Russell", chap. 9 in his *Edwardian Bloomsbury* (Toronto, 1987); and Caroline Moorehead, *Bertrand Russell, A Life* (New York, 1992). See also *Russell: The Journal of the Bertrand Russell Archive*, vols. 1–14 (1980–). Bertrand Russell wrote about 70 books and 2,500 articles; they are being reprinted as *The Collected Papers of Bertrand Russell*, in 28 volumes, of which 7 volumes are already available.

5. On this event see note 4 above; see also G. H. Hardy, *Bertrand Russell and Trinity* (Cambridge, 1970).

6. Tait, *My Father*, 51.

7. Ibid.

8. Bertrand Russell to Stanley Unwin (his publisher), March 20, 1923 (copyright with Russell Archive, McMaster University), in which they discuss the authorship of the book *The Prospects of an Industrial Civilization* (1923).

9. See Crystal Eastman, "Who is Dora Black?" *Equal Rights*, June 5, 1926; reprinted in Blanche Wiesen Cook, *Crystal Eastman on Women and Revolution* (New York, 1978), 114–118.

10. *Hypathia* (1925), *The Right to be Happy* (1927), *In Defense of Children* (1932), and *Children: Why Do We Have Them?* (1933), all published in London by Unwin.

11. Dora Russell, *The Tamarisk Tree*, 80.

12. Ibid., 294. See also Bertrand Russell, *Marriage and Morals* (London, 1929).

13. Tait, *My Father*, 98.

14. For overviews of Alva and Gunnar Myrdals' lives, see Elise Boulding, "Alva Myrdal, 31 January 1902–1 February 1986," in *American Philosophical Society Yearbook* (Philadelphia, 1986), 157–159; Charles P. Kindleberger, "Gunnar Myrdal, December 6, 1898–May 18, 1987," in ibid., 1989, 255–264. See also the books by their children: Sissela Bok, *Alva Myrdal: A Daughter's Memoir* (Boston, 1991), and Jan Myrdal, *Childhood* (New York, 1991); and see also the review of these two books by Sue Halpern, *The New York Review of Books*, March 5, 1992, 30–31. Especially insightful on the Myrdals' work in the context of American race relations is Walter A. Jackson, *Gunnar Myrdal and America's Conscience: Social Engineering and Racial Liberalism, 1938–1987* (Chapel Hill, N.C., and London, 1991). See also Immanuel Wallerstein, "The Myrdal Legacy: Racism and Underdevelopment as Dilemmas," in his *Unthinking Social Science* (Cambridge, 1991), 80–103.

15. For the Myrdals' fellowship records, see Rockefeller Archive Center, North Tarrytown, N.Y.; see also Earlene Craver, "Patronage and the Directions of Research in Economics: The Rockefeller Foundation in Europe, 1924–38", *Minerva* 16 (1988): 217–239.

16. The full extent of the Myrdals' pioneering work in the 1930s on issues of gender and race relations remains to be explored, especially now that their archives have become available. I thank Susan Hoecker-Drysdale for drawing my attention to the fact that the Myrdals were elaborating on the analogy between race and gender, an idea originating in the work of Harriet Martineau.

17. On Mead and Bateson's relationship, as viewed by their only child, see Mary Catherine Bateson, *With a Daughter's Eye: A Memoir of Margaret Mead and Gregory Bateson* (New York, 1984). On Mead, see D. L. Olmstead, ed., "In Memoriam: Margaret Mead, 1901–1978," *American Anthropologist*, 82, no. 2 (1980); Clifford Geertz, "Margaret Mead, 1901–1978," in *Biographical Memoirs of Members of the National Academy of Sciences*, vol. 39 (Washington, D.C., 1979), 329–354; Edward Rice, *Margaret Mead, a Portrait* (New York, 1979); Jane Howard, *Margaret Mead, a Life* (New York, 1984); and Phyllis Grosskurth, *Margaret Mead: A Life of Controversy* (New York, 1988). See also Margaret Mead's *Letters from the Field, 1925–1975* (New York, 1977). On Bateson, see Robert I. Levi and Roy Rappaport, "Obituary: Gregory Bateson, 1904–1980," *American Anthropologist*, 84, no. 2 (1982): 379–394; and David Lipset, *Gregory Bateson: The Legacy of a Scientist* (Englewood Cliffs, N.J., 1980). See also Steve J. Heims, *The Cybernetics Group* (Cambridge, Mass., 1991).

18. On Ruth Benedict see Judith S. Modell, *Ruth Benedict: Patterns of a Life* (Boston, 1983). Mead has also written a biography, *Ruth Benedict* (New York, 1974), and prepared a collection of essays, *An Anthropologist at Work: Writings of Ruth Benedict* (Boston, 1959). Reo Fortune's work at the time of his encounter with Mead and Bateson was published as *Manus Religion* (Philadelphia, 1935).

19. Mead became a celebrity following the publication of her *Coming of Age in Samoa: A Psychological Study of Primitive Youth for Western Civilization* (New York, 1928). Prior to her meeting Bateson in 1932, she had also published *Growing Up in New Guinea: A Comparative Study of Primitive Education* (New York, 1930).

20. See note 1 above.

21. On the social and political conditions affecting transdisciplinary innovation in twentieth-century science, see Pnina G. Abir-Am, "The Biotheoretical Gathering, Transdisciplinary Authority, and the Incipient Legitimation of Molecular Biology in the 1930s," *History of Science* 25 (1987): 1–70; and idem, "From Multidisciplinary Collaboration to Transnational Objectivity: International Space as Constitutive of Mo-

lecular Biology, 1930–1970," in E. Crawford, T. Shinn, and S. Sorlin, eds., *Denationalizing Science: The Contexts of International Scientific Practice* (Dordrecht, 1993), 153–186. On transdisciplinary innovation and gender, see Pnina G. Abir-Am, "Women in Research Schools," in Seymour H. Mauskopf, ed., *Chemical Sciences in the Modern World* (Philadelphia, 1993), 375–392.

List of Contributors

The Editors

HELENA M. PYCIOR is a professor of history at the University of Wisconsin-Milwaukee. Her articles on the history of algebra have appeared in *Historia Mathematica, Isis,* the *Journal of the History of Ideas,* and *Victorian Studies.* For over a decade her second focus of published research has been the history of women in science, particularly a reexamination of Marie and Pierre Curie. She teaches a wide range of courses, from introductory history of science to graduate seminars on Darwin and Darwinism and on gender, race, science, and medicine. Her recent writing includes a book on the history of British algebra through 1750 (Cambridge, forthcoming). She is presently working on a dual biography of the Curies.

NANCY G. SLACK is professor of biology at Russell Sage College, Troy, New York, where she teaches courses in ecology, the history of science, and the history of women in science and medicine. Her research interests are in plant community ecology, women in science, and the history of American botany and ecology. She has held an AAUW fellowship for research in Sweden and was visiting research scholar in the Program of History of Medicine and Life Sciences at Yale University in 1990–1991. In addition to ecological publications, including books published in 1993 and 1995, she has written a chapter on American women botanists in *Uneasy Careers and Intimate Lives: Women in Science, 1789–1979,* ed. Pnina G. Abir-Am and Dorinda Outram (New Brunswick, N. J., 1987), and several papers on the professionalization of botany in the United States and on the men and women, including couples, involved in the botanical exploration of the American West. She is currently writing a biography of the ecologist G. Evelyn Hutchinson.

PNINA G. ABIR-AM is a resident fellow at the Dibner Institute for the History of Science and Technology in Cambridge, Massachusetts. She has published widely in the history of science, including her award-winning contribution to *Uneasy Careers and Intimate Lives: Women in Science, 1789–1979* (Rutgers University Press, 1987). She is founding editor of "Lives of Women in Science," a series started by Rutgers University Press in 1989. She is now working on a dual biography of Dorothy Wrinch and Dorothy Hodgkin, and on a comparative history of research schools of molecular biology in the United States, United Kingdom, and France. She is co-editor of the forthcoming *La mise en mémoire de la science: Commémoration et célébration* (Paris, 1995).

Contributors

MARIANNE GOSZTONYI AINLEY is professor and chair, Women's Studies/ Gender Studies, University of Northern British Columbia. She worked as an "invisible" chemist for twenty years before becoming a historian of science. She is the author of *Restless Energy: A Biography of William Rowan* (Montreal, 1993) and the editor of *Despite the Odds: Essays on Canadian Women and Science* (Montreal, 1990). She has published book chapters and articles on women and science and the history of Canadian ornithology. Her research program includes both issue-oriented and biographical studies of Canadian women scientists and engineers, and collaborative work on First Nations women and environmental knowledge.

BARBARA J. BECKER is a senior research associate at the Southwest Regional Laboratory, Los Alamitos, California. She is currently developing an introductory physical science curriculum that integrates history in the teaching of science.

BERNADETTE BENSAUDE-VINCENT teaches philosophy of science at the University of Paris, X (Nanterre), and coordinates an international project on scientific popularization at the Centre de recherche en histoire des sciences et des techniques, La Villette. She is the author of *Langevin, science et vigilance* (Paris, 1987), *Lavoisier, mémoires d'une révolution* (Paris, 1993), and numerous articles on the history and philosophy of French chemistry and related topics. Her joint book with Isabelle Stengers, *Histoire de la chimie* (Paris, 1993), received the Prix Rostand of the French Association for the Advancement of Science.

MILDRED COHN spent fourteen years, 1946–1960, as a member of Carl Cori's department at Washington University, where she began as a research associate and ended as an associate professor. During that time she had the rewarding experience of friendship with Gerty and Carl Cori, which lasted until their respective deaths. Since 1960, she has been at the University of Pennsylvania (with a brief interlude at the Institute for Cancer Research at Fox Chase, 1982–1985), and she is now the Benjamin Rush Professor of Physiological Chemistry Emeritus. During her career Dr. Cohn has published about 150 papers, including one with Gerty Cori and one with Carl Cori. One paper, coauthored with T. R. Hughes in 1962, is a citation classic. She has received numerous awards, including the National Medal of Science and nine honorary degrees. She is a member of various professional societies, the American Academy of Arts and Sciences, the National Academy of Sciences, and the American Philosophical Society.

JANET BELL GARBER, a retired biology professor from Los Angeles Harbor College, received a Ph.D. in history of science from the University of California, Los Angeles, in 1989, with a dissertation entitled "Darwin as a Laboratory Director." She wrote a chapter about Darwin's correspondents in the Pacific for *Darwin's Laboratory: Evolutionary Theory and Natural History in the Pacific,* ed. Roy MacLeod and Philip F. Rehbock (Honolulu, 1994), and is currently editing *Naturalists: A Biographical Encyclopedia* for Garland Publishing. She represented the city of Los Angeles as a commissioner of the Southern California Coastal Water Research Authority for eighteen years and published a paper, "Maintaining Scientific Integrity under Pressure," on this experience.

CHRISTIANE GROEBEN is archivist at the Stazione Zoologica "Anton Dohrn" and is book review editor for the journal *History and Philosophy of the Life Sciences.* She has recently edited "Karl Ernst von Baer-Anton Dohrn Correspondence" (*Transactions of the American Philosophical Society,* vol. 83, part 3 [1993]), a collection of letters illuminating the role of the Naples Station in nineteenth-century biology. She is particularly interested in European women biologists and marine stations.

JOY HARVEY has a Ph.D. in history of science from Harvard University. She has taught history of science at Harvard, Skidmore, Sarah Lawrence, and Virginia Tech. Her publications include articles and book chapters on French nineteenth- and twentieth-century science and medicine. Her book on the French nineteenth-century Darwin translator, feminist, and philosopher of science Clémence Royer will soon appear from Rutgers University Press. She has been a Rockefeller fellow in the Department of History of Science at the University of Oklahoma and is currently an associate editor with the Darwin Correspondence, Cambridge University Library.

PAMELA M. HENSON is director of the Institutional History Division of the Office of Smithsonian Institution Archives, where she is responsible for research on the history of the Institution and for the Oral History Program. She received her Ph.D. in the history and philosophy of science from the University of Maryland with a dissertation on the impact of evolutionary theory on insect taxonomy. Her research interests concentrate on the history of natural history, the history of museums, and the role of women in science. Her recent publications include "The Comstock Research School in Evolutionary Entomology," *Osiris* 8 (1993): 159–177; with Marc E. Epstein, "Digging for Dyar: The Man Behind the Myth," *American Entomologist* 38 (fall 1992): 148–169; and with Terri A. Schorzman, "Videohistory: Focusing on the American Past," *Journal of American History* 78 (1991): 618–627.

SUSAN HOECKER-DRYSDALE is associate professor of sociology and chair of the Department of Sociology and Anthropology at Concordia University, Montreal. Her research and teaching center on classical sociological theory and the history of women's contributions in the discipline of sociology. She recently published *Harriet Martineau: First Woman Sociologist* (Oxford, 1992) and is currently completing another book manuscript, "The Sociology of Harriet Martineau." She has written on the careers of women in the first generation of Canadian sociologists.

MAUREEN M. JULIAN, adjunct professor and senior research scientist, Department of Geological Sciences, Virginia Polytechnic Institute and State University, is currently doing theoretical calculations on diamond and silicon nitride and completing a biography of Kathleen Lonsdale. Her latest publications are "Kathleen Yardley Lonsdale," in *Women in Chemistry and Physics: A Biobibliographic Sourcebook*, ed. Louise S. Grinstein, Rose K. Rose, and Miriam H. Rafailovich (Westport, Conn., 1993), 329–336, and "*Ab initio* Modeling of Constrained Systems: The Tetrahedron," *Journal of Physical Chemistry* (in press). She did postdoctoral work at University College, London, with Kathleen Lonsdale.

SYLVIA W. MCGRATH is an American historian specializing in the history of American science and history of women. Since 1968 she has taught a wide array of undergraduate and graduate courses at Stephen F. Austin State University in Nacogdoches, Texas, where she introduced the history of women in the mid-1970s and where her husband serves as the forest pathologist. In 1994–1995 she was named a regents professor at the university. Her publications include a book, *Charles Kenneth Leith: Scientific Adviser* (Madison, Wis., 1971); "Scientific Foundations, Societies, and Museums," in *100 Years of Science and Technology in Texas* (Houston, Tex., 1986); and many encyclopedia entries.

MARILYN BAILEY OGILVIE is curator of the History of Science Collections, professor of bibliography, and adjunct professor of the history of science at the University of Oklahoma. She is the author of *Women in Science, Antiquity through the Nineteenth Century: A Biographical Dictionary with Annotated Bibliography* (Cambridge, Mass., 1986). She is currently working on an annotated bibliography of the history of women in science and is continuing to work in the areas of the history of astronomy and biology.

JOHN STACHEL is professor of physics and director of the Center for Einstein Studies at Boston University. He was editor of the first two volumes of *The Collected Papers of Albert Einstein* (Princeton, 1987, 1989) and is the author of numerous articles on theoretical physics, philosophy of science, and history of science, including a number of studies of the life and work of Albert Einstein.

LINDA TUCKER is a curator of biological sciences in the Smithsonian Institution's National Museum of American History. She is currently completing a doctoral dissertation on science in selected African-American educational institutions in the nineteenth century.

SYLVIA WIEGAND, professor of mathematics at the University of Nebraska, received a Ph.D. from the University of Wisconsin in 1972. She comes from a long tradition of mathematicians, including William Henry and Grace Chisholm Young (her grandparents), Lawrence Chisholm Young (her father), and R.C.H. (Young) Tanner (her aunt). Roger Wiegand, her husband, is also a mathematician. Professor Wiegand has written numerous articles and given many invited talks about her research specialty, commutative algebra. She is an editor of the journal *Communications in Algebra,* of the *Rocky Mountain Journal of Mathematics,* and of the *Collected Works of W. H. Young and G. C. Young* (forthcoming). Professor Wiegand was elected to the Council of the American Mathematical Society in 1994, and she is chair of its Policy Committee on Meetings and Conferences.

Index